Lecture Notes in Computer Science

Vol. 1: GI-Gesellschaft für Informatik e.V. 3. Jahrestagung, Hamburg, 8.–10. Oktober 1973. Herausgegeben im Auftrag der Gesellschaft für Informatik von W. Brauer. XI, 508 Seiten. 1973.

Vol. 2: GI-Gesellschaft für Informatik e.V. 1. Fachtagung über Automatentheorie und Formale Sprachen, Bonn, 9.–12. Juli 1973. Herausgegeben im Auftrag der Gesellschaft für Informatik von K.-H. Böhling und K. Indermark. VII, 322 Seiten. 1973.

Vol. 3: 5th Conference on Optimization Techniques, Part I. (Series: I.F.I.P. TC7 Optimization Conferences.) Edited by R. Conti and A. Ruberti. XIII, 565 pages. 1973.

Vol. 4: 5th Conference on Optimization Techniques, Part II. (Series: I.F.I.P. TC7 Optimization Conferences.) Edited by R. Conti and A. Ruberti. XIII, 389 pages. 1973.

Vol. 5: International Symposium on Theoretical Programming. Edited by A. Ershov and V. A. Nepomniaschy. VI, 407 pages. 1974.

Vol. 6: B. T. Smith, J. M. Boyle, B. S. Garbow, Y. Ikebe, V. C. Klema, and C. B. Moler, Matrix Eigensystem Routines – EISPACK Guide. X, 387 pages. 1974.

Vol. 7: 3. Fachtagung über Programmiersprachen, Kiel, 5.–7. März 1974. Herausgegeben von B. Schlender und W. Frielinghaus. VI, 225 Seiten. 1974.

Vol. 8: GI-NTG Fachtagung über Struktur und Betrieb von Rechensystemen, Braunschweig, 20.–22. März 1974. Herausgegeben im Auftrag der GI und der NTG von H.-O. Leilich. VI, 340 Seiten. 1974.

Vol. 9: GI-BIFOA Internationale Fachtagung: Informationszentren in Wirtschaft und Verwaltung. Köln, 17./18. Sept. 1973. Herausgegeben im Auftrag der GI und dem BIFOA von P. Schmitz. VI, 259 Seiten. 1974.

Vol. 10: Computing Methods in Applied Sciences and Engineering, Part 1. International Symposium, Versailles, December 17–21, 1973. Edited by R. Glowinski and J. L. Lions. X, 497 pages. 1974.

Vol. 11: Computing Methods in Applied Sciences and Engineering, Part 2. International Symposium, Versailles, December 17–21, 1973. Edited by R. Glowinski and J. L. Lions. X, 434 pages. 1974.

Vol. 12: GFK-GI-GMR Fachtagung Prozessrechner 1974. Karlsruhe, 10.–11. Juni 1974. Herausgegeben von G. Krüger und R. Friehmelt. XI, 620 Seiten. 1974.

Vol. 13: Rechnerstrukturen und Betriebsprogrammierung, Erlangen, 1970. (GI-Gesellschaft für Informatik e.V.) Herausgegeben von W. Händler und P. P. Spies. VII, 333 Seiten. 1974.

Vol. 14: Automata, Languages and Programming – 2nd Colloquium, University of Saarbrücken, July 29–August 2, 1974. Edited by J. Loeckx. VIII, 611 pages. 1974.

Vol. 15: L Systems. Edited by A. Salomaa and G. Rozenberg. VI, 338 pages. 1974.

Vol. 16: Operating Systems, International Symposium, Rocquencourt 1974. Edited by E. Gelenbe and C. Kaiser. VIII, 310 pages. 1974.

Vol. 17: Rechner-Gestützter Unterricht RGU '74, Fachtagung, Hamburg, 12.–14. August 1974, ACU-Arbeitskreis Computer-Unterstützter Unterricht. Herausgegeben im Auftrag der GI von K. Brunnstein, K. Haefner und W. Händler. X, 417 Seiten. 1974.

Vol. 18: K. Jensen and N. Wirth, PASCAL – User Manual and Report. VI, 167 pages. 2nd Edition 1975.

Vol. 19: Programming Symposium. Proceeding, Colloque sur la Programmation, Paris, April 9–11, 1974. V, 425 pages. 1974.

Vol. 20: J. Engelfriet, Simple Program Schemes and Formal Languages. VII, 254 pages. 1974.

Vol. 21: Compiler Construction, An Advanced Course. Edited by F. L. Bauer and J. Eickel. XIV. 621 pages. 1974.

Vol. 22: Formal Aspects of Cognitive Processes. Proceedings, Interdisciplinary Conference, Ann Arbor, March 1972. Edited by T. Storer and D. Winter. V, 214 pages. 1975.

Vol. 23: Programming Methodology. 4th Informatik Symposium, IBM Germany Wildbad, September 25–27, 1974. Edited by C. E. Hackl. VI, 501 pages. 1975.

Vol. 24: Parallel Processing. Proceedings of the Sagamore Computer Conference, August 20–23, 1974. Edited by T. Feng. VI, 433 pages. 1975.

Vol. 25: Category Theory Applied to Computation and Control. Proceedings of the First International Symposium, San Francisco, February 25–26, 1974. Edited by E. G. Manes. X, 245 pages. 1975.

Vol. 26: GI-4. Jahrestagung, Berlin, 9.–12. Oktober 1974. Herausgegeben im Auftrag der GI von D. Siefkes. IX, 748 Seiten. 1975.

Vol. 27: Optimization Techniques. IFIP Technical Conference. Novosibirsk, July 1–7, 1974. (Series: I.F.I.P. TC7 Optimization Conferences.) Edited by G. I. Marchuk. VIII, 507 pages. 1975.

Vol. 28: Mathematical Foundations of Computer Science. 3rd Symposium at Jadwisin near Warsaw, June 17–22, 1974. Edited by A Blikle. VIII, 484 pages. 1975.

Vol. 29: Interval Mathematics. Proceedings of the International Symposium, Karlsruhe, West Germany, May 20–24, 1975. Edited by K. Nickel. VI, 331 pages. 1975.

Vol. 30: Software Engineering. An Advanced Course. Edited by F. L. Bauer. (Formerly published 1973 as Lecture Notes in Economics and Mathematical Systems, Vol. 81) XII, 545 pages. 1975.

Vol. 31: S. H. Fuller, Analysis of Drum and Disk Storage Units. IX, 283 pages. 1975.

Vol. 32: Mathematical Foundations of Computer Science 1975. Proceedings 1975. Edited by J. Bečvář. X, 476 pages. 1975.

Vol. 33: Automata Theory and Formal Languages, Kaiserslautern, May 20–23, 1975. Herausgegeben im Auftrag der GI von H. Brakhage. II, 292 Seiten. 1975.

Vol. 34: GI – 5. Jahrestagung, Dortmund 8.–10. Oktober 1975. Herausgegeben im Auftrag der GI von J. Mühlbacher. X, 755 Seiten. 1975.

Vol. 35: W. Everling, Exercises in Computer Systems Analysis. (Formerly published 1972 as Lecture Notes in Economics and Mathematical Systems, Vol. 65) VIII, 184 pages. 1975.

Vol. 36: S. A. Greibach, Theory of Program Structures: Schemes, Semantics, Verification. XVIII, 364 pages. 1975.

Vol. 37: C. Böhm, λ-Calculus and Computer Science Theorie, Proceedings of the Symposium Held in Rome. March 25–27. XII, 370 pages. 1975.

Lecture Notes in Computer Science

Edited by G. Goos and J. Hartmanis

37

λ-Calculus and Computer Science Theory

Proceedings of the Symposium Held in Rome
March 25–27, 1975

IAC – CNR
Istituto per le Applicazioni del Calcolo
"Mauro Picone"
of the Consiglio Nazionale delle Ricerche

Edited by C. Böhm

Springer-Verlag Berlin Heidelberg GmbH 1975

Editor
Prof. C. Böhm
Università di Roma
Istituto Matematico
"Guido Castelnuovo"
00185 Roma/Italia

Library of Congress Cataloging in Publication Data
Main entry under title:

λ[i. e. Lambda]=calculus and computer science theory.

(Lecture notes in computer science ; 37)
English or French.
Bibliography: p.
Includes index.
1. Combinatory logic--Congresses. 2. Programming languages (Electronic computers)--Congresses. I. Böhm, Corrado, 1923- II. Istituto per le applicazioni del calcolo. III. Series.
QA9.5.L35 511'.3 75-33375

AMS Subject Classifications (1970): 00-02, 00A10, 02C20, 02C99, 68A05
CR Subject Classifications (1974): 5.21

ISBN 978-3-540-07416-8 ISBN 978-3-540-37944-7 (eBook)
DOI 10.1007/978-3-540-37944-7

Originally published by Springer-Verlag Berlin Heidelberg New York in 1975.

Offsetdruck: Julius Beltz, Hemsbach/Bergstr.

Dedicated to
Alonzo Church
Haskell B. Curry
Frederic B. Fitch

This Symposium is organised by IAC-CNR under the sponsorship of the Consiglio Nazio nale delle Ricerche, Associazione Italiana di Calcolo Automatico, European Association for Theoretical Computer Science.

Organizers

C. Böhm

I. Galligani

TABLE OF CONTENTS

INTRODUCTION

This volume may be considered as a first attempt at synthesizing the multiple relationships between the λ-Calculus and Computer Science. The volume arose from a Symposium on λ-Calculus and Computer Science Theory that was held in Rome, Italy, March 25-27, 1975, at the Consiglio Nazionale delle Ricerche, organized by the Istituto per le Applicazioni del Calcolo "Mauro Picone".

The first idea of this rather special symposium was born during a session of the European Association for Theoretical Computer Science. The main topics of the Symposium were: 1. λ-calculus models and semantics of programming languages; 2. the Church-Rosser theorem and its applications; 3. algorithms of the metatheory; 4. applicative terms as models of computation; 5. applications of typed λ-calculus.

Other topics treated were algebraic models of computations and relational caculi.

The Symposium was sponsored by the European Association for Theoretical Computer Science, the Consiglio Nazionale delle Ricerche and the Associazione Italiana per il Calcolo Automatico. The organizing committee for the Symposium consisted of C. Böhm, University of Rome and I. Galligani, Director of the Istituto per le Applicazioni del Calcolo "Mauro Picone" (IAC), Rome. The scientific committee consisted of: G. Ausiello, CSSCCA, Rome; H. Barendregt, University of Utrecht (Holland); C. Böhm, University of Rome; M. Dezani Ciancaglini, University of Torino (Italy); E. Engeler, Eidg. Techn. Hochschule, Zürich (Switzerland); G. Huet, IRIA (France); M. Nivat, University of Paris (France); M. Hyland, University of Oxford (Great Britain); D. Scott, University of Oxford (Great Britain); M. Venturini Zilli, IAC, Rome.

Coordination was cared for by M. Dezani Ciancaglini, University of Torino and M. Venturini Zilli, IAC, Rome.

Seventeen papers were presented and discussed at the Conference; three papers were not presented orally but are included in this volume. It was impossible to include the paper by K. Indermark in this volume. The few words that follow are intended to orient the reader who may be familiar with general aspects of the λ-Calculus and Computer Science Theory but not particularly with the specific work of the participants at the Symposium.

In his introductory talk D. Scott defines the nature of the relation between class abstraction and λ-abstraction in models for the λ-Calculus and gives a short critical analysis of the present state of foundational research concerning Combinatory Logic and its relationship with Predicate Logic.

The problems of least fixed points and of the semantics of programming languages are particularly stressed in De Bakker's paper in which it is stated, contrary to the application by Manna of Scott's theoretical results, that both call-by-value and call-by-name mechanisms are strictly related to the least fixed points.

In the context of Scott's work R. Nakajima introduces a generalization of normal forms considered in the framework of infinite λ-expressions.

The aim of M. Hyland's paper is to stress that, generally speaking, a partial order relation on terms of the λ-calculus possesses a characterization in terms of its computational significance and of contexts. A particular theorem is proved by H. Egli relating the meaning of typed and type-free terms in corresponding λ-calculus models over complete partially ordered sets.

The relationship between typed models and type-free models is investigated also in V. Sazonov's paper in which the author shows a particular definition of sequential and parallel functionals for these kinds of models. L. Aiello and M. Aiello seek to describe a reasonable semantics for a programming language in the context of a typed λ-calculus environment. A particular interpretation of λ-calculus in terms of a certain collection of algorithms is presented by L. Nolin through the construction of certain models very near to Scott's classical models and to the URS (Uniformly Reflexive Structures).

In the framework of algebraic languages A. Dubinsky's paper aims at defining the strict connection between the generalized automata theory and the algebraic theory of computation.

G. Ausiello's work is also related to the automata field and rewriting systems. He examines how λ-calculus can be used to describe the behavior of time-varying systems" in particular the problem of synchronization and the link with the developmental languages are analyzed.

On the other hand R. de Vrijer affirms that the abstract term system λλ presented in his paper is strictly connected to the Automath family of languages and may be considered as a simple generalization of AUT-QE.

In relation to the problems of the mechanization of first and second order theories A. Huet presents a unification algorithm for typed λ-calculus; in particular an algorithm which searches for the existence of unifiers in ω-order logic, and gives a proof of its correctness.

Another central point of interest is represented by the analysis of the URS considered as an interesting and clear axiomatization of recursion theory. H. Barendregt shows how to introduce the use of length of computation in the URS by means of the construction of a Normed Uniformly Reflexive Structures that permit us to overcome some of the defects of standard URS and rediscover some of Moschovakis' results concerning the length of computation in recursion theory.

Moreover, some important properties concerning nonterminating computations and terminating ones are presented in M. Venturini's work in which the problem of the shift from the former to the latter is analyzed.

The idea of analyzing the properties of the complexity of computations in Combinatory Logic is at the base of the work of C. Batini and A. Pettorossi who introduce various notions of computational resource and who seek to characterize levels of subrecursiveness in Combinatory Logic. A particular definition of the concept of "subbase" is given and some results about the generative and computational power of subbases are shown.

In this manner these last papers stress the particular relationship between Combinatory Logic and Recursion Theory.

Lastly some aspects of the pure λ-calculus are revisited in order to extract some computational properties or some technical results.

So J.J. Lévy aims to prove by means of introduction of a labelled λ-calculus Welch's conjecture about the completeness, in the reducibility sense, of inside-out reductions. Then the general problem of inside-out reductions and continuous semantics is examined in P.H. Welch's paper in the light of the consideration that the natural meaning of an expression is just the union of "instant meanings".

The work of C. Böhm and M. Dezani Ciancaglini presents a particular partition of the set of λ-terms in $2\omega+1$ classes considered as a natural kind of type introduction in which every λ-term possesses a unique type. This classification can be effectively determined only for λ-terms in normal form.

From another point of view the purpose of G. Jacopini is to define a necessary and sufficient condition by which two combinators can be identified without introducing contradictions with the axioms of

Combinatory Logic.

J.W. Klop gives an easier version of the main lemma of a well-known theorem by H. Barendregt concerning the solvability of λ-I-terms.

In his final talk D. Scott outlines an informal survey of the wide variety of points of view issuing from the papers presented at the Symposium, in an attempt to establish, on a philosophical basis, some common measure of critical comparison without any pretence to furnish a definitive philosophy of combinators.

At the end of the volume Barendregt presents, as a stimulating challenge, a set of open problems with his critical comments.

Particular thanks are due to A. Faedo, President of the Consiglio Nazionale delle Ricerche, for providing assistance which contributed in large measure to the realization of the meeting. Special thanks go to I. Galligani, Director of the IAC, which provided the financial support, the clerical staff and all other organizing facilities.

I would also like to thank the members of the Scientific Committee for their help in the management of the congress. In particular I would like to express my gratitude to M. Venturini Zilli for her invaluable help and intelligent suggestions.

Corrado Böhm

Roma, June 1975

COMBINATORS AND CLASSES

by

Dana Scott

Oxford University

Abstract. The paper tries to answer the question: What is the relation between class abstraction and λ-abstraction in models for the λ-calculus?

Introduction. It seems fair to say that one of the original motivations for the study of λ-calculus and the combinators was to give a *foundation* for logic. The pure combinators were to be very general operators which would provide an analysis of "substitution" and the behaviour of variables. (An all-pervading confusion between use and mention haunts the literature on λ-calculus. Perhaps it would make more sense to say that the purpose of the combinators is to analyze the notion of functional *dependence* by producing a few basic combinators from which the others could be explicitly defined. And, besides this expressive power, it was also necessary to uncover the laws - usually identities - that hold among the combinators.) Adjoined to these general operators, we were then to have the quantifiers and logical connectives to be used in the analysis of propositions. All of this was to turn out to be a Fregian paradise of "type-free" functions. Alas, the first workers hardly had time to savour the forbidden fruit before they were turned out of this paradise by the discovery of the usual paradoxes. A cruel fate, but in retrospect, it is not a very surprising one.

In one way or the other we have all been trying to get back into this lost paradise, though some have been content to pause along the way to play with the non-paradoxical, pure λ-calculus part of the subject. It is a pity, however, to have a study without real motivation; and without the propositional component of the theory it is not all that easy to explain the interest in the combinators. This is not to say that they are not fun; but aside from formal amusement, what is the point? It is no good pointing to combinatory arithmetic, because there are far easier ways of explaining recursive number-theoretic functions.

There are two good answers which establish the mathematical value of the combinators. The first was given by Kleene when he defined in number theory $\{e\}(n)$, which certainly is a kind of type-free application. (We should also note that Kleene also defined $\{\alpha\}(\beta)$ for arbitrary number-theoretic functions α and β (see the Kleene-Vesley book). This application too has an interesting theory.) Kleene introduced many techniques from λ-calculus into recursive function theory by defining whatever he needed in terms of Gödel numbers. More recently we speak of the URS of Wagner-Strong which gives an abstract version of Kleene's idea (see the paper in this volume by Barendregt). But these structures have not been adequately related to "traditional" λ-calculus, because a URS is only a <u>partial</u> algebra in that the application $\{e\}(n)$ is not always defined.

Models in which application is always meaningful were discovered by the present author. The first were the D_∞ models, and the second kind (building on an idea of Plotkin) were the graph models, like $P\omega$. We shall not need any detail about the construction of such models in this paper except to note that they all satisfy these basic axioms:

$$(\alpha) \qquad \lambda x.\tau = \lambda y.\tau[y/x]$$

$$(\beta) \qquad (\lambda x.\tau)(y) = \tau[y/x]$$

$$(\xi) \qquad \lambda x.\tau = \lambda x.\sigma \leftrightarrow \forall x.\tau = \sigma \quad .$$

We can call these the <u>axioms of extensional</u> λ-<u>calculus</u>. The fact that the specific models mentioned above satisfy much more than what is implied by (α), (β), (ξ) is not relevant here, as everything we want to say can be done in the context of this rather weak system. (That term models for these axioms were already known as a consequence of the Church-Rosser Theorem is also irrelevant for the reasons pointed out in the author's other paper in this volume.)

Before going further, we should stress that the kind of truth definition that we shall propose in this paper could also be done for the URS. We stick to λ-calculus of the traditional sort, because it is more familiar and it illustrates a general idea neatly.

What was learned from the construction of the λ-calculus models is that there are many coherent notions of function and functional application which admit the closure properties demanded by the combinators. In all of these examples paradox is avoided because the functions used are of a <u>limited</u> sort and are far from arbitrary (e.g. the space of <u>continuous</u> functions rather than the full function space in the usual sense of set theory). Conflict is resolved by eliminating operators of an infinitary nature (like the quantifiers!) The trick was to retain enough to have the (pure) combinators. It was possible; and it was good for λ-calculus, but bad for logic, because it was just the propositional notions that had been eliminated. There was a gain for recursion theory, since the extensional combinators are less messy than Kleene's and ideas could be applied more abstractly. But still the original motivation was definitely not regained.

The one person who had climbed farthest back into "paradise" was Fitch. For one reason or the other his ideas are not very well known, partly because his presentations are highly formal and rather complicated. There is no claim here to have correctly interpreted his program, but his method was the direct inspiration. In somewhat different contexts both Feferman and Aczel have used the plan to advantage and have found connections with other theories, but their papers are not yet published. (As the method is very closely related to iterated truth definitions similar to Kleene's hyperarithmetic hierarchy, they may not feel that their inspiration can be traced directly to Fitch, though Feferman comments on Fitch's system and indicates how his own has advantages.) Earlier, in work with Fitch, Myhill had pusued the idea and mentioned it in conversation. The author at this point cannot remember whether Myhill came to any definite conclusions and cannot recall what he published about it. But the problem of priority is not very acute: Fitch started in the mid-thirties and gets the main credit. There are many variations, and we have to try to judge which of the roads back into paradise (if any!)are not a dead ends.Maybe they all are. The purpose of this paper is to encourage more exploration.

It comes to mind that Chuch mentions in his little booklet on λ-calculus that there were some lecture notes on a hierarchy of quantifiers to be added to his system. These notes were not published; and though he spent some time at Princeton, the author does not remember seeing them and does not know whether they are relevant to the present discussion. By the time Church published his monograph one feels that he had lost interest in giving any foundation *via* λ-calculus of the untyped kind. No one seems to have tried to follow up Church's ideas. As all the principals are still alive (except for Turing, of course), someone should perhaps do some historical investigation. It is not always so easy to deduce from the writings what was intended. In particular the recent publication by Fitch of "Elements of Combinatory Logic" is very disappointing in that he did not try to make a uniform exposition of his papers in the JSL which contain the details of the consistency proof (truth definition). He proposed over the years several different systems, and we could have hoped to see a complete, final version.

One conclusion we might reach in this paper is that the terminology "combinatory logic" is still premature dispite all the works of Curry and Fitch. We shall certainly establish connections with the usual kind of predicate logic, but it seems to this author that much remains to be done to determine whether these are the right connections or even especially useful ones. And the question of whether we have foundations in this way should also remain open.

§ 1. Syntax. In the background we are assuming that we have any non-trivial model for (α), (β), (ξ). "Non-trivial" here means a domain of individuals (or objects) of at least two elements. The axioms are all schematic, and the terms used (the τ and σ) are just pure λ-terms built up from variables by application and λ-abstraction.

We shall have to make a very conscious effort to avoid confusion between use and mention because we are going to formalize the syntax in the model. We shall not be quite as rigorous as Quine in keeping the distinction, but we shall be rigorous enough. The point is that certain objects of the model are going to represent formulas somewhat in the style of Gödel numbers. As we have an abstract model, however, we do not speak of its elements as numbers. Nevertheless, certain elements can be chosen to represent numbers, and we regard 0,1,2,3,4 as distinct elements of our model. (The choice of representation is not important. This is a standard construction as in the Curry volumes.)

Aside from the numbers we need combinators to form tuples like $\langle a,b,c \rangle$. Again, just how the tuple is defined is not important, though to save notation we will assume that a construction is effected whereby:

$$\langle a,b,c \rangle (0) = a \ ,$$

$$\langle a,b,c \rangle (1) = b \ ,$$

$$\langle a,b,c \rangle (2) = c \ ,$$

and similarly for other size tuples. We shall often write "u_k" for "u(k)", especially for numerical subscripts.

<u>Definition</u>. The <u>primitive</u> formula constructs are represented as follows:

$$a = b \quad = \quad \langle 0,a,b \rangle$$

$$\forall x . \varphi \quad = \quad \langle 1,\lambda x . \varphi \rangle$$

$$\sim \varphi \quad = \quad \langle 2,\varphi \rangle$$

$$\varphi \wedge \psi \quad = \quad \langle 3,\varphi,\psi \rangle$$

$$\varphi \times\!\!\times \psi \quad = \quad \langle 4,\varphi,\psi \rangle$$

We have tried to choose as few primitives as possible without being incomprehensible. Still we need various defined operations as well in order to make formulas look normal and familiar.

<u>Definition</u>. The <u>defined</u> formula constructs are represented as follows:

$$a \in b \quad = \quad b(a)$$

$$\exists x . \varphi \quad = \quad \sim \forall x . \sim \varphi$$

$$\varphi \vee \psi \quad = \quad \sim [\sim \varphi \wedge \sim \psi]$$

$$\varphi \rightarrow \psi \quad = \quad \sim \varphi \vee \psi$$

$$\varphi \leftrightarrow \psi \quad = \quad [\varphi \wedge \psi] \vee [\sim \varphi \wedge \sim \psi]$$

$$T \quad = \quad 0 = 0$$

$$F \quad = \quad \sim T$$

$$* \quad = \quad T \times\!\!\times F$$

By way of example consider the formula:

$$\exists x . \ a(x) = x \ .$$

In the model this is represented by the element:

$$< 2,<1,\lambda x.<2,<0,a(x),x >>>> .$$

A little odd looking, but it is a perfectly good λ-term. (As we know, this is a _true_ formula; but we do not get the truth definition until the next section.)

The point to keep in mind here is that we are doing _syntax_ (except for the case of $a \in b$, but more of that later). At the moment we are concerned with _form_ and not with meaning. What we have done is to make it possible to assign to every logical formula (containing possibly _constants_ for elements of the model) a λ-term which, of course, denotes an element of the model. This element represents the formula itself, not its interpretation. And, as long as we do not use the ϵ-notation, the representation of formulas by elements is _unique_, because the whole system is based on tuples. (If we wanted full uniqueness, we could make the ϵ-combination a primitive - say $<5,a,b>$ - and save the $b(a)$-part for the truth definition.) We have not tried to respect the distinction between use and mention by going to the metalanguage and then defining a _mapping_ from formulas to objects. Instead we have shown the effect of the mapping by regarding the logical connectives as combinators. But note - and this is very important - the approach is intensional. Our definition contributes almost nothing toward meaning.

A smaller point we should keep in mind is that we are not saying _which_ elements represent formulas. All the definition does is to give laws of composition whereby new formulas can be obtained from old. Since $a = b$ is at once a formula (better: an element that _represents_ a formula), we have something to start with. We just have no need to say how far we want to _iterate_ formula construction. Whatever they are, they are elements of the model; and we shall find we do not need to be more definite than that.

Part of the trick of the definition (aside from the obvious use of tuples to give us an "abstract" syntax) is the use of λ-abstraction in the quantifier. It would not be unreasonable to say that the formulas represented in the model are those without free variables. A formula with free variables could be taken as the corresponding λ-abstract. Or if we like: a formula with a free variable is a mapping from constants to the corresponding substitution instance. In this way we eliminate all fuss with variables in the formalization - the use of λ does all the work behind the scenes. This may not be quite the idea that Curry had in mind in the begining of the subject, but it does not seem like such a bad idea. (Thus, if we choose to regard $\langle 1,u \rangle$ as a quantified formula, then for any constant a , the element u(a) is the corresponding "substitution" instance.) At least we can say we are putting λ-calculus to work.

§ 2. Semantics. We can now pass on to the truth definition, but we shall find it is a transfinite one. This is something that Fitch has never made especially clear in his writings, but the method is actually very well known in logic. (For references see the recent book by Moschovakis.) We need only take a little care to formulate the definition in such a way that we will know - on the grounds of general principles - that the truth predicate actually exists. To do this it is helpful to define the predicate of falsehood at the same time as truth in a mutually recursive way. This will be found to be only a very slight complication in concept which makes it easier to formulate the clauses of the definition. We will see that the definition is very much like that of first-order truth in model theory, but since we made syntax part of the model there is extra "feed back" that drives us to transfinite lengths. The important question will be: how can we make use of this feed back? It is not likely that we can give a very convincing answer at once.

<u>Definition</u>. The subsets $\mathfrak{T}$ and $\mathfrak{F}$ are the least subsets of the model such that the following equivalences hold for all elements u :

$$\begin{array}{llll}
\mathfrak{T}u & \text{iff either} & u = \langle 0,u_1,u_2\rangle & \text{and } u_1 = u_2 \\
 & \text{or} & u = \langle 1,u_1\rangle & \text{and } \mathfrak{T}u_1(x) \text{ for all } x \\
 & \text{or} & u = \langle 2,u_1\rangle & \text{and } \mathfrak{F}u_1 \\
 & \text{or} & u = \langle 3,u_1,u_2\rangle & \text{and } \mathfrak{T}u_1 \text{ and } \mathfrak{T}u_2 \\
 & \text{or} & u = \langle 4,u_1,u_2\rangle & \text{and } \mathfrak{T}u_1 \text{ and } \mathfrak{T}u_2
\end{array}$$

$$\begin{array}{llll}
\mathfrak{F}u & \text{iff either} & u = \langle 0,u_1,u_2\rangle & \text{and } u_1 \neq u_2 \\
 & \text{or} & u = \langle 1,u_1\rangle & \text{and } \mathfrak{F}u_1(x) \text{ for some } x \\
 & \text{or} & u = \langle 2,u_1\rangle & \text{and } \mathfrak{T}u_1 \\
 & \text{or} & u = \langle 3,u_1,u_2\rangle & \text{and } \mathfrak{F}u_1 \text{ or } \mathfrak{F}u_2 \\
 & \text{or} & u = \langle 4,u_1,u_2\rangle & \text{and } \mathfrak{F}u_1 \text{ and } \mathfrak{F}u_2
\end{array}$$

Since we went to all the trouble to formalize the syntax <u>within</u> the model, the reader will notice that we have relapsed into English in the metalanguage. We have written "$\mathfrak{T}u$" to mean the same as "u belongs to the subset $\mathfrak{T}$". We will want te read "$\mathfrak{T}u$" as "u is <u>true</u>", but first we should see why the definition is a proper one.

We remark first that negation is never applied to the predicates $\mathfrak{T}$ and $\mathfrak{F}$ in the definition. So, by transfinite recursion, let us begin with $\mathfrak{T}_0$ and $\mathfrak{F}_0$ as the <u>empty</u> subsets. At each ordinal stage α in the recursion, put $\mathfrak{T}_\alpha$ and $\mathfrak{F}_\alpha$ on the right hand sides of the above, thereby defining by these equivalences new predicates $\mathfrak{T}_{\alpha+1}$ and $\mathfrak{F}_{\alpha+1}$. At a limit stage we take unions:

$$\mathfrak{T}_\alpha = \bigcup_{\beta<\alpha} \mathfrak{T}_\beta$$

$$\mathfrak{F}_\alpha = \bigcup_{\beta<\alpha} \mathfrak{F}_\beta$$

Because negation was avoided, we can prove by transfinite induction that if $\beta < \alpha$, then $\mathfrak{T}_\beta \subseteq \mathfrak{T}_\alpha$ and $\mathfrak{F}_\beta \subseteq \mathfrak{F}_\alpha$; that is, we have two chains of subsets. But the model is a <u>set</u> (that is, of limited cardinality), thus there must exist a stage α where the definition "closes off" in the sense that:

$$\mathfrak{T}_{\alpha+1} = \mathfrak{T}_\alpha \quad \text{and} \quad \mathfrak{F}_{\alpha+1} = \mathfrak{F}_\alpha \ .$$

These are the desired predicates $\mathfrak{T}$ and $\mathfrak{F}$, and we call the least such α the <u>ordinal</u> of the model. (Question: Which ordinals are ordinals of models of λ-calculus?).

We also note that thanks to the separation of cases (by 0,1,2,3,4) and by the careful choice of the clauses, we can prove by transfinite induction that $\mathfrak{T}_\alpha$ and $\mathfrak{F}_\alpha$ are always <u>disjoint</u>. This means that no formula (element of the model) can be <u>both</u> true and false at the same time.

We note, too, that if we restrict attention to the usual first-order formulas (that is, start with equations between λ-terms, and then use only $\forall, \sim, \wedge$), then the definition is exactly the usual truth definition - word for word, except for our use of the formalized syntax. To check these points it is helpful to separate out cases making use of a readable notation:

<u>Lemma</u>. The following equivalences hold:

$\mathfrak{T}a = b$	iff	$a = b$	$\mathfrak{F}a = b$	iff	$a \neq b$
$\mathfrak{T}a \in b$	iff	$\mathfrak{T}b(a)$	$\mathfrak{F}a \in b$	iff	$\mathfrak{F}b(a)$
$\mathfrak{T}\forall x.\varphi$	iff	$\mathfrak{T}\varphi[a/x]$ all a	$\mathfrak{F}\forall x.\varphi$	iff	$\mathfrak{F}\varphi[a/x]$ some a

$\mathfrak{T}\exists x.\varphi$	iff	$\mathfrak{T}\varphi[a/x]$			$\mathfrak{F}\exists x.\varphi$	iff	$\mathfrak{F}\varphi[a/x]$		
		some a					all a		
$\mathfrak{T} \sim \varphi$	iff	$\mathfrak{F}\varphi$			$\mathfrak{F} \sim \varphi$	iff	$\mathfrak{T}\varphi$		
$\mathfrak{T}\varphi \wedge \psi$	iff	$\mathfrak{T}\varphi$	and	$\mathfrak{T}\psi$	$\mathfrak{F}\varphi \wedge \psi$	iff	$\mathfrak{F}\varphi$	or	$\mathfrak{F}\psi$
$\mathfrak{T}\varphi \vee \psi$	iff	$\mathfrak{T}\varphi$	or	$\mathfrak{T}\psi$	$\mathfrak{F}\varphi \vee \psi$	iff	$\mathfrak{F}\varphi$	and	$\mathfrak{F}\psi$
$\mathfrak{T}\varphi \times\!\!\times \psi$	iff	$\mathfrak{T}\varphi$	and	$\mathfrak{T}\psi$	$\mathfrak{F}\varphi \times\!\!\times \psi$	iff	$\mathfrak{F}\varphi$	and	$\mathfrak{F}\psi$

We do not bother to include the other connectives, except to remark that T is true, F is false, and * is <u>neither</u>. Thus some formulas <u>lack</u> truth values.

This last remark seems very arbitrary indeed. Why put in that stupid connective $\times\!\!\times$ at all? There is an answer to this, but it was not just to find truth-value gaps. These are forced on us in any case by the Russel Paradox. Consider the abstract:

$$r = \lambda x.\ \sim x \in x$$

We ask as usual for the truth value of $r \in r$. If it is true, it is false; if false, then true. But true and false are exclusive; hence, $r \in r$ has no truth value.

Is this really the Russell Paradox? (Actually there is no paradox unless you thought every likely looking formula should have a truth value.) The answer is <u>yes</u>, because for any abstract we have:

$$\mathfrak{T}a \in \lambda x.\varphi \quad \text{iff} \quad \mathfrak{T}\varphi[a/x]$$
$$\mathfrak{F}a \in \lambda x.\varphi \quad \text{iff} \quad \mathfrak{F}\varphi[a/x]\ .$$

That is to say, λ-abstraction in conjunction with the truth definition works just like <u>class</u> abstraction. All we have to do is to interpret membership by functional application. This is something we have always wanted to do clear back to the genesis of the subject when Schoenfinkel created the combinators. We have now done it, but at a certain loss of innocence.

The price we have had to pay is the tax of intensionality. When we write:

$$\lambda x.\varphi = \lambda x.\psi \quad ,$$

we <u>cannot</u> mean that the two classes are extensionally equivalent. The way the truth definition is set up, what we mean is, for each constant a , that $\varphi[a/x]$ is the <u>same formula</u> as $\psi[a/x]$. This is a much stricter relationship than that we thought we were promised at the base of the tree of knowledge. Is the price worth the outcome? Fitch certainly thinks so, but one must say that in his book he does not explain all that well just what rules of extensionality he really wants. Maybe he will find the choice made here <u>too strict</u>; but if so, the alternatives will have to be spelled out more simply.

Since we see now as a consequence of allowing self-application that we are stuck with truth-value gaps, we realize that the "logic" of this truth definition is <u>three valued</u>. Using the symbols to stand for the truth "values", we have these tables:

~	
T	F
*	*
F	T

∧	T	*	F
T	T	*	F
*	*	*	F
F	F	F	F

∨	T	*	F
T	T	T	T
*	T	*	*
F	T	*	F

⋈	T	*	F
T	T	*	*
*	*	*	*
F	*	*	F

We of course read "*" as "having no value". And now we can see the reason for the notation "⋈": this connective gives the common part of the <u>and</u> and the <u>or</u>. Since "∨" is often called the wedge , we call "⋈" the <u>squadge</u>, since it is formed by squeezing symbols together making an ideogram. What is the point of this new connective?

The idea of ⋈ may well have been thought of before, but it was discovered as necessary by Stephen Blamey of Oxford in his study of presuppositions and truth-value gaps. Without going into the details of Blamey's motivation, we can say that considering the truth values as partially ordered in this natural pattern:

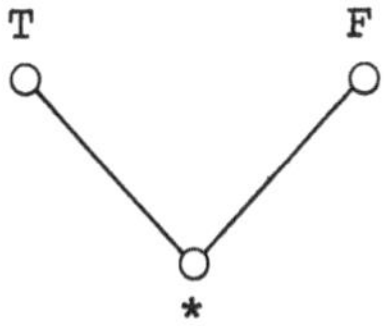

Then the xx-connective is just what you need to adjoin to $\sim, \wedge$ and $\vee$ to have a system <u>complete</u> for defining all <u>monotone</u> three-valued propositional operators (that is, monotone in the partial ordering). These are the connectives whose $\mathfrak{T}$-$\mathfrak{F}$ conditions can be defined <u>positively</u> in terms of $\mathfrak{T}$ and $\mathfrak{F}$. We could use no others.

(We might remark that if this method were applied to a URS, then in the clauses for $a = b$ we would only assign a truth value when <u>both</u> sides are defined, and in the quantifier clause the variable should range over just the defined elements. It does not seem reasonable to this author to treat the undefined element of a URS as a real element. It seems especially unnatural to let it enter into decidable equations (equations with truth values).)

A good question to ask is whether we need any other primitive quantifiers besides $\forall$, but the author does not have enough experience with the model theory of this three-valued logic to answer the question.

<u>§ 3. Proof Theory.</u> The sketch here will be quite brief; for the case of pure (three-valued) first-order logic, more details will be found in Blamey's Oxford thesis. It is useful to see something of what is achieved by the truth definition, however, in order to evaluate further the claims that there are connections between λ-calculus and logic.

Definition. The consequence relation $\Gamma \vdash \Delta$, which stands between two subsets of the model, holds if and only if we have these two conditions:

$$\text{(i)} \quad \text{if} \quad \Gamma \subseteq \mathfrak{T} \;, \quad \text{then} \quad \Delta \cap \mathfrak{T} \neq \emptyset \; ; \quad \text{and}$$

$$\text{(ii)} \quad \text{if} \quad \Delta \subseteq \mathfrak{F} \;, \quad \text{then} \quad \Gamma \cap \mathfrak{F} \neq \emptyset \; .$$

We have taken care to make the definition of consequence symmetric in the true and the false. In words $\Gamma \vdash \Delta$ means that whenever all formulas in Γ are true, then at least one in Δ must be true also; and whenever all formulas in Δ are false, then at least one in Γ is false also. In two-valued logic condition (ii) follows from (i), but in the three-valued case we must be more explicit. The definition was formulated for arbitrary subsets, but in the sequel we consider only finite Γ and Δ. In expressing various laws, we use a comma to indicate union of sets:

$$\Gamma, \Gamma' = \Gamma \cup \Gamma' \quad \text{and}$$

$$\Gamma, \varphi = \Gamma \cup \{\varphi\} \; .$$

We also write: $\sim \Gamma = \{\sim \varphi \mid \varphi \in \Gamma\}$

The Laws of Consequence. The following general principles hold for all subsets:

$$\text{(R)} \quad \Gamma \vdash \Delta \quad \text{provided} \quad \Gamma \cap \Delta \neq \emptyset$$

$$\text{(M)} \quad \frac{\Gamma \vdash \Delta}{\Gamma, \Gamma' \vdash \Delta, \Delta'}$$

$$\text{(T)} \quad \frac{\Gamma, \varphi \vdash \Delta \qquad \Gamma \vdash \varphi, \Delta}{\Gamma \vdash \Delta}$$

The consequence relation is a kind of multiary partial ordering, and these general laws are the reflexive, monotone, and transitive properties of $\vdash$. In logic the transitive law (T) is usually called the <u>cut rule</u>. It is a general form of <u>modus ponens</u>.

<u>The Laws of the Truth Values</u>.

$$(T * F) \quad \vdash T \qquad F \vdash \qquad * \vdash \sim * \qquad \sim * \vdash *$$

We want T to be true, F to be false, and $*$ to have no value. We have not quite said the latter; rather the law gives $*$ the same truth value as its negation. We could write this more shortly as:

$$* \dashv\vdash \sim *$$

as an abbreviation for the two laws. After we exhibit the laws of negation we will discuss how $*$ fails to have a truth value.

<u>The Laws of Negation</u>.

$$(\sim) \quad \frac{\Gamma \vdash \Delta}{\sim\Delta \vdash \sim\Gamma} \qquad \varphi \dashv\vdash \sim\sim \varphi \qquad \varphi, \sim \varphi \vdash *$$

The first principle captures part of the symmetry between the true and the false; while the second, the law of double negation, provides the rest. In particular, the two together - with the aid of cut - give:

$$\frac{\sim \Gamma \vdash \sim \Delta}{\Delta \vdash \Gamma}$$

and other obvious variations. The last law is that non-contradiction: if φ and $\sim \varphi$ were both true, then something without a truth value would have one. But why does $*$ lack a truth value?

The answer to the question is that $*$ could indeed be given a truth value if we would want to be perverse, but there is no need to be so. In giving a completeness proof for (a suitable portion of) our logic, we would consider the situation where $\Gamma \vdash \Delta$ was not derivable. By cut, either $\Gamma \vdash *, \Delta$ or $\Gamma, * \vdash \Delta$ would fail. In the latter case by all the various laws involving negation, we would equivalently say that $\sim \Delta \vdash *, \sim \Gamma$ fails; and this is now parallel to $\Gamma \vdash *, \Delta$.

When an entailment fails in such a formal system, we can usually argue that there is a <u>total</u> valuation v defined on formulas where all on the left of the failing $\vdash$ have the value T, while all on the right have the value F. In either case above we are making sure that $v(*) = F$. Now this equation <u>does not</u> mean that $*$ has the truth value <u>false</u>; rather it means that it is <u>not</u> true. The truth sets corresponding to v are defined by:

$$\mathfrak{T}_v = \{\varphi \mid v(\varphi) = T\}$$

$$\mathfrak{F}_v = \{\varphi \mid v(\sim\varphi) = T\}$$

Now we can see the reasoning behind the rules: we want $\mathfrak{T}_v$ and $\mathfrak{F}_v$ to have the biconditionals of a truth definition as done in the previous section. And we also see that $*$ belongs neither to $\mathfrak{T}_v$ nor to $\mathfrak{F}_v$, as was desired. (We are of course not giving the details of any completeness proof here, and these remarks correspond to just a few steps in such a proof.)

Inumerable writers have discussed three-valued logic. Feferman rejected it in his study an not practical, but the reasons do not seem just. Anderson/Belnap (partly influenced by Fitch one would guess) have mentioned it - also the four-valued case without the law of non-contradiction. But to the author's best knowledge, no paper has made any really serious contribution to a general study of the model theory. Perhaps the very rich λ-calculus models will provide some incentive.

<u>The Laws of the Connectives</u>.

($\wedge$)	$\varphi, \psi \vdash \varphi\wedge\psi$	($\vee$)	$\varphi\vee\psi \vdash \varphi, \psi$
	$\varphi\wedge\psi \vdash \varphi$		$\varphi \vdash \varphi\vee\psi$
	$\varphi\wedge\psi \vdash \psi$		$\psi \vdash \varphi\vee\psi$

$(\times\!\!\times)$	$\varphi, \psi \vdash *, \varphi \times\!\!\times \psi$	$\varphi \times\!\!\times \psi, * \vdash \varphi, \psi$
	$\varphi \times\!\!\times \psi \vdash *, \varphi$	$\varphi, * \vdash \varphi \times\!\!\times \psi$
	$\varphi \times\!\!\times \psi \vdash *, \psi$	$\psi, * \vdash \varphi \times\!\!\times \psi$

What has been done here is just to transcribe the three-valued truth tables for these connectives. Note how useful $*$ is in blocking the symmetry of the rules when necessary. Note too that the various De Morgan laws follow easily using laws of negation:

$$\sim [\varphi \wedge \psi] \vdash\dashv \sim \varphi \vee \sim \psi$$

$$\sim [\varphi \vee \psi] \vdash\dashv \sim \varphi \wedge \sim \psi$$

$$\sim [\varphi \times\!\!\times \psi] \vdash\dashv \sim \varphi \times\!\!\times \sim \psi$$

<u>The Laws of the Quantifiers</u>.

$(\forall)$	$\forall x . \varphi \vdash \varphi[a/x]$	$(\exists)$	$\varphi[a/x] \vdash \exists x . \varphi$
	$\dfrac{\Gamma \vdash \varphi, \Delta}{\Gamma \vdash \forall x . \varphi, \Delta}$		$\dfrac{\Delta, \varphi \vdash \Gamma}{\Delta, \exists x . \varphi \vdash \Gamma}$

<u>provided that</u> x <u>is not free in</u> Γ .

These laws are the standard ones and they work just as well in the three-valued logic.

<u>The Laws of Equality</u>.

$$(=) \qquad \vdash a = a \qquad\qquad a = b,\ \varphi[a/x] \vdash \varphi[b/x]$$

Transitivity and symmetry of $=$ of course follow. If we use φ as $\sim x = b$, we have:

$$a = b\ ,\ \sim a = b \vdash \sim b = b \quad .$$

From the first law of equality and the laws of negation we derive:

$$\vdash a = b\ ,\ \sim a = b\ ,$$

that is, the law of the excluded middle for equality. As remarked earlier, we might not wish to assume this for all systems like those based on a URS.

Remember that formulas are elements of the model. Thus a special case of (=) above is the odd looking:

$$\varphi = \psi \text{ , } \varphi \vdash \psi$$

In our type-free logic this make quite good sense, once you think it over. In his book (§ 5.5 pp. 43-44) Fitch seems to suggest the following law of "extensionality":

$$\frac{\varphi \dashv\vdash \psi}{\vdash \varphi = \psi}$$

The present author is quite unable to see why Fitch wants this. Perhaps, for some very restricted set of φ and ψ it should hold, but even this seems intuitively doubtful. Experience based on some of Feferman's arguments suggests that almost any form of extensionality is risky in such systems. Fitch of course claims a consistency proof, but his rules are so numerous (four double columns of index!) and the exceptions are so well hidden in the text, it is next to impossible to judge his claim. (In any case he does not give the proof in the book.) Certainly the very intensional view of formulas in our model makes this rule invalid (for example $* \neq \sim *$ even though $* \dashv\vdash \sim *$ holds). Possibly a further study of extensionality would be worthwhile, but a way out is to be satisfied with equivalence relations rather than identities. Equivalences and congruences are necessary sooner or later any way, and Fitch's rule of extensionality seems to have no very special merit, since we can assume all the equations we need in a more standard format. For example:

The Laws of λ-Calculus.

(α) $\vdash \lambda x.\tau = \lambda y.\tau[y/x]$ (y not free in τ)

(β) $\vdash (\lambda x.\tau)(y) = \tau[y/x]$

(ξ) $\lambda x.\tau = \lambda y.\sigma \vdash\dashv \forall x.\tau = \sigma$

The only difference with what we said at the beginning of this paper is the use of "$\vdash$" and "$\vdash\dashv$" and the view that the "$\forall x$" is now formalized. It all fits together quite neatly.

Because "ϵ" is a defined symbol, the (β) rule along with the rules of equality give us at once what might be called "Church's Rule":

$$y \in \lambda x.\varphi \vdash\dashv \varphi[y/x] \quad .$$

By using laws (and definitions!) of the connectives and the quantifiers, we can show:

$$\forall x[\varphi \vee \sim \varphi] \vdash \exists a \ \forall x[x \in a \leftrightarrow \varphi] \ ,$$

where a is not free in φ . Thus a <u>version</u> of naive set theory does come out, but it is one protected from contradiction by the three-valued logic (the basic idea of Fitch). The question remains: is this theory of classes any good? What-ever we do it will remain somewhat messy owing to the failure of extensionality.

The laws that have been stated in this section refer to a fixed but arbitrary model onto which we have grafted a fairly natural truth definition. (Truth is of course not directly definable in the original model without the external transfinite recursion. Note, however, that truth is definable <u>in terms of truth</u>:

$$\varphi \in \lambda x.x \vdash\dashv \varphi$$

That is, the combinator $\lambda x.x$ defines the class $\mathfrak{T}$ while $\lambda x.\sim x$ defines $\mathfrak{F}$.) Once formal laws have been isolated, though, they take on a life of their own - unfortunately. Even we cannot refrain from asking some formal questions: are the laws as formulated complete with respect to the intended interpretations? The author is afraid that the proposed truth definition takes on something of the character of truth in higher-order logic and is therefore not axiomatizable. But he does not see the answer off hand. Even if the rules are not complete for "standard" models, the formal system _might_ be interesting. Can there be a _cut-free_ version of this logic with a normalization theorem, or do the λ-terms without normal forms spoil things? The author simply does not know enough proof theory to answer this question.

It might be useful to look at other variations of this logic using, for instance, the lattice or cpo models of λ-calculus. For these, we should replace $=$ as a primitive by $\subseteq$ and include laws of a partial ordering (even: lattice). Remember too that the paradoxical combinator Y may have special properties in such models, so one may also wish to formalize what is called "Scott's Induction Rule":

$$\frac{\Gamma,\ a(x) \subseteq b(x) \vdash a(u(x)) \subseteq b(u(x))}{\Gamma,\ a(\bot) \subseteq b(\bot) \vdash a(Y(u)) \subseteq b(Y(u))}$$

provided x is not free in Γ.

§ 4. Applications. There is no question now that we have well-grounded _connections_ between logic and λ-calculus: the truth definition can be faulted for its infinitistic character, but it cannot be ignored. We simply leave aside in this paper the question of whether using these ideas the tables can be turned so that we could argue that we have a _foundation_ for logic. Until we have the applications more firmly in hand, this question is unimportant. We understand set theory and model theory and truth definitions well enough to be able to investigate λ-calculus from this point of view more thoroughly without going all mystical or becoming militant type-libers.

One application that has in effect already been proposed is the investigation of particular models by trying to formalize more fully the rules that are valid for them. This activity is interesting only in-so-far as the models themselves are interesting. But there may be some point in connecting the models with the logic. Mc Carthy has suggested many times adjoining quantifiers to the LCF system of Milner. There is a serious question of whether a quantifier calculus is ever practical for any kind of automatic theorem proving or even theorem checking program; but if we put this question aside, the proposal of this paper does just what is asked for. Indeed the feed back generated by the syntax formalization goes far beyond what was expected. If we can show that the expressive power of the proposed system is of independent interest, then a further look at automated rules may be called for. Some preliminary hints about the expressive power will now be given.

Go back and look at the form of the truth definition in Section 2. We have a monotone, first-order, inductive definition. We are using (in the metalanguage) (1) equations and the given algebraic structure of the λ-calculus model; (2) positive connectives like and and or; (3) quantifiers over the model. (The fact that we had to define two predicates $\mathfrak{T}$ and $\mathfrak{F}$ is irrelevant.) Consider any other such definition defining, say, a one-place predicate $\mathfrak{P}$. The definition is done in the metalanguage remember, but we can formalize it by replacing $\mathfrak{P}\,x$ everywhere by $x \in p$ and by using the formal connectives and quantifiers. Having done so, we find corresponding to the right-hand side of the definition a formula φ with p and x as the only free variables. By the fixed point theorem of the λ-calculus, we can find a specific element p of the model such that the equation

$$p = \lambda x.\,\varphi$$

is true. By itself this is not the definition of $\mathfrak{P}$, but only a "formal" solution. We get the required solution by invoking the truth definition, for we can prove by transfinite induction that

$$\mathfrak{P}x \quad \text{iff} \quad \mathfrak{T}x \in p$$

for all x in the model. Thus the element p does in fact represent the predicate (or class) $\mathfrak{P}$. Another way of stating this result is to say that the $\mathfrak{T}$-predicate is universal for all such recursively defined predicates, since a direct formalization of their definitions provides a <u>reduction</u>. This makes it clear that the classes represented by the combinators are very, very complicated (and that the ordinal of the model is quite large).

As usual with recursive definitions, the law of the excluded middle fades. It is interesting to ask about those classes for which it holds. We define:

$$V = \lambda a.\ \forall x\,[x \in a \vee \sim x \in a]\ .$$

Note that the formula $a \in V$ can never be <u>false</u>; it often fails to be true, but that is something else. Those a where $a \in V$ is true are called <u>definite</u>: for each x the truth definition has decided membership one way or the other. There are many definite a because we made = a definite relation. Thus

$$a = (\lambda x.\ u(x) = v(x))$$

is definite for any choice of u and v , and the author has shown that this can be a very extensive family of classes. Quite generally, however, we can show that V enjoys pleasant closure properties.

Let us define some new "logical" combinators:

$$\mathrm{Fab} = \lambda f.\ \forall x[\sim x \in a \vee f(x) \in b]$$

$$\Pi ab = \lambda f.\ \forall x[\sim x \in a \vee f(x) \in b(x)]$$

$$\Sigma ab = \lambda u.\ \exists x\ \exists y[x \in a \wedge y \in b(x) \wedge u = \langle x,y\rangle\ ]$$

We want to relate F to Curry's ideas on functionality and Π and Σ to de Bruijn's and Martin-Löf's ideas on types.

As regards F , we can easily establish the validity of these rules:

$$f \in \mathrm{Fab}\ ,\ x \in a \vdash *\ ,\ f(x) \in b$$

$$\frac{\Gamma,\ x \in a \vdash *,\ f(x) \in b}{\Gamma,\ a \in V \vdash *,\ f \in \mathrm{Fab}}$$

provided x is not free in Γ . In other words, when a is definite, we can read $f \in \mathrm{Fab}$ as meaning that f maps a into b . This seems to formalize in our system what Curry had in mind about functionality, but he was never very definite about what it means to be definite.

A very important point about F and V is the closure condition:

$$a \in V\ ,\ b \in V \vdash \mathrm{Fab} \in V\ \ .$$

This is already indication enough that V is a big universe since so many different definite classes can be represented by F-expressions. We should note too the functionality laws of the favourite combinators:

$$a \in V \vdash *\ ,\ I \in \mathrm{Faa}$$

$$a \in V\ ,\ b \in V \vdash *\ ,\ K \in \mathrm{Fa\ Fba}$$

$$a \in V\ ,\ b \in V\ ,\ c \in V \vdash *\ ,\ S \in \mathrm{FFaFbcFFabFac}$$

In the case of I we can even write:

$$a \in V \vdash\!\dashv I \in \mathrm{Faa}\ ,$$

but this does not seem too important.

The combinator F gives a function-space construction, while Π makes a cartesian product. The notation $\Pi a\ \lambda x.b(x)$ would correspond to the notation $[x,a]b(x)$ of de Bruijn (see the paper by de Vrijer from this symposium). We have the rules:

$$f \in \Pi ab\ ,\ x \in a \vdash *\ ,\ f(x) \in b(x)$$

$$\frac{\Gamma,\ x \in a \vdash *\ ,\ f(x) \in b(x)}{\Gamma,\ a \in V \vdash *\ ,\ f \in \Pi ab}$$

provided x is not free in Γ . The relevant closure condition reads:

$$a \in V\ ,\ b \in FaV \vdash \Pi ab \in V\ .$$

This is a principle of "type inclusion" (see § 1.1.3 of de Vrijer's paper), but we are not at all identifying b with Πab which would be very confusing. The rules for the Σ-combinator can be left to the reader.

We are thinking of V as the type of all types. Of course V is not really a type because $V \in V$ is not true; however, it is a semi-type, and what we wrote about Π and V above is quite coherent. Thus V represents a very interesting class (which gives some kind of a model for some kind of type theory) and we would like to do some recursive definitions on it. We cannot go very far, however, because V is not definite. Well, why not make it definite? We can, but this involves a new recursive truth definition. (The author learned this idea from Aczel at the Kiel Logic Summer School.) In fact we shall make a whole hierarchy of truth definitions in a way that would seem to incorporate an old idea of Fitch.

We add to our syntactical primitives a new equation:

$$\varphi^{(i)} = \langle\ 5+i, \varphi\ \rangle$$

Instead of one pair $\mathfrak{T}, \mathfrak{F}$ we have an infinite number $\mathfrak{T}^{(n)}, \mathfrak{F}^{(n)}$ where the recursion is such that the $n^{\underline{th}}$ pair is defined in terms of the pairs for $i < n$. Indeed we change the clauses of the truth definition to be able to adjoin these equivalences:

$$\mathfrak{T}^{(n)}\varphi^{(i)} \quad \text{iff} \quad i < n \quad \text{and} \quad \mathfrak{T}^{(i)}\varphi$$

$$\mathfrak{F}^{(n)}\varphi^{(i)} \quad \text{iff} \quad i < n \quad \text{and } \underline{\text{not}} \quad \mathfrak{T}^{(i)}\varphi \ .$$

We see at once that this is not a monotone definition because negation is being used. However, the pair $\mathfrak{T}^{(0)}, \mathfrak{F}^{(0)}$ is just our original $\mathfrak{T}, \mathfrak{F}$; so we know they exist. When we pass then to $\mathfrak{T}^{(1)}, \mathfrak{F}^{(1)}$ we are taking $\mathfrak{T}^{(0)}$ as given (that is to say, definite). As far as the recursion goes we are not defining the infinite number of predicates together, but one after the other. (The ordinals will get really large now.) From the infinitistic, set-theoretical point of view there is nothing wrong in this, and there is no reason why we cannot iterate it, as suggested.

Unless the author has made an oversight, we have the consistency conditions:

$$\mathfrak{T}^{(n)} \subseteq \mathfrak{T}^{(n+1)} \quad \text{and} \quad \mathfrak{F}^{(n)} \subseteq \mathfrak{F}^{(n+1)} \ ,$$

since each successive truth definition changes nothing of the previous ones but just makes more formulas "meaningful". Thus we can take the unions, obtaining $\mathfrak{T}^{(\infty)}, \mathfrak{F}^{(\infty)}$. (We could also iterate this passage, but enough is enough.) Using $\mathfrak{T}^{(\infty)}$ and $\mathfrak{F}^{(\infty)}$ for our logic, we can now employ a whole range of universes (somewhat in the way proposed by Martin-Löf), which are defined by:

$$V^{(i)} = \lambda a.\ \forall x[x \in a \vee \sim x \in a]^{(i)}$$

Now $V^{(0)}$ represents our old V; but since we changed the truth sets, the symbol V has a new meaning (that is, it represents a different class). In fact,

$$\vdash V^{(n)} \in V^{(n+1)}$$

and

$$\vdash V^{(n)} \in V \ .$$

The various $V^{(n)}$ will have very extensive closure properties.

Fitch once suggested a hierarchy of stronger and stronger negations.

The author is only guessing, but why not look at formulas $(\sim \varphi)^{(i)}$?
We would have

$$(\sim \varphi)^{(i)} \vdash_{*} , (\sim \varphi)^{(i+1)}$$

and

$$\vdash (\sim \varphi)^{(i)} \text{ iff } \mathfrak{F}^{(i)}\varphi .$$

In other words, if we are right, Fitch introduced the series of $\mathfrak{F}$-classes. But since $(\sim \sim \varphi)^{(i)}$ is just as good as $\varphi^{(i)}$, the two ideas come to the same thing. Clearly this plan of making certain classes definite could be carried out in many different ways.

This would seem to be enough to show that there are applications of the truth definition to connect with other ideas. So the author now leaves it to someone else to make the next move.

Acknowledgement. Thanks are due to Oxford University for giving the author leave for one term and to the ETH, Zürich, whose very generous hospitality provided the excellent working conditions under which this paper could be written and very efficiently typed.

LEAST FIXED POINTS REVISITED

J.W. de Bakker *

Mathematical Centre, Amsterdam

ABSTRACT

Parameter mechanisms for recursive procedures are investigated. Contrary to the view of Manna et al., it is argued that both call-by-value and call-by-name mechanisms yield the least fixed points of the functionals determined by the bodies of the procedures concerned. These functionals differ, however, according to the mechanism chosen. A careful and detailed presentation of this result is given, along the lines of a simple typed lambda calculus, with interpretation rules modelling program execution in such a way that call-by-value determines a change in the environment and call-by-name a textual substitution in the procedure body.

KEY WORDS AND PHRASES: *Semantics, recursion, least fixed points, parameter mechanisms, call-by-value, call-by-name, lambda calculus.*

* Author's address
Mathematical Centre
2e Boerhaavestraat 49
Amsterdam
Netherlands

NOTATION

Section 2

$s,t,t_i,t',\ldots$	individual terms
$\sigma,\tau,\tau_j,\tau',\ldots$	function terms
$S,T,\ldots$	functional terms
$p,p',\ldots$	boolean individual terms
$\pi,\ldots$	boolean function terms
$P,\ldots$	boolean functional terms
$a \in A,\ \alpha \in \mathcal{A}$	constants
$b \in B,\ \beta \in \mathcal{B}$	
$x,y,z,u \in X,\ \xi,\eta \in \mathcal{X}$	variables
$q \in Q,\ \chi \in \mathcal{Q}$	
$\phi \in \mathcal{F},\ \psi \in \mathcal{G}$	procedure symbols
$\tau(t_1,\ldots,t_n),\ \pi(t_1,\ldots,t_n)$	application
$T(\tau_1,\ldots,\tau_r),\ P(\pi_1,\ldots,\pi_r)$	
$\nu x_1\ldots x_\ell \lambda y_1\ldots y_m.t,$	abstraction
$\nu x_1\ldots x_\ell \lambda y_1\ldots y_m.p,$	
$\lambda\xi_1\ldots\xi_r.\tau,\ \lambda\chi_1\ldots\chi_r.\pi$	
<u>if</u> p <u>then</u> t' <u>else</u> t"	selection
<u>if</u> p <u>then</u> p' <u>else</u> p"	
$n(\tau),\ \langle n(T),r(T)\rangle$	rank
$1 \le i \le n,\ 1 \le j \le r,$	indices
$1 \le h \le \ell,\ 1 \le k \le m$	
$\bar{t},\bar{\tau},\bar{x},\bar{\xi},\ldots$	vector notation
$\equiv$	syntactic identity between terms

Section 3

$s[t/x],\ s[\tau/\xi],\ldots$	substitution
$\mathcal{E}^{x\xi}_{vf},\ \mathcal{E}^{\bar{x}}_{\bar{v}},\ldots$	change of environment

Section 4

$\mathcal{V} = \mathcal{V}_0 \cup \{\bot\}$	domain
$\mathcal{C}$	constant-interpretation
$\mathcal{E}$	variable-interpretation (environment)
$\mathcal{D}$	declarations
$\mathcal{J} = <\mathcal{V},\mathcal{C},\mathcal{E};\mathcal{D}>$	interpretation
$v,w \in \mathcal{V},\ f \in \mathcal{V}^n \to \mathcal{V},$ $F \in (\mathcal{V}^n \to \mathcal{V})^r \to (\mathcal{V}^n \to \mathcal{V})$	elements of (higher) domains
$valt(t,\mathcal{J},N)$ $val(t,\mathcal{J}),\ val(t,\mathcal{E})$	evaluation functions

Section 5

$v_1 \sqsubseteq v_2,\ f_1 \sqsubseteq f_2,\ F_1 \sqsubseteq F_2$	partial orderings
$val(\tau,\mathcal{J}),\ val(T,\mathcal{J})$	extension of *val*
$t_1 \sqsubseteq t_2,\ \tau_1 \sqsubseteq \tau_2,\ T_1 \sqsubseteq T_2$	atomic formulae
Φ,Ψ	sets of atomic formulae
ρ	element of Φ
$\Phi \models_{\mathcal{D}} \Psi$	assertion
$\Phi[\mathcal{V},\mathcal{C},\mathcal{E};\mathcal{D}]$	$<\mathcal{V},\mathcal{C},\mathcal{E};\mathcal{D}>$ satisfies Φ
$\Phi \models \Psi,\ \models \Psi,\ \models t_1 = t_2,\ldots$	simplified forms of assertions

Section 7

ϕ_0	the nowhere defined procedure
ω	$\phi_0(\bar{z})$
Ω	$\nu\bar{x}\lambda\bar{y}.\omega$
$\mathit{0}$	$\lambda\bar{\xi}.\Omega$
$\tilde{t},\tilde{\tau},\tilde{T}$	replacing procedure symbols by new variables
$t^{(i)},\tau^{(i)},T^{(i)}$	approximants to t,τ,T

1. INTRODUCTION

1.1. MOTIVATION

The fixed point approach in the semantics of programming languages has gained considerable popularity in recent years. The basic programming notions of recursion and iteration have found a satisfactory *mathematical* treatment in terms of least fixed points, as opposed to the *operational* methods, where the emphasis is on techniques using stacks, displays and the like. (For a discussion on the distinction between mathematical and operational semantics one should consult the works of SCOTT and STRACHEY, such as [20].) In order to explain our motivation for the present paper, a brief sketch of the history of the subject is needed.

KLEENE's first recursion theorem [6] already gave a characterization of recursive functions (albeit restricted to integer functions with parameters called-by-value, see below) in terms of least fixed points. For some time, applications in programming theory remained tentative, however. We refer for example to the results of McCARTHY [10], and, in particular, to the early work of LANDIN (e.g. [7]) where CURRY's Y combinator was used to deal with recursion in such a way that the fact that the fixed points concerned are *least* with respect to a suitable partial ordering remained implicit. As another important predecessor we mention MORRIS [12]. In 1969, a number of people arrived independently at some methods and results causing a revival of the fixed point approach, viz. BEKIC, PARK [15] and SCOTT and DE BAKKER [19]. To be more specific, by "fixed point approach" we refer to the whole of techniques for proving properties of programs which take as starting point the fact that the function defined by a recursive procedure can, in a sense to be made precise presently, be viewed as the least fixed point of a functional which is associated in a rather natural way with the body of the procedure declaration. The paper [19] also contained the first statement of an important rule of proof, SCOTT's induction rule, which has found a variety of applications in the next few years [1,2,3,4,8,9,11]. On the theoretical side, the invention of SCOTT's models of the lambda calculus, where the relationship between the least fixed point result and the Y combinator could be settled, has added to the success of the method.

As to the applications, we are in particular interested in the papers by MANNA and his colleagues - then at Stanford - which contain both a long list of examples, and a discussion of the relationship between SCOTT's induction (computation induction, as they call it), and other methods such as MORRIS' truncation induction [13]. Moreover, their work drew attention to a problem which inspired the present paper. This problem is stated e.g. in MANNA & VUILLEMIN [9] (p.529): "many programming languages use implementations (such as call-by-value) which do not necessarily lead to the least fixed point", or in MANNA, NESS & VUILLEMIN [8] (p.496): "we are interested in computation rules that yield the least fixed point ... we call such computation rules fixed point computation rules ... the left-most innermost rule is not a fixed point rule ...". In both of these papers, the work of MORRIS [12] is mentioned in support of the quoted statements. The present author believes that it is advantegeous to take a different approach to these matters: We view the main assertion of Morris to be that the function determined by a procedure with parameters called-by-value (f_v, say) may be properly included in the function determined by the same procedure with parameters called-by-name (f_n, say) (A partial function f is said to be included in a partial function g ($f \subseteq g$) iff whenever f is defined, g is defined with the same value.) However, it may well be that f_v and f_n, though different, are *both* least fixed points, albeit of different functionals. In order to explain this, consider the example of a recursive procedure (from MANNA & VUILLEMIN [9]):

(1.1) $\phi(x,y) \Leftarrow \underline{\text{if}}\ x = 0\ \underline{\text{then}}\ 0\ \underline{\text{else}}\ \phi(x-1,\phi(x,y))$

Suppose we consider ϕ for integer x,y. Then, if ϕ has parameters called-by-value, we obtain for the function f determined by (1.1), say f_v:

$$f_v(x,y) = \underline{\text{if}}\ x = 0\ \underline{\text{then}}\ 0\ \underline{\text{else}}\ \text{undefined}$$

whereas, when the parameters are called-by-name, we obtain f_n:

$$f_n(x,y) = \underline{\text{if}}\ x \geq 0\ \underline{\text{then}}\ 0\ \underline{\text{else}}\ \text{undefined.}$$

Clearly, $f_v \subsetneq f_n$. Writing

(1.2) $\phi(x,y) \Leftarrow \Phi(\phi)(x,y)$

as short-hand for (1.1), with Φ determining a functional F, it is now certainly impossible that f_v and f_n are both *the* least fixed point of F. *However, the notation in (1.2) leaves out the important distinction between the two parameter mechanisms used,* since one same Φ is used for both cases. What is needed is a treatment of (1.2) such that the functional term Φ carries the information about the parameters along: We then have *two* declarations: $\phi(x,y) \Leftarrow \Phi_v(\phi)(x,y)$, and $\phi(x,y) \Leftarrow \Phi_n(\phi)(x,y)$, which determine functions f_v and f_n, and functionals F_v and F_n, such that f_v is the least fixed point of F_v, and f_n is the least fixed point of F_n (and that $f_v \subseteq f_n$, with the possibility that $f_v \subsetneq f_n$).

Now we can state the goal of our paper: We want to make the above considerations precise, and to prove, in careful detail, the least fixed point result for both parameter mechanisms. As we understand the literature, this result has been obtained before for call-by-name only, and by using quite different proof methods, viz. in Cadiou's Stanford Ph.D. thesis, in ROSEN [17] and NIVAT [14]. An attempt at clarification of the same issues was made by DE ROEVER [16], who also emphasizes that different parameter mechanisms give rise to different transformations, but elaborates this idea in the framework of axiomatized relations. An elementary exposition, which does not involve the somewhat advanced logical and algebraic tools of Nivat and Rosen, but which has certainly benefited from the ideas of the lambda calculus, may be of interest. No applications are dealt with; we do not even prove the continuity of the functionals, nor do we give a justification of Scott's induction rule. (Observe, however, that our results imply that the rule is valid for (each combination of) call-by-value *and* call-by-name; this should be contrasted with the position taken in [8,9].) On the contrary, we concentrate solely on the stated problem. We shall use a rather extensive formalism for this purpose, and spend much attention to a detailed development of the argument. We feel that this may be justified in a situation where wedeal with an intricate issue which has led to somewhat diverging opinions.

1.2. OUTLINE OF THE PAPER

In section 2 we define the syntax of a formal language centered around the notions of *application, abstraction, recursion* and *selection*. In spirit, the syntax is very much like a typed lambda calculus, with two major differences

- the explicit addition of recursion by procedure declarations and - calls (as opposed to the implicit recursion via the Y-combinator in the non-typed lambda calculus)
- the explicit notational distinction between call-by-value and call-by-name parameters

and one minor difference: the number of types is restricted to three: individual, function, and functional.

In section 3 we give the standard definitions of free and bound (occurrences of) variables, and of substitution.

Section 4 is of central importance, giving the semantics of the constructs in our language. Terms are provided with an interpretation: select a domain V, map constants and variables to elements of V (and of the derived domains of higher type), and, moreover, fix a set of declarations for the procedure symbols. This being done, a process of evaluation is prescribed: The well-known extension of V with an extra element to provide a "value" in the case of non-terminating evaluations is used, and, for terminating evaluations, the number of steps needed is carried along (this being of importance in a later proof which uses induction on this number). Recursion is defined by body replacement, call-by-value parameters by changing the environment, and call-by-name parameters by substitution.

Section 5 leads up to the formalism to state facts about our terms which hold for all interpretations (so-called valid assertions). First a partial order on the domains is introduced, and various properties of the *val*-function are derived, which are of technical importance for the proof of the monotonicity theorem in section 6.

Section 7 introduces the notion of approximating a term by procedure-free approximants, and develops a precise notation for this, resulting in the lemma that, for each term t, its value in a given interpretation is also the value of an approximant to t.

Section 8, finally, brings the proof of the least fixed point theorem, which relies heavily on the result of section 7.

ACKNOWLEDGMENT. *Our interest in applying a formalism near to the lambda calculus in investigating semantics has been stimulated by lecture notes of Robin Milner, explaining work of Peter Landin and Dana Scott.*

2. SYNTAX

We introduce a formal language which contains the main programming concepts relevant for our purpose. Starting from some initial classes of expressions, construction rules are provided to build up more complex expressions. These rules correspond to the following programming concepts:

- *application* apply a function to one or more arguments
- *abstraction* an expression may be "parametrized", yielding a function of one or more arguments (such abstraction is part of the mechanism usually invoked at procedure declaration)
- *recursion* a mechanism for "declaring" and "calling" (possibly recursive) procedures is introduced
- *selection* this gives the usual conditional construct in programming

A subset of the expressions of our language, with elements called *terms*, is intended as the class of "abstract programs". In section 4, a method will be given to *interpret* these terms by means of an evaluation mechanism yielding *individuals, functions* and *functionals* as values. The evaluation mechanism is, of course, designed in such a way that it models program execution, in sofar as this is concerned with the programming concepts just mentioned. Anticipating the precise definitions, we already indicate that an interpretation will start with the choice of a domain V, such that the terms of our language will, in this interpretation, have values according to the following table:

Class of terms	Denotation	Intended interpretation
individual	$s,t,t_0,t_i,t',\ldots$	$\in V$
function	$\sigma,\tau,\tau_0,\tau_j,\tau',\ldots$	$\in V^n \to V$
functional	$S,T,T',\ldots$	$\in (V^n \to V)^r \to (V^n \to V)$
boolean	$p,p',\ldots$	$\in \{0,1\}$
boolean function	$\pi,\ldots$	$\in V^n \to \{0,1\}$
boolean functional	$P,\ldots$	$\in (V^n \to \{0,1\})^r \to (V^n \to \{0,1\})$

Figure 1. Intended interpretation of terms

The terms are made up by means of the construction rules mentioned above starting from certain given symbol classes, of *constants*, *variables*, and *procedure symbols*.

Type of class		Notation	
		Class	Element
Individual	constants	A	a
Function	constants	$\mathcal{A}$	α
Boolean	constants	B	b
Boolean function	constants	$\mathcal{B}$	β
Individual	variables	X	x,y,z,u
Function	variables	$\mathcal{X}$	ξ,η
Boolean	variables	Q	q
Boolean function	variables	$\mathcal{Q}$	χ
Procedure	symbols	$\mathcal{F}$	ϕ
Boolean procedure	symbols	$\mathcal{G}$	ψ

Figure 2. The initial classes of symbols

The syntax of our language is now given in the following two tables (explanatory remarks follow the definition):

term \ operation	*constant*	*variable*	*application*
t	$a \in A$	$x \in X$	$\tau(t_1,\ldots,t_n)$
(individual)			
p	$b \in B$	$q \in Q$	$\pi(t_1,\ldots,t_n)$
τ	$\alpha \in \mathcal{A}$	$\xi \in \mathcal{X}$	$T(\tau_1,\ldots,\tau_r)$
(function)			
π	$\beta \in \mathcal{B}$	$\chi \in \mathcal{Q}$	$P(\pi_1,\ldots,\pi_r)$
T	-	-	-
(functional)			
P	-	-	-

Figure 3a. Syntax, first part

term \ operation	*abstraction*	*selection*	*recursion*
t	-	<u>if</u> p <u>then</u> t' <u>else</u> t"	-
(individual)			
p	-	<u>if</u> p <u>then</u> p' <u>else</u> p"	-
τ	$\nu x_1 \ldots x_\ell \lambda y_1 \ldots y_m . t$	-	$\phi \in F$
(function)	$(\ell+m>0)$		
π	$\nu x_1 \ldots x_\ell \lambda y_1 \ldots y_m . p$	-	$\psi \in G$
T	$\lambda \xi_1 \ldots \xi_r . \tau$	-	
(functional)	$(r>0)$		
P	$\lambda \chi_1 \ldots \chi_r . \pi$	-	

Figure 3b. Syntax, second part

We assume - without bothering to justify this - that each term can be uniquely parsed. Moreover, parentheses will be used freely to enhance readability.

Syntactic identity between terms will be denoted by "$\equiv$". Some examples of terms are:

1. Individual terms:

   a, x, $\phi(x_1,x_2)$, $(\lambda x.x)(a)$, <u>if</u> p <u>then</u> a <u>else</u> $\phi(\alpha(x_1), \phi(x_1,x_2))$

2. Function terms:

   α, ξ, $(\lambda\xi.\alpha)(\phi)$, $\nu x_1 x_2$. <u>if</u> p <u>then</u> a <u>else</u> $\phi(\alpha(x_1), \phi(x_1,x_2))$

   $\lambda y_1 y_2$. <u>if</u> p <u>then</u> a <u>else</u> $\phi(\alpha(y_1), \phi(y_1,y_2))$

   (we adopt here the obvious conventions for $\ell = 0$ or $m = 0$)

3. Functional terms:

   $\lambda\xi.\ \nu x_1 x_2$. <u>if</u> p <u>then</u> a <u>else</u> $\xi(\alpha(x_1), \xi(x_1,x_2))$

   $\lambda\xi.\ \lambda y_1 y_2$. <u>if</u> p <u>then</u> a <u>else</u> $\xi(\alpha(y_1), \xi(y_1,y_2))$

   (for suitable interpretation of p,a and α, these two functional terms correspond to the Φ_v and Φ_n of the introduction, cf. remark 2 below).

The following remarks will help the reader in reading and understanding the syntactic definitions in figures 3a, 3b.

1. (Reading the tables). Consider, e.g., the first line (after t in fig. 3a). This should be read as:
   - Each individual constant or individual variable is an individual term.
   - If τ is a function term, and each t_i, $i=1,\ldots,n$, is an individual term, then $\tau(t_1,\ldots,t_n)$ is an individual term (obtained by the construction rule of application).
2. (Abstraction). Most of the tables should now be readable, apart from the *abstraction*-column, which needs further explanation. Consider the construct $\tau \equiv \nu x_1\ldots x_\ell \lambda y_1\ldots y_m.t$. For each individual term t, τ denotes a function term which, in the interpretation to be given presently, will obtain as meaning a function with
   - the $\ell \geq 0$ arguments (formal parameters) $x_1,\ldots,x_\ell$ *called-by-value*
   - the $m \geq 0$ arguments (formal parameters) $y_1,\ldots,y_m$ *called-by-name*.

   In other words, the ν-abstraction is intended to model call-by-value parametrization, the λ-abstraction call-by-name parametrization. (The reader should not confuse this statement of intention, to be made precise in section 4, with the "normal" λ-abstraction in the lambda calculus, where the conversion order is (very much) left open.) Again anticipating, call-by-value parameters will be evaluated by changing the environment, i.e. the variable-value correspondence, and call-by-name parameters by textual substitution. For these definitions to make sense, we assume from now on that the x_n, $1 \leq h \leq \ell$, y_k, $1 \leq k \leq m$, are all different variables. An analogous requirement is imposed upon the ξ_j, $j=1,\ldots,r$, in the formation of $\lambda\xi_1\ldots\xi_r.\tau$.

3. (Functionals). For functional terms, we do not need the call-by-value type of arguments, and we restrict ourselves to the usual λ-abstraction, thus turning function terms into functional terms. It will be noticed that only very limited means are provided to construct functional terms. As a matter of fact, they are, strictly speaking, unnecessary, and are only introduced to obtain eventually a more appealing form of the least fixed point theorem. Note also that we introduce functionals of a rather restricted format: Instead of elements of $[(V^{n_1}\to V) \times (V^{n_2}\to V) \times\ldots\times (V^{n_r}\to V)] \to$ $\to (V^n\to V)$, we have the simple form as given. This is a restriction imposed for convenience sake only. All the results of the paper go through for the more general case, but we did not want to add an extra burden to the already rather heavy formalism.
4. (Recursion). The recursion column is as yet rather meagre: no declarations are given yet, only the procedure symbols. We find it more convenient to introduce declarations as part of the interpreting mechanism, though an appraoch which brings in declarations at an earlier stage might also have been adopted.
5. (Rank and arity). The syntax tables are not very explicit on the role of the integers n and r. The following supplementary information is in order: Each function term τ has a certain so-called rank n(τ), each functional term T has a rank-pair <n(T),r(T)>. The rank is initially given for the α, ξ and ϕ, and for the other constructs it is defined as follows
   - if $\tau \equiv \nu x_1\ldots x_\ell \lambda y_1\ldots y_m.t$, then $n(\tau) = \ell + m$
   - if $\tau \equiv T(\tau_1,\ldots,\tau_r)$, and $n(\tau_i) = n(i=1,\ldots,r)$, then $n(\tau) = n$
   - if $T \equiv \lambda\xi_1\ldots\xi_r.\tau$, and $n(\tau) = n$, then $n(T) = n$ and $r(T) = r$
   - similar definitions hold for the boolean case.

   Furthermore, we require
   - if $t \equiv \tau(t_1,\ldots,t_n)$, then $n(\tau) = n$
   - if $\tau \equiv T(\tau_1,\ldots,\tau_r)$, then $n(T) = n(\tau_i)$, $i=1,\ldots,r$, and $r(T) = r$.

   This system is, of course, designed in this way in order that
   - each τ of rank $n = n(\tau)$ is to be interpreted as a function of arity

n: $V^n \to V$

- each T of rank <n,r> is to be interpreted as a functional: $(V^n \to V)^r \to (V^n \to V)$.
- similarly for the boolean case.

We are aware of the fact that we have not adopted here the most general solution. We have envisaged a system with terms t of rank $n(t) \geq 0$, with abstraction restricted to one variable, such that, for $t \equiv \lambda x.t'$ or $t \equiv \nu x.t'$, $n(t) = n(t')+1$, and $n(t) = 0$ indicating that t is to be interpreted as an individual ($\in V$), and with application restricted to the one-argument case. For our present purpose, the syntax as in figs. 3a,b was thought to be preferable. The restriction of our presentation to terms of three levels (individual, function, functional) is closer to the concepts as they appear in programming than a system with an infinite hierarchy, and our main goal - clarification of the two parameter mechanisms - seems, after some experiments with the mathematically more elegant approach just sketched, to be achieved in a better way.

One last remark on notation: We sometimes use a "vector"-notation, and write $\tau(\bar{t})$ for $\tau(t_1,\ldots,t_n)$, $T(\bar{\tau})$ for $T(\tau_1,\ldots,\tau_r)$, $\bar{x}$ for $(x_1,\ldots,x_\ell)$, etc.

3. SUBSTITUTION

In the interpretation of terms to be given in section 4, we shall define the evaluation of call-by-name parameters by a process of textual replacement of "formal" by "actual" parameters, i.e., by means of *substitution*. Therefore, we devote this section to a precise definition of this operation. We do this by a restatement of standard techniques, see e.g. [5], as adopted to our present goals. At the end of the section a notation for changing the environment - which is used to model call-by-value evaluation - is given.

First we introduce the notion of a *variable occurring in a term* in

<u>DEFINITION</u> 3.1 (Occurrences).

1. x occurs in t iff
   1.1. $t \equiv x$
   1.2. $t \equiv \tau(t_1,\dots,t_n)$, and x occurs in τ or any of the t_i, $i=1,\dots,n$.
   1.3. $t \equiv$ <u>if</u> p <u>then</u> t' <u>else</u> t", and x occurs in p, t' or t".

2. x occurs in τ iff
   2.1. $\tau \equiv \nu\bar{x}\lambda\bar{y}.t$, and x occurs in t
   2.2. $\tau \equiv T(\tau_1,\dots,\tau_r)$, and x occurs in T or any of the τ_j, $j=1,\dots,r$.

3. x occurs in T iff
   3.1. $T \equiv \lambda\bar{\xi}.\tau$, and x occurs in τ.

4. ξ occurs in t iff
   4.1. $t \equiv \tau(t_1,\dots,t_n)$, and ξ occurs in τ or any of the t_i, $i=1,\dots,n$.
   4.2. $t \equiv$ <u>if</u> p <u>then</u> t' <u>else</u> t", and ξ occurs in p, t' or t".

5. ξ occurs in τ iff
   5.1. $\tau \equiv \xi$
   5.2. $\tau \equiv \nu\bar{x}\lambda\bar{y}.t$, and ξ occurs in t
   5.3. $\tau \equiv T(\tau_1,\dots,\tau_r)$, and ξ occurs in T or any of the τ_j, $j=1,\dots,r$.

6. ξ occurs in T iff
   6.1. $T \equiv \lambda\bar{\xi}.\tau$, and ξ occurs in τ.

7. The definitions for p, π or P are similar.

Observe that x does not occur in $\lambda x.a$. Next, we need the notions of *bound* and *free* occurrences of a variable in a term.

<u>DEFINITION</u> 3.2 (Bound and free occurrences).

1. An occurrence of a variable x in a term is bound iff x occurs in a part of that term of the form $\nu x_1\dots x_\ell\lambda y_1\dots y_m.t$, with $x \equiv x_h$, for some h, $1 \le h \le \ell$, or $x \equiv y_k$, for some k, $1 \le k \le m$.

2. An occurrence of a variable x in a term is free, otherwise.

3. An occurrence of a variable ξ in a term is bound iff ξ occurs in a part of that term of the form $\lambda\xi_1,\dots,\xi_r.\tau$, with $\xi \equiv \xi_j$, for some j, $1 \le j \le r$.

4. An occurrence of a variable ξ in a term is free, otherwise.

EXAMPLES:

- x occurs bound in $\lambda x.x$, free in $\alpha(x)$, and bound *and* free in $(\lambda x.x)(\alpha(x))$.
- both ξ and x occur bound in $\lambda\xi.\lambda x.\alpha_1((\lambda\eta.\xi)(\eta)(x))$, whereas η occurs free in that same term.

We now define the important notion of substitution. A term t (or τ) may be substituted for (i.e. replace all free occurrences of) a variable x (or ξ) in any term s,σ,S,p,π or P. The results are denoted by $s[t/x],\ldots$ $\ldots,P[t/x]$, $s[\tau/\xi],\ldots,P[\tau/\xi]$. The process of substitution is defined by induction on the complexity of the terms involved:

DEFINITION 3.3 (Substitution).

1. $s[t/x]$
   1.1. $a[t/x] \equiv a$, $x[t/x] \equiv t$, $y[t/x] \equiv y (y \not\equiv x)$.
   1.2. $\sigma(\bar{s})[t/x] \equiv \sigma[t/x](\bar{s}[t/x])$.
   1.3. $(\underline{\text{if}}\ p\ \underline{\text{then}}\ s'\ \underline{\text{else}}\ s'')[t/x] \equiv \underline{\text{if}}\ p[t/x]\ \underline{\text{then}}\ s'[t/x]\ \underline{\text{else}}\ s''[t/x]$.
2. $\sigma[t/x]$
   2.1. $\alpha[t/x] \equiv \alpha$, $\xi[t/x] \equiv \xi$, $\phi[t/x] \equiv \phi$.
   2.2. $(\nu\bar{x}\lambda\bar{y}.s)[t/x]$
   $\equiv \nu\bar{x}\lambda\bar{y}.s$, if $x \equiv x_h$, for some h, $1 \le h \le \ell$, or $x \equiv y_k$, for some k, $1 \le k \le m$; otherwise
   $\equiv \nu\bar{x}\lambda\bar{y}.s[t/x]$, if none of the $\bar{x}$ or $\bar{y}$ occurs free in t; otherwise
   $\equiv \nu\bar{z}\lambda\bar{u}.s[\bar{z}/\bar{x}][\bar{u}/\bar{y}][t/x]$ where the $\bar{z},\bar{u}$ are *new* variables
   2.3. $S(\bar{\sigma})[t/x] \equiv S[t/x](\bar{\sigma}[t/x])$
3. $S[t/x]$
   3.1. $(\lambda\bar{\xi}.\tau)[t/x] \equiv \lambda\bar{\xi}.\tau[t/x]$
4. $s[\tau/\xi]$, $\sigma[\tau/\xi]$, $S[\tau/\xi]$. We only give the central cases, the remaining ones then follow as in 1-3.
   4.1. $\xi[\tau/\xi] \equiv \tau$, $\eta[\tau/\xi] \equiv \eta\ (\eta \not\equiv \xi)$
   4.2. $(\lambda\bar{\xi}.\sigma)[\tau/\xi]$
   $\equiv \lambda\bar{\xi}.\sigma$, if $\xi = \xi_j$, for some j, $1 \le j \le r$; otherwise
   $\equiv \lambda\bar{\xi}.\sigma[\tau/\xi]$, if none of the $\bar{\xi}_j$ occurs free in τ; otherwise
   $\equiv \lambda\bar{\eta}.\sigma[\bar{\eta}/\bar{\xi}][\tau/\xi]$ where the $\bar{\eta}$ are *new* variables
5. Substitution in boolean terms is defined similarly.

REMARK: The precautions in 2.2 and 4.2 have the usual reason: Without them, variables free in t (or τ) would be turned into bound variables (*then*, e.g., $(\lambda y.x)[y/x] \equiv \lambda y.y$), and substitution would not be "meaning preserving" (e.g., in the intended interpretation, $\lambda y.x$ determines a function which, for each argument, yields the value of x as a result, whereas $\lambda y.y$ determines the identity function).

EXAMPLES OF SUBSITUTION:

1. $(\lambda\xi.\lambda x.\alpha_1((\lambda\eta.\xi)(\eta)(x)))[\lambda z.\xi(a)/\eta] \equiv \lambda\zeta.\lambda x.\alpha_1((\lambda\eta.\zeta)((\lambda z.\xi(a))(x))$
2. $((\lambda y.x)(\phi(x)))[\xi(y)/x] \equiv (\lambda z.\xi(y))(\phi(\xi(y)))$

The following lemma states a number of basic properties of substitutions to be used in later sections:

LEMMA 3.4.

1. *If* y *is not free in* s, *and neither* y *nor any variable free in* t *is bound in* s, *then* $s[y/x][t/y] \equiv s[t/x]$.
2. *If* $y \not\equiv x$, *and* y *is not free in* t", *and no variable free in* t' *or* t" *is bound in* s, *then* $s[t'/y][t''/x] \equiv s[t''/x][t'[t''/x]/y]$.

PROOF. See HINDLEY, LERCHER & SELDIN [5]. □

Call-by-name parameters are dealt with by means of substitution, call-by-value parameters by changing the environment. A definition of this follows in the next section, but we already introduce a notation designed for this purpose.

Let, for the moment, E be any function mapping arguments x to values v, and arguments ξ to values f. Then $E^{x\xi}_{vf}$ is a function which satisfies:

1. $(E^{x\xi}_{vf})(x) = v$, $(E^{x\xi}_{vf})(\xi) = f$.
2. $(E^{x\xi}_{vf})(y) = E(y)$, for each $y \not\equiv x$.

   $(E^{x\xi}_{vf})(\eta) = E(\eta)$, for each $\eta \not\equiv \xi$.

Extension of the notation to the vector case: $E^{\bar{x}\bar{\xi}}_{\bar{v}\bar{f}}$, and restriction of it to cases such as E^{x}_{v}, should be clear. Observe that a notation such as $E^{x\xi}_{vf}$ has nothing to do with substitution which is a notion making sense only for *linguistic* entities.

4. SEMANTICS

The variety of terms as introduced in section 2 are now provided with a meaning. We define a process of interpretation of terms, in which the notion of their evaluation - by means of the function *val* - plays a central part.

An interpretation $\mathcal{J} = \langle \mathcal{V},\mathcal{C},\mathcal{E};\mathcal{D}\rangle$ has the following components:

1. A domain (non-empty set) $\mathcal{V}$.
2. $\mathcal{C}$ (dealing with the *constants*) maps A to $\mathcal{V}$, B to $\{0,1\}$, $\mathcal{A}$ to $\mathcal{V}^n \to \mathcal{V}$, $\mathcal{B}$ to $\mathcal{V}^n \to \{0,1\}$ (with, for $\alpha \in \mathcal{A}$, $n = n(\alpha)$, etc.).
3. $\mathcal{E}$ (dealing with the *variables*) maps X to $\mathcal{V}$, Q to $\{0,1\}$, $\mathcal{X}$ to $\mathcal{V}^n \to \mathcal{V}$ and $\mathcal{Q}$ to $\mathcal{V}^n \to \{0,1\}$ (with, for $\xi \in \mathcal{X}$, $n = n(\xi)$, etc.) (The variable-value mapping established by $\mathcal{E}$ is often referred to as the *environment*.)
4. $\mathcal{D}$ (dealing with the *declarations*) maps procedure symbols ϕ (or boolean procedure symbols ψ) of rank n to function terms τ (or boolean function terms π) of the same rank.

We now discuss the way in which the interpretations $\mathcal{J}$ are used to obtain values of terms: A certain computational process is defined, which is intended to model the semantics of the programming concepts concerned - such as described e.g. in the ALGOL 60 report - and which, for each of the pairs $\langle t,\mathcal{J}\rangle$ and $\langle p,\mathcal{J}\rangle$ will yield a value in $\mathcal{V}$ or $\{0,1\}$, respectively, as their *value*. However, this gives rise to an important point: We know that some computations in a programming language with recursion do not terminate, and, hence, that our function *val* will have to be *partial*: for some terms, no value will be delivered. In order to deal with this problem, the domain $\mathcal{V}$ is *extended* with one special element $\perp$, which is not an element of $\mathcal{V}$, and which stands for "undefined". From now on, $\mathcal{V}$ will refer to this extended set, and the subset of all "defined" elements of $\mathcal{V}$ will be called $\mathcal{V}_0$; i.e., we have $\mathcal{V} = \mathcal{V}_0 \cup \{\perp\}$. This extension is seemingly a trick which does not do away with any of the essential problems stemming from possibly unending computations. However, it will turn out to lead to a streamlining of much of the ensuing argument, and may be compared to some extent with the introduction of ∞ in the calculus.

We shall next define the function *val*, with $val(t,\mathcal{J})$ yielding $v \in \mathcal{V}$, according to the following scheme: First we introduce the *partial* function *valt*(t,J,N) (*valt* standing for terminating evaluation), where N is an integer which tells us how many computation steps are needed in order to arrive at the result v. Then we define the *total* function *val*(t,J) in terms of *valt*(t,J,N).

DEFINITION 4.1 (Terminating evaluations).
Let $J = \langle V,C,E;D\rangle$ be an interpretation, and t an individual term. *valt*(t,J,N) is defined by the following inductive definition:

1. $t \equiv a \in A$.
   If $C(a) = v \in V_0$, then $valt(a,J,1) = v$.
2. $t \equiv x \in X$.
   If $E(x) = v \in V_0$, then $valt(x,J,1) = v$.
3. $t \equiv \tau(t_1,\ldots,t_n) \equiv \tau(\bar{t})$.
   3.1. $\tau \equiv \alpha \in A$
   If $valt(t_i,J,N_i) = v_i \in V_0$, $i=1,\ldots,n$, and $C(\alpha)(\bar{v}) = v \in V_0$, then
   $$valt(\alpha(\bar{t}),J,(\textstyle\sum_i N_i)+1) = v.$$
   3.2. $\tau \equiv \xi \in X$
   If $valt(t_i,J,N_i) = v_i \in V_0$, $i=1,\ldots,n$, and $E(\xi)(\bar{v}) = v \in V_0$, then
   $$valt(\xi(\bar{t}),J,(\textstyle\sum_i N_i)+1) = v.$$
   3.3. (The central case). $\tau \equiv \nu\bar{x}\lambda\bar{y}.t$
   If $valt(t_h,J,N_h) = v_h \in V_0$, $h=1,\ldots,\ell$,
   and
   $$valt(t[z_h/x_h]_{h=1}^{\ell}[t_{\ell+k}/y_k]_{k=1}^{m},\ \langle V,C,E_{\bar{v}}^{\bar{z}};D\rangle,N) = v$$
   where the $\bar{z} = (z_1,\ldots,z_\ell)$ are *new* variables,
   then
   $$valt((\nu\bar{x}\lambda\bar{y}.t)(\bar{t}),J,(\textstyle\sum_h N_h)+N) = v.$$
   3.4. $\tau \equiv T(\tau_1,\ldots,\tau_r) = T(\bar{\tau})$, where $T \equiv \lambda\bar{\xi}.\tau_0$.
   If $valt(\tau_0[\bar{\tau}/\bar{\xi}](\bar{t}),J,N) = v$
   then
   $$valt(T(\bar{\tau})(\bar{t}),J,N+1) = v.$$

3.5. $\tau \equiv \phi \in F$

If $valt(\mathcal{D}(\phi)(\bar{t}),J,N) = v$

then

$valt(\phi(\bar{t}),J,N+1) = v$.

4. $\tau \equiv$ <u>if</u> p <u>then</u> t' <u>else</u> t".

If $valt(p,J,N) = 1$ and $valt(t',J,N') = v$, then

$valt($<u>if</u> p <u>then</u> t' <u>else</u> t"$,J,N+N') = v$.

If $valt(p,J,N) = 0$ and $valt(t'',J,N'') = v$, then

$valt($<u>if</u> p <u>then</u> t' <u>else</u> t"$,J,N+N'') = v$.

5. The definition of $valt(p,J,N)$ is completely analogous to 1-4 and omitted.

<u>DEFINITION</u> 4.2 (Evaluations).

1. $val(t,J) = v$ if there exists N such that $valt(t,J,N) = v$.
2. $val(t,J) = \perp$, otherwise.
3. Similarly for $val(p,J)$.

The following remarks have to be made on these definitions:

1. Observe that, if $valt(t,J,N) = v$, then $v \in V_0$. This follows by induction on the complexity of t. All the "terminal" cases in the inductive definition explicitly require that v be an element of V_0, and this property is inherited by the "non-terminal" cases.
2. (Clauses 1,2 of def. 4.1). The case that t is a constant or variable are clear.
3. (Clauses 3.1, 3.2 of def. 4.1). Let $t = \tau(\bar{t})$ be an application, with τ a constant or variable. Here we require that, for *valt* to be defined, v and each of the v_i, $i=1,\ldots,n$, be in V_0. Observe that, otherwise, $val(t,J)$ will, by definition 4.2, be set to $\perp$. The requirement that v_i be in V_0 is justified by our desire to have that our *basic* functions (i.e., the functions that are not defined via our language) satisfy the property that a function value be undefined when one of its arguments is undefined. That $v \in V_0$ fits in with our scheme that *valt* defines only terminating computations.
4. (The central case). Let $t \equiv \tau(\bar{t})$, with τ an abstraction. Here we observe that

   a. The value parameters t_h, $h=1,\ldots,\ell$, are evaluated first. The fact that the v_h are an outcome of *valt* guarantees that these evaluations terminate (cf. remark 1). Note that, if *valt* would not terminate for some t_h, then the attempt at defining *valt* for t would fail, and clause 2 of definition 4.2 would apply.

b. The environment is changed to a new environment which links new variables z_h to the v_h obtained above. The need for this change of variables from $\bar{x}$ to $\bar{z}$ is explained by the possibility that the $\bar{x}$ occur free in the $t_{\ell+k}$.

c. For the name parameters, no evaluation takes place, but a process of substituting the actual parameters $t_{\ell+k}$ for the formal parameters y_k, $k=1,\ldots,m$, is instead applied.

5. (Clause 3.4 of def. 4.1). This case is dealt with only for completeness sake.
6. (Clause 3.5 of def. 4.1). Here we find the rule of body replacement, which gives the standard meaning to recursion: The procedure symbol ϕ is replaced by the term $\mathcal{D}(\phi)$ which forms its body, and, next, the evaluation is continued.
7. (Clause 4 of def. 4.1). This defines the standard meaning of conditionals.

5. ASSERTIONS

Before we introduce the formalism to assert that certain facts hold for our terms under all interpretations, we need some preparatory concepts and lemma's.

Firstly, we introduce a partial ordering on our domains:

DEFINITION 5.1.

1. For $v_1, v_2 \in V$,
   $v_1 \subseteq v_2$ iff $v_1 = \perp$ or $v_1 = v_2$.
2. For $f_1, f_2 \in V^n \to V$,
   $f_1 \subseteq f_2$ iff, for all $\bar{v} \in V^n$, $f_1(\bar{v}) \subseteq f_2(\bar{v})$.
3. For $F_1, F_2 \in (V^n \to V)^r \to (V^n \to V)$,
   $F_1 \subseteq F_2$ iff, for all $\bar{f} \in (V^n \to V)^r$, $F_1(\bar{f}) \subseteq F_2(\bar{f})$.

Clearly, "$\subseteq$" is indeed a partial ordering. Thus $v_1 = v_2$ iff $v_1 \subseteq v_2$ and $v_2 \subseteq v_1$, etc. Anticipating again, now that $\subseteq$ is defined on V, we know what $val(t,J) \subseteq val(t',J)$ means, viz. that, for this J, either the evaluation of t does not terminate, or, t and t' have the same value in V_0. If the inclusion and its reverse hold for all J, we shall call t and t' semantically equivalent. Details follow. (We have made here the first step towards the extensive lattice-theoretic treatment in the more advanced theory of SCOTT, see e.g. [18]. The development of this is not necessary for our present purpose.)

As the next step, we extend the *val* function to terms τ and T. In this definition (and many of the subsequent formulations) we write $val(t,E)$, etc., instead of $val(t,J)$, etc., since it is only the E-component of J which interests us, the other components remaining fixed throughout.

<u>DEFINITION</u> 5.2.

1. $val(\tau,E)$ is defined as that function f: $V^n \to V$ which satisfies:
   $f(\bar{v}) = v$ iff, for *new* $\bar{x}$, $val(\tau(\bar{x}),E^{\bar{x}}_{\bar{v}}) = v$.
2. $val(T,E)$ is defined as that functional F: $(V^n \to V)^r \to (V^n \to V)$, which satisfies:
   $F(\bar{f}) = f$ iff, for *new* $\bar{\xi}$, $val(T(\bar{\xi}),E^{\bar{\xi}}_{\bar{f}}) = f$.

As the first lemma about the extended *val* we state

<u>LEMMA</u> 5.3.

1. *If none of the* $\bar{z},\bar{u}$ *occur free in* s,
   $val(\nu\bar{x}\lambda\bar{y}.s,E) = val(\nu\bar{z}\lambda\bar{u}.s[\bar{z}/\bar{x}][\bar{u}/\bar{y}],E)$.
2. *If none of the* $\bar{\eta}$ *occur free in* σ,
   $val(\lambda\bar{\xi}.\sigma,E) = val(\lambda\bar{\eta}.\sigma[\bar{\eta}/\bar{\xi}],E)$.

<u>PROOF</u>. Follows from the definitions and lemma 3.3. □

<u>REMARK</u>. This lemma is, clearly, the analogue of the rule of α-conversion in the ordinary lambda calculus. It allows us a rewriting of bound variables where this is convenient.

The next lemma is of considerable technical importance in our development. It is the main tool in the proof of the monotonicity theorem of the next section, which, in turn, plays an important part in the proof of the least fixed point theorem.

LEMMA 5.4.

Assume

1. $val(t,E) = v$, $val(\tau,E) = f$

2.1. $val(\sigma,E) \subseteq val(\sigma',E')$

2.2. $val(S,E) \subseteq val(S',E')$

3.1. $val(s_i,E) \subseteq val(s_i',E')$, $i=1,\dots,n$

3.2 $val(\sigma_j,E) \subseteq val(\sigma_j',E')$, $j=1,\dots,r$

4. $v' \subseteq v''$, $f' \subseteq f''$.

Then

1.1. $val(s[t/x],E) = val(s,E_v^x)$

1.2. $val(\sigma[t/x],E) = val(\sigma,E_v^x)$

1.3. $val(S[t/x],E) = val(S,E_v^x)$

1.4. $val(s[\tau/\xi],E) = val(s,E_f^\xi)$

1.5. $val(\sigma[\tau/\xi],E) = val(\sigma,E_f^\xi)$

1.6. $val(S[\tau/\xi],E) = val(S,E_f^\xi)$

2.1. $val(\sigma(s_1,\dots,s_n),E) \subseteq val(\sigma'(s_1',\dots,s_n'),E')$

2.2. $val(S(\sigma_1,\dots,\sigma_r),E) \subseteq val(S'(\sigma_1',\dots,\sigma_r'),E')$

3.1. $val(s,E_{v'f'}^{x\,\xi}) \subseteq val(s,E_{v''f''}^{x\,\xi})$

3.2. $val(\sigma,E_{v'f'}^{x\,\xi}) \subseteq val(\sigma,E_{v''f''}^{x\,\xi})$

3.3. $val(S,E_{v'f'}^{x\,\xi}) \subseteq val(S,E_{v''f''}^{x\,\xi})$

PROOF. By simultaneous induction on the complexity of the terms. We prove a few selected cases:

1.1. $s \equiv a$: $val(a[t/x],E) = val(a,E) = val(a,E^x_v)$.

$s \equiv x$: $val(x[t/x],E) = val(t,E) = v = val(x,E^x_v)$.

$s \equiv y \not\equiv x$: $val(y[t/x],E) = val(y,E) = val(y,E^x_v)$.

$s \equiv \sigma(\bar{s})$: $val(\sigma(\bar{s})[t/x],E) = val(\sigma[t/x](\bar{s}[t/x]),E)$.

We have, by conclusions 1.2, 1.1, and induction:

$$val(\sigma[t/x],E) = val(\sigma,E^x_v)$$

$$val(\bar{s}[t/x],E) = val(\bar{s},E^x_v)$$

Hence, by conclusion 2.1 applied twice, and induction,

$$val(\sigma[t/x](\bar{s}[t/x]),E) = val(\sigma(\bar{s}),E^x_v)$$

$s \equiv$ <u>if</u> p <u>then</u> s' <u>else</u> s". For the reader.

1.2. The cases that σ is a constant, variable or application are clear. Now let $\sigma \equiv \nu\bar{x}\lambda\bar{y}.s$. By suitably rewriting of bound variables (lemma 5.3) we may assume that none of the $\bar{x}$ or $\bar{y}$ occurs free in s or t. Then $val((\nu\bar{x}\lambda\bar{y}.s)[t/x],E) = val(\nu\bar{x}\lambda\bar{y}.s[t/x],E)$. We have, by definition 5.2, $val(\nu\bar{x}\lambda\bar{y}.s[t/x],E) = val(\nu\bar{x}\lambda\bar{y}.s,E^x_v)$ iff, for new $\bar{z}$ and arbitrary $\bar{w}$, $val((\nu\bar{x}\lambda\bar{y}.s[t/x])(\bar{z}),E^{\bar{z}}_{\bar{w}}) = val((\nu\bar{x}\lambda\bar{y}.s)(\bar{z}),E^{x\bar{z}}_{v\bar{w}})$, or, by definition of val and since the $\bar{z}$ are new, $val(s[t/x][\bar{z}/\bar{y}],E^{\bar{z}}_{\bar{w}}) = val(s[\bar{z}/\bar{y}],E^{x\bar{z}}_{v\bar{w}})$ (we assume that none of the $w_h = \perp$; otherwise, the result is obvious) or, by lemma 3.3, $val(s[\bar{z}/\bar{y}][t/x],E^{\bar{z}}_{\bar{w}}) = val(s[\bar{z}/\bar{y}],E^{\bar{z}x}_{\bar{w}v})$, and this holds by conclusion 1.1 of the lemma and induction.

1.3, 1.4, 1.5, 1.6. For the reader

2.1. By assumption 2.1, $val(\sigma,E) \subseteq val(\sigma',E')$, i.e. by definition 5.2, $val(\sigma(\bar{x}),E^{\bar{x}}_{\bar{v}}) \subseteq val(\sigma(\bar{x}),E^{\bar{x}}_{\bar{v}})$, for new $\bar{x}$ and arbitrary $\bar{v}$. By conclusion 3.1 and induction, $val(\sigma'(\bar{x}),E'^{\bar{x}}_{\bar{v}}) \subseteq val(\sigma'(\bar{x}),E'^{\bar{x}}_{\bar{v}'})$, if $\bar{v} \subseteq \bar{v}'$. Now choose $\bar{v} = val(\bar{s},E)$, $\bar{v}' = val(\bar{s}',E')$. Then $\bar{v} \subseteq \bar{v}'$ by assumption 3.1. We then have

$val(\sigma(\bar{x}),E^{\bar{x}}_{\bar{v}}) \subseteq val(\sigma'(\bar{x}),E'^{\bar{x}}_{\bar{v}'})$ (derived above)

$val(\sigma(\bar{x})[\bar{s}/\bar{x}],E) = val(\sigma(\bar{x}),E^{\bar{x}}_{\bar{v}})$ (by conclusion 1.1 and induction)

$val(\sigma'(\bar{x})[\bar{s}'/\bar{x}],E') = val(\sigma'(\bar{x}),E'^{\bar{x}}_{\bar{v}'})$ (similarly)

Hence, $val(\sigma(\bar{s}),E) \subseteq val(\sigma'(\bar{s}'),E')$, as was to be shown.

2.2 Omitted.

3.1. The cases that s is a constant or variable are clear. Now let $s \equiv \sigma(\bar{s})$. We have $val(\sigma, E^{x\ \xi}_{v'f'}) \subseteq val(\sigma, E^{x\ \xi}_{v''f''})$, by conclusion 3.2 and induction, $val(\bar{s}, E^{x\ \xi}_{v'f'}) \subseteq val(\bar{s}, E^{x\ \xi}_{v''f''})$, by conclusion 3.1 and induction; the result then follows by two applications of conclusion 2.1 and induction. The case that s is a selection is, once more, for the reader.

3.2, 3.3. Omitted. □

As the first consequence of lemma 5.4 we have lemma 5.5, which shows that our extension of the definition of *val* as given in definition 5.1, is consistent.

LEMMA 5.5.

1. $val(\tau(\bar{t}),E) = val(\tau,E)(val(\bar{t},E))$.
2. $val(T(\bar{\tau}),E) = val(T,E)(val(\bar{\tau},E))$.

PROOF.

1. Let $val(\bar{t},E) = \bar{v}$. Then $val(\tau,E)(val(\bar{t},E)) = val(\tau,E)(\bar{v}) =$
   $=$ (df. 5.2)$val(\tau(\bar{x}),E^{\bar{x}}_{\bar{v}}) =$ (lemma 5.4, part 1.1)$val(\tau(\bar{x})[\bar{t}/\bar{x}],E) =$
   $= val(\tau(\bar{t}),E)$.
2. Similar. □

We are also in the position to show that changing call-by-value to call-by-name for one or more of the parameters yields a possibly extended function:

LEMMA 5.6. $val(\nu x_1 \ldots x_{\ell+1} \lambda x_{\ell+2} \ldots x_n . t, E) \subseteq val(\nu x_1 \ldots x_\ell \lambda x_{\ell+1} \ldots x_n . t, E)$.

PROOF. We show that, for new $\bar{z} = (z_1,\ldots,z_n)$
$val((\nu x_1 \ldots x_{\ell+1} \lambda x_{\ell+2} \ldots x_n . t)(\bar{z}), E^{\bar{z}}_{\bar{v}}) \subseteq val((\nu x_1 \ldots x_\ell \lambda x_{\ell+1} \ldots x_n . t)(\bar{z}), E^{\bar{z}}_{\bar{v}})$.
If any of the z_n, $1 \le h \le \ell+1$, equals $\bot$, the left-hand side of this inclusion has $\bot$ as value, and we are done. Otherwise, since the $\bar{z}$ are new,
$val((\nu x_1 \ldots x_{\ell+1} \lambda x_{\ell+2} \ldots x_n . t)(\bar{z}), E^{\bar{z}}_{\bar{v}}) =$ (def. 4.1)
$val(t[z_h/x_h]^{\ell+1}_{h=1}[z_k/x_k]^{n}_{k=\ell+2}, E^{\bar{z}}_{\bar{v}}) = val(t[z_h/x_h]^{\ell}_{h=1}[z_k/x_k]^{n}_{k=\ell+1}, E^{\bar{z}}_{\bar{v}}) =$
$val((\nu x_1 \ldots x_\ell \lambda x_{\ell+1} \ldots x_n . t)(\bar{z}), E^{\bar{z}}_{\bar{v}})$. □

We can now, at last, introduce the formalism in which we shall below state the main theorem of our paper.

We are interested in proving *assertions* about *formulae* $\Phi, \Psi \ldots$. A formulae is a *set* of *atomic* formulae, and an atomic formula is an inclusion of one of the forms $t_1 \subseteq t_2$, $\tau_1 \subseteq \tau_2$, or $T_1 \subseteq T_2$. An assertion has the form

$$\Phi \models_{\mathcal{D}} \Psi$$

For assertions in this format we introduce the notion of *validity* in

DEFINITION 5.6 (Validity of assertions).
Let $\mathcal{D}$ be a given mapping from (boolean) procedure symbols to (boolean) function terms.

1. Let $t_1 \subseteq t_2$ be an atomic formula. We call this formula *satisfied* by an interpretation J, iff $val(t_1, J) \subseteq val(t_2, J)$. Similarly for $\tau_1 \subseteq \tau_2$ and $T_1 \subseteq T_2$.
2. J satisfies a formula Φ iff it satisfies each element of the set Φ. If $J = \langle V, C, E; \mathcal{D} \rangle$ satisfies Φ, we also say that "$\Phi[V,C,E;\mathcal{D}]$ holds".
3. An assertion $\Phi \models_{\mathcal{D}} \psi$ is called *valid* iff:
   For all V,C, whenever, for all E, $\Phi[V,C,E;\mathcal{D}]$ holds, then, for all E, $\Psi[V,C,E;\mathcal{D}]$ holds.

Note carefully the structure of clause 3 in the definition. Firstly, $\mathcal{D}$ will remain fixed in the application we have in mind, and is not subject to quantification. However, we emphasize the difference between the role of the V and C on the one hand, and E on the other hand: We define $\ldots \models \ldots$ as a statement of the form $\forall V,C[\forall E\ldots \supset \forall E\ldots]$, and not of the form $\forall V,C,E[\ldots \supset \ldots]$. In order to explain this, consider for instance the desired monotonicity property, which includes as special case: $\{x \subseteq y\} \models_{\mathcal{D}} \{x[t/x] \subseteq y[t/x]\}$. Now, according to the second (rejected) definition, this means that for all E, if $E(x) \subseteq E(y)$ then $val(t,E) \subseteq E(y)$, which is clearly absurd. According to the first (adopted) definition, all it states is that, if for all E, $E(x) \subseteq E(y)$, then for all E, $val(t,E) \subseteq E(y)$, and this implication does hold since its antecedent is false.

The section is concluded with some additional pieces of notation:

1. For $\Phi \models_{\mathcal{D}} \Psi$ we write, in the case that $\mathcal{D}$ is understood, just $\Phi \models \Psi$.
2. For $\phi \models \Psi$, with ϕ the empty set, we write $\models \Psi$.
3. When confusion is improbable, we omit the {} around a collection $\Phi = \{\rho_1,\rho_2,\ldots\}$ of atomic formulae.
4. For $\models t_1 \subseteq t_2, t_2 \subseteq t_1$ we write $\models t_1 = t_2$, and similarly with $\models \tau_1 = \tau_2$ and $\models T_1 = T_2$.

6. MONOTONICITY

The first theorem of the paper states the monotonicity of our terms: The syntactical constructions of substitution, application, abstraction and selection all preserve the semantic ordering "$\subseteq$" between terms. (Recursion also preserves "$\subseteq$", but this can be proved only *after* the least fixed point result has been established. No further attention will be paid to this; the reader will have no problem to adapt the proof e.g. in [1] to the present formalism.)

<u>THEOREM</u> 6.1. (Monotonicity).

$s \subseteq s'$, $t \subseteq t'$, $t_i \subseteq t_i'$, $i=1,\ldots,n$, $p \subseteq p'$
$\sigma \subseteq \sigma'$, $\tau \subseteq \tau'$, $\tau_j \subseteq \tau_j'$, $j=1,\ldots,r$
$S \subseteq S'$,

$\models$

$s[t/x] \subseteq s'[t'/x]$,
$\sigma[t/x] \subseteq \sigma'[t'/x]$, (Substitution,1)
$S[t/x] \subseteq S'[t'/x]$,

$s[\tau/\xi] \subseteq s'[\tau'/\xi]$,
$\sigma[\tau/\xi] \subseteq \sigma'[\tau'/\xi]$, (Substitution,2)
$S[\tau/\xi] \subseteq S'[\tau'/\xi]$,

$\sigma(t_1,\ldots,t_n) \subseteq \sigma'(t_1',\ldots,t_n')$, (Application)
$S(\tau_1,\ldots,\tau_r) \subseteq S'(\tau_1',\ldots,\tau_r')$,

$\nu x_1 \ldots x_\ell \lambda y_1 \ldots y_m . t \subseteq \nu x_1 \ldots x_\ell \lambda y_1 \ldots y_m . t'$, (Abstraction)
$\lambda \xi_1 \ldots \xi_r . \tau \subseteq \lambda \xi_1 \ldots \xi_r . \tau'$,

if p then s else t $\subseteq$ (Selection)
if p' then s' else t'.

PROOF.

1. (Substitution). $s[t/x] \subseteq s'[t'/x]$. Choose any E, and let $val(t,E) = v$, $val(t',E) = v'$. Then $val(s[t/x],E) = val(s,E^x_v) \subseteq val(s,E^x_{v'}) \subseteq val(s',E^x_{v'}) = val(s'[t'/x],E)$, by lemma 5.4, part 1.1, lemma 5.4, part 3.1, the assumption, and lemma 5.4, part 1.1. The other cases for substitution are similar.

2. (Application). Direct from lemma 5.4, part 2.1.

3. (Abstraction). Let $\bar{z}$ be *new* variables, and let $\bar{v}$ be an arbitrary element in V^n. We shall show $val((\nu\bar{x}\lambda\bar{y}.t)(\bar{z}),E^{\bar{z}}_{\bar{v}}) \subseteq val((\nu\bar{x}\lambda\bar{y}.t')(\bar{z}),E^{\bar{z}}_{\bar{v}})$. If any of the v_h, $1 \le h \le \ell$, equals $\perp$, the whole evaluation on the left-hand side yields $\perp$, and we are done. Otherwise, we argue as follows: we apply the definition of *val*, the fact that $val(\bar{z},E^{\bar{z}}_{\bar{v}}) = \bar{v}$, and the fact that the $\bar{z}$ do not occur in $\nu\bar{x}\lambda\bar{y}.t$ or $\nu\bar{x}\lambda\bar{y}.t'$, and obtain successively:
$val((\nu\bar{x}\lambda\bar{y}.t)(\bar{z}),E^{\bar{z}}_{\bar{v}})$ = (def. 4.1)
$val(t[\bar{z}/\bar{x}][\bar{z}/\bar{y}],E^{\bar{z}}_{\bar{v}})$ = (lemma 5.4, part 1.1)
$val(t,E^{\bar{z}\bar{x}\bar{y}}_{\bar{v}\bar{v}\bar{v}}) \subseteq$ (assumption)
$val(t',E^{\bar{z}\bar{x}\bar{y}}_{\bar{v}\bar{v}\bar{v}}) = \ldots = val((\nu\bar{x}\lambda\bar{y}.t')(\bar{z}),E^{\bar{z}}_{\bar{v}})$, as was to be shown.
The functional case is left to the reader, as is

4. (Selection). □

Statement and proof of the monotonicity theorem for the various boolean cases are omitted.

7. APPROXIMATIONS

We arrive at our last body of definitions and preparatory lemmas. We assume from now on that we deal with one fixed $\mathcal{D}$, defined on each of

$\phi_1,\ldots,\phi_r$, with $\mathcal{D}(\phi_j) \equiv \tau_j$, j=1,...,r. The proof to be given presently has to have available terms which, for all interpretations, have non-terminating evaluations. This is the reason for the following conventions: We extend our set of procedure symbols with the symbol ϕ_0, with declaration $\mathcal{D}(\phi_0) \equiv \phi_0$. Hence, ϕ_0 is a procedure which, when evaluated, causes nothing but a call upon itself. Clearly, therefore, the evaluation of ϕ_0 terminates for no argument.

Let us write ω, Ω, $\mathcal{O}$ for the "nowhere defined" individual-, function- and functional terms defined by:

$$\omega \equiv \phi_0(\bar{z}), \text{ with } \bar{z} \text{ new variables,}$$
$$\Omega \equiv \nu\bar{x}\lambda\bar{y}.\omega,$$
$$\mathcal{O} \equiv \lambda\bar{\xi}.\Omega.$$

It is left to the reader to verify that, for all J, $val(\omega,J) = \bot$, $val(\Omega,J) = f_0$, where $f_0(\bar{v}) = \bot$ for all $v \in V^n$, and $val(\mathcal{O},J) = F_0$, where $F_0(\bar{f}) = f_0$ for all $\bar{f} \in (V^n \to V)^r$.

The reason we are interested in these constructs is the following: We want to define a process of *approximation* to our terms. An intuitive explanation is given first. Consider an individual term t. In general, t contains one or more occurrences ("calls") of the (recursive) procedure symbols $\phi_1,\ldots,\phi_r$, and evaluation of t will result in a, more or less elaborate, "calling tree" for the ϕ, where, in general, some ϕ_{j1} may call a ϕ_{j2}, this calls ϕ_{j3}, etc., with the possibility that $\phi_{j_i} \equiv \phi_{j_k}$ for $i \neq k$. However complicated this process may be, we always have that, if the evaluation of t terminates with value $v \in V_0$, then the calling tree is finite. It is then possible to obtain the same v as value of a new term, which is derived from t by suitable *finite* replacement of procedure symbols by the bodies of their declarations, where the procedures at the innermost level are not called any more (in general as a result of selection choosing another branch). These innermost occurrences of procedure symbols may then be replaced by whatever term we like, without changing the outcome. We now choose for this the undefined term Ω just introduced, since this choice guarantees a convenient ordering of the approximations, as will be seen soon. To be somewhat more specific, we shall prove that, for each term t and interpretation J, there exists a term $t^{(i)}$ (with i an integer which is derived from the size of the calling tree determined by t, and, therefore, depending upon J) such that $val(t,J) = val(t^{(i)},J)$, and, moreover, $t^{(i)}$ contains no occurrences of any procedure symbol.

The first step towards the precise formulation of this idea is the introduction of one more syntactic operation on terms t, τ and T. The operation is denoted by "$\sim$", and defined *with respect to* the collection $\{\phi_1,\dots,\phi_r\}$. It amounts to the replacement, in the term at hand, of each occurrence of a procedure symbol ϕ_j by a *new* variable ξ_j, for $j=1,\dots,r$. In other words:

a. $\tilde{x} \equiv x$, $\tilde{a} \equiv a$, $\tau(\bar{t})^{\sim} \equiv \tilde{\tau}(\tilde{\bar{t}})$,
   $(\underline{\text{if}}\ p\ \underline{\text{then}}\ t'\ \underline{\text{else}}\ t'')^{\sim} \equiv \underline{\text{if}}\ \tilde{p}\ \underline{\text{then}}\ \tilde{t}'\ \underline{\text{else}}\ \tilde{t}''$.

b. $\tilde{\alpha} \equiv \alpha$, $\tilde{\xi} \equiv \xi$, $\tilde{\phi}_j \equiv \xi_j$, $T(\bar{\tau})^{\sim} \equiv \tilde{T}(\tilde{\bar{\tau}})$,
   $(\nu\bar{x}\lambda\bar{y}.t)^{\sim} \equiv \nu\bar{x}\lambda\bar{y}.\tilde{t}$.

c. $(\lambda\bar{\xi}.\tau)^{\sim} \equiv \lambda\bar{\xi}.\tilde{\tau}$.

The approximations are now defined in

<u>DEFINITION</u> 7.1 (Approximations).

1. $t^{(0)} \equiv \omega$, $\tau^{(0)} \equiv \Omega$, $T^{(0)} \equiv \mathcal{O}$.
2. $t^{(i+1)} \equiv \tilde{t}[\tau_j^{(i)}/\xi_j]_{j=1}^r$, $i=0,1,\dots$
   $\tau^{(i+1)} \equiv \tilde{\tau}[\tau_j^{(i)}/\xi_j]_{j=1}^r$, $i=0,1,\dots$
   $T^{(i+1)} \equiv \tilde{T}[\tau_j^{(i)}/\xi_j]_{j=1}^r$, $i=0,1,\dots$

It should be observed here that the τ_j in this definition are the bodies of the procedures ϕ_j, and that the ξ_j are the new variables introduced in the definition of the $\sim$-operation. In words, the zero-th approximation to τ is Ω, the i+1-st approximation is a term resulting from τ by
- replacing all procedure symbols by new variables
- substituting for these new variables the i-th approximations to the τ_j

(This somewhat roundabout process is necessary since we cannot substitute for procedure symbols.)

As first lemma on these new constructs we have

LEMMA 7.2.

1. $\phi_j^{(i+1)} \equiv \tau_j^{(i)}$ $(\equiv(\mathcal{D}(\phi_j))^{(i)})$, $j=1,\ldots,r$, $i=0,1,\ldots$

2. $\models \tau = (\lambda\bar{\xi}.\tilde{\tau})(\phi_1,\ldots,\phi_r)$

3. $\models t^{(i)} \subseteq t$, $\tau^{(i)} \subseteq \tau$, $T^{(i)} \subseteq T$, $i=0,1,\ldots$

4. $s \subseteq t$, $\sigma \subseteq \tau$, $S \subseteq T$

$\models$

$s^{(i)} \subseteq t^{(i)}$, $\sigma^{(i)} \subseteq \tau^{(i)}$, $S^{(i)} \subseteq T^{(i)}$, $i=0,1,\ldots$

PROOF. Straightforward from the definitions. □

The next lemma is the key result for the proof of the least fixed point theorem.

LEMMA 7.3. *For all* t *and* J, *if* $val(t,J) = v$, *then there exists some* i *such that* $val(t^{(i)},J) = v$.

PROOF. If $v = \perp$, take $i = 0$. Otherwise, $v \in V_0$, and $valt(t,J,N) = v$, for some N. We prove, by induction on N, the following statement: If $valt(t,J,N) = v$, then, for some i and M, $valt(t^{(i)},J,M) = v$.

1. $t \equiv a$ or $t \equiv x$. Take $i = 1$.
2. $t \equiv \tau(\bar{t})$.
   a. $\tau \equiv \phi_j$. We have successively
      $valt(\phi_j(\bar{t}),J,N) = v$, or, by definition 4.1,
      $valt(\tau_j(\bar{t}),J,N-1) = v$, or, by induction, for some i_0 and M,
      $valt((\tau_j(\bar{t}))^{(i_0)},J,M) = v$, or, by definition 3.3 and 7.1,
      $valt(\tau_j^{(i_0)}(\bar{t}^{(i_0)}),J,M) = v$, or, by lemma 7.2, part 1,
      $valt(\phi_j^{(i_0+1)}(\bar{t}^{(i_0)}),J,M) = v$, or, by monotonicity,
      $valt(\phi_j^{(i_0+1)}(\bar{t}^{(i_0+1)}),J,M) = v$, or, as above,
      $valt((\phi_j(\bar{t}))^{(i_0+1)},J,M) = v$.
      Taking $i = i_0+1$ thus proves this case.
   b. $\tau \equiv \alpha$. Let $valt(\alpha(\bar{t}),J,N) = v$. Then there exist v_k such that $valt(t_k,J,N_k) = v_k$, with $N_k < N$, for $k=1,\ldots,n$. Thus, by induction,

$valt(t_k^{(i_k)}, ,M_k) = v_k$. Now taking $i = \max(i_1,\ldots,i_n)$ settles this case.

c. $\tau \equiv \xi$ or, $\tau \equiv T(\bar{\tau})$, or $\tau \equiv \nu\bar{x}\lambda\bar{y}.t$. These cases are proven similarly using induction and monotonicity.

3. $t \equiv$ if p then t' else t". For the reader. □

REMARK. As pointed out by the referee, it should be observed that this lemma - though sufficient for our present purposes - needs additional assumptions in order to be generalized to domains with a richer partial ordering than the one imposed on our V by def. 5.1.1.

8. LEAST FIXED POINTS

The time for the payoff of our labour has arrived. We state and prove

THEOREM 8.1 (The least fixed point theorem).
Let $\phi_1,\ldots,\phi_r$ *be procedure symbols with* $\mathcal{D}(\phi_j) \equiv \tau_j$, $j=1,\ldots,r$. *Let us put* $T_j \equiv \lambda\bar{\xi}.\tilde{\tau}_j$. *Then*

1. $\models \{T_j(\phi_1,\ldots,\phi_r) = \phi_j\}_{j=1}^r$
2. $\{T_j(\sigma_1,\ldots,\sigma_r) = \sigma_j\}_{j=1}^r \models \{\phi_j \subseteq \sigma_j\}_{j=1}^r$

REMARK

1. The first statement tells us that the ϕ's *are* fixed points of the T's; the second that they are *least* w.r.t. "$\subseteq$".
2. Observe that the first statement is nothing but $\models\{\tau_j = \phi_j\}_{j=1}^r$. However, this formulation does not bring out the fixed point aspect, and we have taken no inconsiderable trouble to provide a notation - with all the extra's for functionals - which does emphasize this.

PROOF.

1. We show that $\models\{\tau_j = \phi_j\}_{j=1}^r$.
   a. $\subseteq$: Let $val(\tau_j(\bar{z}),E_{\bar{v}}^{\bar{x}}) = v$. If $v = \bot$, we are done. Otherwise, there exists N such that $valt(\tau_j(\bar{z}),E_{\bar{v}}^{\bar{z}},N) = v$. Then, by definition 4.1, $valt(\phi_j(\bar{z}),E_{\bar{v}}^{\bar{z}}, N+1) = v$, hence $val(\phi_j(\bar{z}),E_{\bar{v}}^{\bar{z}}) = v$. This proves "$\subseteq$".
   b. $\supseteq$: Reverse the argument of part a.
2. We first show

(8.1) $\{T_j(\sigma_1,\ldots,\sigma_r) \subseteq \sigma_j\}_{j=1}^r \models \{\phi_j^{(i)} \subseteq \sigma_j\}_{j=1}^r$, i=0,1,...

by induction on i.

a. i = 0. Immediate from the definitions.

b. Assume the result for some i:

(8.2) $\{T_j(\sigma_1,\ldots,\sigma_r) \subseteq \sigma_j\}_{j=1}^r \models \{\phi_j^{(i)} \subseteq \sigma_j\}_{j=1}^r$.

By lemma 7.2 and the definitions,

$$\models \phi_j^{(i+1)} = \tau_j^{(i)}, \text{ and}$$

$$\models \tau_j^{(i)} = (T_j(\phi_1,\ldots,\phi_r))^{(i)} \quad (\equiv T_j^{(i)}(\bar{\phi}^{(i)})).$$

Now $T_j^{(i)} \equiv (\lambda\bar{\xi}.\tilde{\tau}_j)^{(i)} \equiv (\lambda\bar{\xi}.\tilde{\tau}_j)^{\sim}[\tau_j^{(i-1)}/\xi_j]_{j=1}^r \equiv (\lambda\bar{\xi}.\tilde{\tilde{\tau}})[\tau_j^{(i-1)}/\xi_j]_{j=1}^r \equiv$ $\equiv \lambda\bar{\xi}.\tilde{\tilde{\tau}} \equiv \lambda\bar{\xi}.\tilde{\tau} \equiv T_j$, as follows from the definitions of "$\sim$" and of substitution. Thus we obtain

(8.3) $\models \phi_j^{(i+1)} = T_j(\phi_1^{(i)},\ldots,\phi_r^{(i)})$.

By monotonicity we have

(8.4) $\{\phi_j^{(i)} \subseteq \sigma_j\}_{j=1}^r \models \{T_j(\phi_1^{(i)},\ldots,\phi_r^{(i)}) \subseteq T_j(\sigma_1,\ldots,\sigma_r)\}_{j=1}^r$.

From (8.2), (8.3) and (8.4) we conclude

$$\{T_j(\sigma_1,\ldots,\sigma_r) \subseteq \sigma_j\}_{j=1}^r \models \{\phi_j^{(i+1)} \subseteq \sigma_j\}_{j=1}^r$$

thus completing the inductive proof of (8.1) (or, in fact, of a stronger version with "$\subseteq$" instead of "=" in its antecedent). Next, we choose some V and C, and assume that for all E, $val(T_j(\sigma_1,\ldots,\sigma_r),J) \subseteq val(\sigma_j,J)$, for j=1,...,r and with $J = \langle V,C,E;\mathcal{D}\rangle$. We show that then, for all E (using the notation with V and C suppressed), $val(\phi_j,E) \subseteq val(\sigma_j,E)$, i.e., for new $\bar{z}$ and arbitrary $\bar{v}$, $val(\phi_j(\bar{z}),E_{\bar{v}}^{\bar{z}}) \subseteq val(\sigma_j(\bar{z}),E_{\bar{v}}^{\bar{z}})$, j=1,...,r. Assume

$val(\phi_j(\bar{z}),E_{\bar{v}}^{\bar{z}}) = v$. Then, by lemma 7.3, for some i, $val((\phi_j(\bar{z}))^{(i)},E_{\bar{v}}^{\bar{z}}) =$ $= v$, i.e., $val(\phi_j^{(i)}(\bar{z}),E_{\bar{v}}^{\bar{z}}) = v$. From (8.1) it then follows that $val(\sigma_j(\bar{z}),E_{\bar{v}}^{\bar{z}}) = v$, as was to be shown. □

We have completed the proof of the least fixed point theorem, thus achieving the goal of our paper.

REFERENCES

[1] DE BAKKER, J.W., *Recursive Procedures*, Mathematical Centre Tracts 24, Mathematisch Centrum, Amsterdam, 1971.

[2] DE BAKKER, J.W., *Fixed points in programming theory*, in: Foundations of Computer Science (J.W. de Bakker, ed.), p.1-49, Mathematical Centre Tracts 63, Mathematisch Centrum, 1975.

[3] DE BAKKER, J.W. & W.P. DE ROEVER, *A calculus for recursive program schemes*, in: Automata, Languages and Programming (M. Nivat, ed.), p.167-196, North-Holland, Amsterdam, 1973.

[4] DE BAKKER, J.W. &. L.G.L.T. MEERTENS, *On the completeness of the inductive assertion method*, to appear in J. of Comp. Syst. Sciences.

[5] HINDLEY, J.R., B. LERCHER & J.P. SELDIN, *Introduction to Combinatory Logic*, Cambridge University Press, Cambridge, 1972.

[6] KLEENE, S.C., *Introduction to Metamathematics*, North-Holland, Amsterdam, 1952.

[7] LANDIN, P.J., *The mechanical evaluation of expressions*, Comp. J, 6 (1964), p.308-320.

[8] MANNA, Z., S. NESS & J. VUILLEMIN, *Inductive methods for proving properties of programs*, CACM, 16 (1973), p.491-502.

[9] MANNA, Z. & J. VUILLEMIN, *Fixpoint approach to the theory of computation*, CACM, 15 (1972), p.528-536.

[10] McCARTHY, J., *A basis for a mathematical theory of computation*, in: Computer Programming and Formal Systems (P. Braffort & D. Hirschberg, eds.), p.33-70, North-Holland, Amsterdam, 1963.

[11] MILNER, R., *Implementation and applications of Scott's logic for computable functions*, in: Proc. of an ACM Conference on Proving Assertions about Programs, p.1-6, ACM, 1972.

[12] MORRIS, J.H., *Lambda-calculus models of programming languages*, Ph.D Thesis, M.I.T., 1968.

[13] MORRIS, J.H., *Another recursion induction principle*, CACM, 14 (1971), p.351-354.

[14] NIVAT, M., *On the interpretation of recursive program schemes*, Report A74/09, Saarland University, Saarbrücken, 1974.

[15] PARK, D., *Fixpoint induction and proof of program semantics*, in: Machine Intelligence, Vol. 5 (B. Meltzer & D. Michie, eds.) p.59-78, Edinburgh University Press, 1970.

[16] DE ROEVER, W.P., *Recursion and parameter-mechanisms: an axiomatic approach*, in: Automata, Languages and Programming (J. Loeckx, ed.), p.34-65, Lecture Notes in Computer Science Vol. 14, Springer-Verlag, Berlin etc., 1974.

[17] ROSEN, B.K., *Tree-manipulating systems and Church-Rosser theorems*, J.ACM, 20 (1973), p.160-187.

[18] SCOTT, D., *Outline of a mathematical theory of computation*, in: Proc. of the Fourth Annual Princeton Conference on Information Sciences and Systems, p.169-176, Princeton, 1970.

[19] SCOTT, D. & J.W. DE BAKKER, *A theory of programs*, unpublished notes, IBM Seminar, Vienna, 1969.

[20] SCOTT, D. & C. STRACHEY, *Towards a mathematical semantics for computer languages*, in: Proc. of the Symposium on Computers and Automata (J. Fox, ed.), p.19-46, Polytechnic Inst. of Brooklyn, 1971.

INFINITE NORMAL FORMS FOR THE λ-CALCULUS[†*]

Reiji Nakajima

Computer Science Division
University of California, Berkeley

Abstract

The notion of C-function is introduced to λ-calculus with η-convertibility as a generalization of normal forms. C is a function from the λ-expressions, Λ, onto a partially ordered set, $\mathbb{C}_{fin}$. The D_∞-value of $X \in \Lambda$ is characterized by $C(X) \in \mathbb{C}_{fin}$. Extending the syntactical structure of $\mathbb{C}_{fin}$ into $\mathbb{C}_{inf}$, we generalize Λ to Λ^∞, the infinite λ-expressions. The lattice topology of Λ and Λ^∞ induced by D_∞ is equivalent to the lattice topology of $\mathbb{C}_{inf}$. Since $\mathbb{C}_{inf}$ is deduced from Λ independent of D_∞, $\mathbb{C}_{inf}$ can be said to give a natural lattice structure of Λ.

§1. Preparations

In this paper, the reader is assumed to be familiar with Scott's theory of computation, especially D_∞-model [7-10] and Wadsworth's model theory of λ-calculus in D_∞ [12-13].

We introduce several useful concepts and notations.

(1) $\mathbb{Z}$: set of the integers

$\mathbb{N}$: set of all positive integers

U: denumerable set of the variables denoted by lower case letters, z, u, v, etc.

Λ: set of all λ-expressions generated from U. We give a special name, Ω, to $(\lambda x.xx)(\lambda x.xx) \in \Lambda$. We use capital letters X,Y,... to denote λ-expressions.

Due to Wadworth's theory [12-13], $X \in \Lambda$ is equal to $\perp$ in D_∞ if and only if X has no head normal form. For convenience, we will take Ω as the representative of those expressions without a head normal form.

---

†This work was partially supported by National Science Foundation Grant GJ-34342X.

*This paper is incomplete due to the space limitation. For the complete presentation, refer to [4].

Λ_c: set of all closed λ-expressions (λ-expressions that have no occurrence of free variables)

(2) η-abstraction: the opposite of η-reduction, i.e. for $X \in \Lambda$ and $s \in U$, if s does not occur free in X, $X \xrightarrow[\eta\text{-}ab]{} \lambda s.Xs$. We write $X \xrightarrow[\eta]{} Y$ if Y derives from X by applying η-reductions and/or η-abstractions to some subexpressions of X.

Ω-conversion: For $X, Y \in \Lambda$, $X \xrightarrow[\Omega]{} Y$ if Y derives from X by replacing some expressions of X that have no head normal form by Ω, e.g. $(\lambda x.xxx)(\lambda x.xxx) \xrightarrow[\Omega]{} \Omega$. (Note that this conversion is not effective since the existence of a head normal form is not recursively decidable.)

$\xrightarrow{CNV}$: The reflective and transitive closure of $R_\alpha \cup R_\beta \cup R_\eta$ where $R_\xi = \{(X,Y) \in \Lambda \times \Lambda \mid X \xrightarrow[\xi]{} Y\}$ for $\xi = \alpha, \beta, \eta, \Omega$.

$\approx$: The reflective, transitive and symmetric closure of $R_\alpha \cup R_\beta \cup R_\eta \cup R_\Omega$.

$\xrightarrow[\beta h]{}$: For $X, Y \in \Lambda$, $X \xrightarrow[\beta h]{} Y$ if there is an outermost, leftmost β-reduction sequence: $X = X_1 \xrightarrow[\beta]{} X_2 \xrightarrow[\beta]{} X_3 \xrightarrow[\beta]{} \cdots \xrightarrow[\beta]{} X_n = Y$ where X_i is not in a head normal form for $i < n$ and Y is in a head normal form. (Note that if a λ-expression X has a head normal form, it can be obtained by applying outermost leftmost β-reductions to X. Also note that if $X \xrightarrow[\beta h]{} Y$, Y is unique for X.)

(3) Semantic function: Let $EN = U \to D_\infty$ and $\mathbb{V}: \Lambda \to (EN \to D_\infty)$ be the semantic function as in [10]. For $X, Y \in \Lambda$, we say $X \underset{D_\infty}{\subseteq}^{\dagger} Y$ if and only if $\mathbb{V}[\![X]\!]\rho \subseteq \mathbb{V}[\![Y]\!]\rho$ for all $\rho \in EN$ and $X \underset{D_\infty}{=} Y$ if and only if $\mathbb{V}[\![X]\!]\rho = \mathbb{V}[\![Y]\!]\rho$ for all $\rho \in EN$. Note that $\underset{D_\infty}{\subseteq} \subseteq \Lambda \times \Lambda$ is not anti-symmetric and so is not a partial order.

<u>Note</u>. We should remark that there are two important bases for the theory to be developed in this paper.

(a) D_∞ is an extensional model for λ-calculus, so η-conversions preserve

---

† We use $\subseteq$ and $\cup$ instead of $\sqsubseteq$ and $\sqcup$ for typographical convenience.

D_∞-values. η-convertibility is essential in the arguments of this paper.

(b) Due to Wadsworth's theory, $X \underset{D_\infty}{=} \perp$ for $X \in \Lambda$ if and only if X has no head normal form. Thus, Ω-conversion preserves D_∞-value, too. However, this conversion is not effective. The C-function to be defined in the next section is not computable for this reason, although it is well-defined.

Here we informally state the idea† which is fundamental to the theory in this paper by using flow chart programs as an example. Given the following flow chart A0 where <α> is a Boolean function and S is a statement.

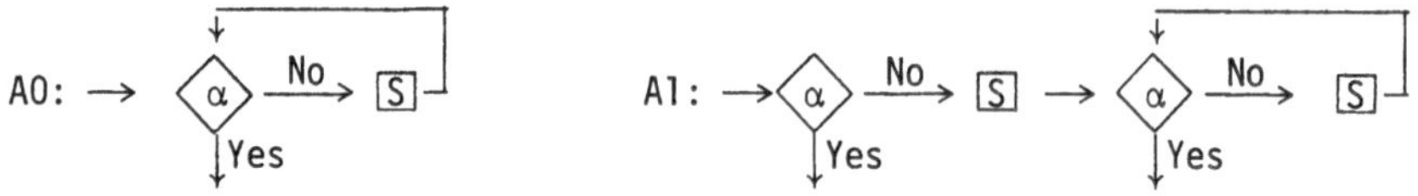

A1 is the result of unwinding the loop of A0 for one time. Also A1 can be said to be the execution of A0 for one time under unspecified input. Applying the same operation, we have

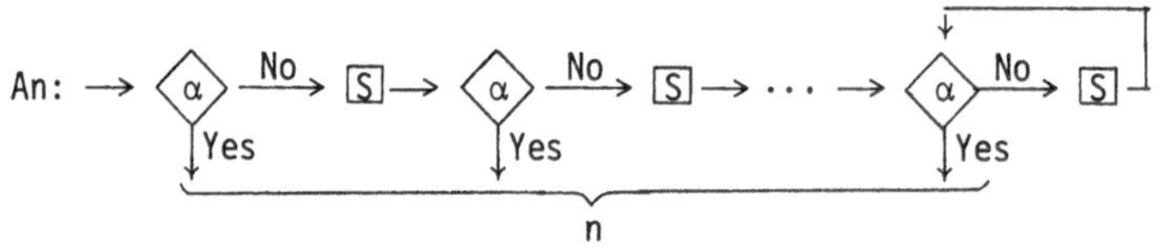

Letting $n \to \infty$,

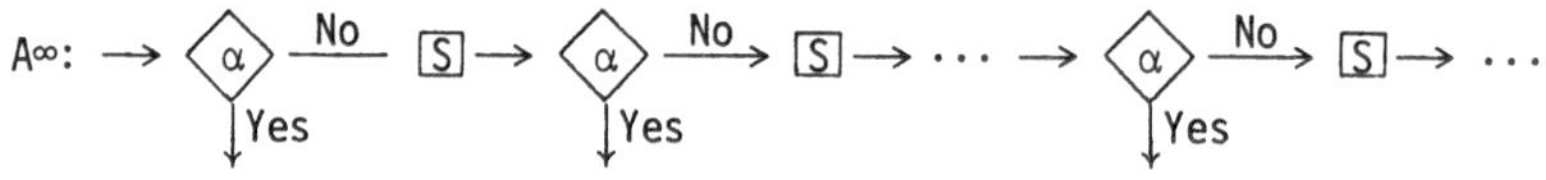

which is an infinite completely sequential flow chart. We formalize this idea of infinite expansion as follows:

Let P be the domain of programs and I be the domain of the infinite expansions of the programs in P. Let $E: P \to I$ be the infinite expansion, i.e., $E: A0 \longrightarrow A\infty$. Then we ask the following questions.

1) How are I and E to be constructed from P?

2) Can the meaning of a program P be determined by E(P)?

For example, is P_1 equivalent to P_2 if and only if $E(P_1) = E(P_2)$?

†Refer to [11] for a comprehensive presentation.

3) What sort of structure does I have? Does it have, for instance, a lattice-like structure?

We shall answer these questions for the λ-expressions as programs.

§2. C-function - Generalized Normal Forms

2.1 Definition. Let $\Delta = \{0\} \cup \{(n_1, n_2, \ldots, n_k) \mid n_i, k \in \mathbb{N}\}$.

a) Partial order $\leq$ in Δ is defined by: For $\delta_1, \delta_2 \in \Delta$, $\delta_1 \leq \delta_2$ if

either $\quad \delta_1 = 0$

or $\quad \delta_1 = (m_1, m_2, \ldots, m_i)$, $\quad \delta_2 = (n_1, n_2, \ldots, n_j)$

where $i \leqq j$ and $m_1 = n_1, m_2 = n_2, \ldots, m_i = n_i$.

b) Given $\delta \in \Delta$ and $i \in \mathbb{N}$, we define $\delta \circ i$ to be:

$$(i) \in \Delta \quad \text{if} \quad \delta = 0$$

$$(n_1, n_2, \ldots, n_j, i) \quad \text{if} \quad \delta = (n_1, n_2, \ldots, n_j) \quad .$$

For simplicity, we will denote $(n_1, n_2, \ldots, n_k)$ as $n_1 n_2 \cdots n_k$, so, for instance, 5 will be used instead of (5).

We will formalize the idea of "expanding λ-expressions". There are four operations involved in this process.

Operation a. β-reduction

Operation b. η-abstraction

Operation c. Ω-conversion

Operation d. Renaming of the bound variables according to their position.

Given $X \in \Lambda$, the C-function of X is the limit of arbitrary times of applications of the four operations to X.

We take two mutually disjoint subsets F and T of U and set $V = F \cup T$, where $F = \{f_i \mid i \in \mathbb{N}\}$ and $T = \{t_\delta \mid \delta \in \Delta - \{0\}\}$.

We assume, in the rest of this paper, that <u>if any given expression has some occurrences of a free variable, it is one of f_i's in</u> F. Our intention as for T is to convert any given expression to one whose bound variables are in T by applying α-conversions.

Let $h\colon \Lambda \to \{0,1\}$ be a non-computable predicate defined as:

$$h(X) = \begin{cases} 0 & \text{if } X \text{ has no head normal form} \\ 1 & \text{if } X \text{ has a head normal form.} \end{cases}$$

2.2 <u>Definition</u> (L-function). L-function is a map $L\colon \Lambda \to (\Delta \to \Lambda)$ defined inductively as follows: Given $X \in \Lambda$, assume that any t_δ in T does not appear in X (by applying α-conversions if necessary).

<u>Step 0</u>.

$$L(X,0) = \begin{cases} \Omega & \text{if } h(X) = 0 \ (\text{: operation c}) \\ \lambda t_1 t_2 \cdots t_m . z X_1 X_2 \cdots X_n & \text{if } h(X) = 1 \text{ and } X \underset{\beta h}{\rightarrow} \lambda s_1 s_2 \cdots s_m . w X'_1 X'_2 \cdots X'_n \\ & \text{and } z_1 X_1 X_2 \cdots X_n = S^{s_1, s_2, \ldots, s_m}_{t_1, t_2, \ldots, t_m} w X'_1 X'_2 \cdots X'_n \\ & (\text{: operations a \& d}) \end{cases}$$

<u>Step δ</u>. Suppose we have defined $L(X,\delta')$ for all $\delta' \in \Delta$ such that $\delta' \leq \delta$. We now define $L(X,\delta \circ i)$ for each $i \in \mathbb{N}$.

<u>Case I</u>. If $L(X,\delta) = \Omega$ then $L(X,\delta\circ i) = \Omega$ for each $i \in \mathbb{N}$.

<u>Case II</u>. If $L(X,\delta) = \lambda t_{\delta\circ 1} t_{\delta\circ 2} \cdots t_{\delta\circ m} . z X_1 X_2 \cdots X_n$.

(i) If $i \leqq n$ then

(a) $L(X,\delta\circ i) = \Omega$ if $h(X_i) = 0$ (: operation c).

(b) $L(X,\delta\circ i) = \lambda t_{\delta\circ i\circ 1} t_{\delta\circ i\circ 2} \cdots t_{\delta\circ i\circ p} . v Y_1 Y_2 \cdots Y_q$

if $X_i \underset{\beta h}{\rightarrow} \lambda r_1 r_2 \cdots r_p . u Y'_1 Y'_2 \cdots Y'_q$

and $v Y_1 Y_2 \cdots Y_q = S^{r_1, r_2, \ldots, r_p}_{t_{\delta\circ i\circ 1}, t_{\delta\circ i\circ 2}, \ldots, t_{\delta\circ i\circ p}} u Y'_1 Y'_2 \cdots Y'_q$

(: operations a & d)

(ii) If $i > n$ then $L(X,\delta\circ i) = t_{\delta\circ(m-n+i)}$ (: operations b & d).

Here note that for each $L(X,\delta)$ such that $L(X,\delta) \neq \Omega$, its head variable is in F if it is free in X. Otherwise it is in T.

2.3 Corollary. If $X \approx Y$, then $L(X,\delta) \approx L(Y,\delta)$ for each $\delta \in \Delta$. □

2.4 Definition (C-function). $C: \Lambda \to (\Delta \to V \cup \{\omega\})$ is defined by:

$$C(X,\delta) = \begin{cases} z & \text{if } L(X,\delta) = \lambda t_{\delta\circ 1} t_{\delta\circ 2} \cdots t_{\delta\circ m}.zX_1X_2\cdots X_n \\ \omega & \text{if } L(X,\delta) = \Omega. \end{cases}$$

The transformation of $L(X)$ into $C(X)$ may look fairly drastic. We discard every information except the head variable.

2.5 Corollary. If $X \approx Y$ for $X, Y \in \Lambda$, $C(X) = C(Y)$.

2.6 Theorem. Given X, Y in Λ,

1) If there exists $\delta \in \Delta$ such that, for different u, v in V,

$$C(X,\delta) = u \quad \text{and} \quad C(Y,\delta) = v \quad ,$$

then, for arbitrarily given a, b in D_∞, we can choose $e_1,e_2,\ldots,e_n \in \Lambda$ and an environment ρ for which

$$\mathbb{V}[\![Xe_1e_2\cdots e_n]\!]\rho = a$$
$$\mathbb{V}[\![Ye_1e_2\cdots e_n]\!]\rho = b \quad .$$

If $a, b \in \Lambda_c$, then we can choose ρ so that $\rho(V) \subseteq \Lambda_c$.

2) If there exists $\delta \in \Delta$ such that $C(X,\delta_0) = C(Y,\delta_0)$ for any δ_0 satisfying $|\delta_0| < |\delta|$ and that $C(X,\delta) = u \in V$, $C(Y,\delta) = \omega$, then, for arbitrarily given a in D_∞, there exist $e_1,e_2,\ldots,e_n$ in Λ and an environment ρ for which

$$\mathbb{V}[\![Xe_1e_2\cdots e_n]\!]\rho = a$$
$$\mathbb{V}[\![Ye_1e_2\cdots e_n]\!]\rho = \bot \quad .$$

If $a \in \Lambda_c$, we can choose ρ so that $\rho(V) \subset \Lambda_c$.

Proof. Refer to [4]. (The proof is similar to that of Böhm's Theorem in [2].) □

Let $\mathbb{C} = \Delta \to V \cup \{\omega\}$. We introduce a partial order $\leq$ over $\mathbb{C}$ as follows: For $c_1, c_2 \in \mathbb{C}$, $c_1 \leq c_2$ if and only if for all $\delta \in \Delta$, $c_1(\delta) = \omega$ or $c_1(\delta) = c_2(\delta)$.

Using this partial order, we have the following corollary to Theorem 2.4.

2.7 Corollary. For X, Y in Λ, if $X \underset{D_\infty}{\subseteq} Y$, then $C(X) \leq C(Y)$.

Proof. Suppose that $C(X) \nleq C(Y)$. Then there must exist $\delta \in \Delta$ such that, for some $u, v \in V$,

$$\text{either} \quad \begin{matrix} C(X,\delta) = u \\ C(Y,\delta) = v \end{matrix} \quad \text{where } u \neq v$$

$$\text{or} \quad \begin{matrix} C(X,\delta) = u \\ C(Y,\delta) = \omega \end{matrix} \quad .$$

In either case, there must be at least one $\delta \in \Delta$ for which the condition of part 1 or 2 of Theorem 2.4 holds. So, by the conclusion of the theorem, there exist $e_1, e_2, \ldots, e_n \in \Lambda$ and an environment ρ such that

$$\text{either} \quad \begin{matrix} \mathbb{V}[\![Xe_1e_2\cdots e_n]\!]\rho = \lambda x\lambda y.x \\ \mathbb{V}[\![Ye_1e_2\cdots e_n]\!]\rho = \lambda x\lambda y.y \end{matrix}$$

$$\text{or} \quad \begin{matrix} \mathbb{V}[\![Xe_1e_2\cdots e_n]\!]\rho = \lambda x\lambda y.x \\ \mathbb{V}[\![Ye_1e_2\cdots e_n]\!]\rho = \bot \end{matrix} \quad .$$

Since $\lambda x\lambda y.x \underset{D_\infty}{\nsubseteq} \lambda x\lambda y.y$ and $\lambda x\lambda y.x \underset{D_\infty}{\neq} \bot$, this contradicts $X \underset{D_\infty}{\subseteq} Y$. □

We translate Theorem 2.6 into one stated in pure λ-calculus language.

2.8 Corollary. Let x, y be in Λ. If $C(X) \neq C(Y)$, then, for any $u, v \in V$, we can choose λ-expressions $e_1, e_2, \ldots, e_n$ in Λ, variables $z_1, z_2, \ldots, z_m$ in V and closed λ-expressions $h_1, h_2, \ldots, h_m$ in Λ_c so that one of the following (1), (2) or (3) holds: Let $X^* = S^{z_1, z_2, \ldots, z_m}_{h_1, h_2, \ldots, h_m}(Xe_1 e_2 \cdots e_n)$ and $Y^* = S^{z_1, z_2, \ldots, z_m}_{h_1, h_2, \ldots, h_m}(Ye_1 e_2 \cdots e_n)$.

(1) $X^* \xrightarrow{CNV} u$ and $Y^* \xrightarrow{CNV} v$.

(2) $X^* \xrightarrow{CNV} u$ and Y^* has no head normal form.

(3) X^* has no head normal form and $Y^* \xrightarrow{CNV} u$. □

This is an extension of Böhm's Theorem [2] by regarding $C(X)$ and $C(Y)$ as <generalized normal forms> of X and Y. The point is that we are no longer concerned with conventional normal forms. Corollary 2.8 is a statement regarding general λ-expressions no matter whether they are normal or not. The opposite of Corollary 2.7 is also true:

2.9 <u>Theorem</u>. For $X, Y \in \Lambda$, if $C(X) \leq C(Y)$ then $X \underset{D_\infty}{\subseteq} Y$.

<u>Proof</u>. Outline of the proof is described as follows: Due to the fact that $C(X) \leq C(Y)$, we can choose sequences of λ-expressions $A_p^1(X), A_p^2(X), \ldots, A_p^n(X), \ldots$ and $A_p^1(Y), A_p^2(Y), \ldots, A_p^n(Y), \ldots$ where the following four conditions hold:

(1) There are X_i and Y_i for each $i \in \mathbb{N}$ such that

$$X \xrightarrow{CNV} X_i$$
$$Y \xrightarrow{CNV} Y_i$$

where $A_p^i(X)$ matches X_i except at occurrences of Ω in $A_p^i(X)$ and $A_p^i(Y)$ matches Y_i except at occurrences of Ω in $A_p^i(Y)$.

(2) For each $i \in \mathbb{N}$, $A_p^i(X)$ matches $A_p^{i+1}(X)$ except at occurrences of Ω in $A_p^i(X)$ and $A_p^i(Y)$ matches $A_p^{i+1}(Y)$ except at occurrences of Ω in $A_p^i(Y)$.

(3) For each $i \in \mathbb{N}$, $A_p^i(X)$ matches $A_p^i(Y)$ except at occurrences of Ω in $A_p^i(X)$.

(4) $X \underset{D_\infty}{=} \bigcup_{i=1}^{\infty} A_p^i(X)$ and $Y \underset{D_\infty}{=} \bigcup_{i=1}^{\infty} A_p^i(Y)$.

Since $A_p^i(X) \underset{D_\infty}{\subseteq} A_p^i(Y)$ by (3), we conclude that $X \underset{D_\infty}{\subseteq} Y$ by (4). For the complete proof refer to [4]. Also see Lemma 4.6. □

By Corollary 2.5 and Theorem 2.7, we have:

2.10 Theorem.† For X, Y ∈ Λ, $X \underset{D_\infty}{\subseteq} Y$ if and only if $C(X) \leq C(Y)$. So, $X \underset{D_\infty}{=} Y$ if and only if $C(X) = C(Y)$. □

2.11 Example (Wadsworth). Let $I = \lambda x.x$ and $J = Y(\lambda f\lambda x\lambda y.x(fy))$. It is easy to see that $C(I) = C(J)$ by applying η-abstractions to I and β-reductions to J. So $I \underset{D_\infty}{=} J$. Note that I is normal while J is not.

The following fact is interesting in relation to ω-completeness discussions in [1,6].

2.12 Theorem. Let X, Y be in Λ. If $XW \underset{D_\infty}{=} YW$ for all closed λ-expressions W in Λ_c, then $X \underset{D_\infty}{=} Y$.

Proof. Assume that $X \underset{D_\infty}{\neq} Y$. By Theorem 2.10, $C(X) \neq C(Y)$. So by Theorem 2.4, there exist $e_1, e_2, \ldots, e_n \in \Lambda$ and an environment ρ such that

$$\mathbb{V}[\![Xe_1e_2\cdots e_n]\!]\rho \neq \mathbb{V}[\![Ye_1e_2\cdots e_n]\!]\rho \quad . \tag{*}$$

Especially, (*) can be realized with both sides being in Λ_c. So, again by Theorem 2.4, we can choose ρ so that $\rho(V) \subset \Lambda_c$. Let

$$\bar{X} = S^{u_1,u_2,\ldots,u_p}_{\rho(u_1),\rho(u_2),\ldots,\rho(u_p)} X , \qquad \bar{Y} = S^{v_1,v_2,\ldots,v_q}_{\rho(v_1),\rho(v_2),\ldots,\rho(v_q)} Y$$

and

$$\bar{e}_i = S^{w_1^i,w_2^i,\ldots,w_{m(i)}^i}_{\rho(w_1^i),\rho(w_2^i),\ldots,\rho(w_{m(i)}^i)} e_i$$

where $u_1, u_2, \ldots, u_p$ are the free variables in X, $v_1, v_2, \ldots, v_q$ are the free

†Refer to [14] for an alternative characterization of $\underset{D_\infty}{\sqsubseteq}$.

variables in Y and $w_1^i, w_2^i, \dots, w_{m(i)}^i$ are the free variables in e_i for $i = 1,2,\dots,n$.

Now the inequality (*) can be written as:

$$\bar{X}\bar{e}_1\bar{e}_2\cdots\bar{e}_n \underset{D_\infty}{\neq} \bar{Y}\bar{e}_1\bar{e}_2\cdots\bar{e}_n$$

where $\bar{X}$, $\bar{Y}$ and $\bar{e}_i \in \Lambda_c$. By extensionality of D_∞, we conclude that

$$\bar{X}\bar{e}_1\bar{e}_2\cdots\bar{e}_{n-1} \underset{D_\infty}{\neq} \bar{Y}\bar{e}_1\bar{e}_2\cdots\bar{e}_{n-1}$$

and so

$$\bar{X}\bar{e}_1\bar{e}_2\cdots\bar{e}_{n-2} \underset{D_\infty}{\neq} \bar{Y}\bar{e}_1\bar{e}_2\cdots\bar{e}_{n-2}$$

$$\vdots$$

$$\bar{X}\bar{e}_1 \underset{D_\infty}{\neq} \bar{Y}\bar{e}_1 \quad .$$

Thus we have shown that if $X \underset{D_\infty}{\neq} Y$, there exists $\bar{e}_1 \in \Lambda_c$ such that

$$X\bar{e}_1 \underset{D_\infty}{\neq} Y\bar{e}_1 \quad . \quad \square$$

2.13 Corollary. Let X, Y be in Λ. If $C(XW) = C(YW)$ for all closed λ-expressions W, then $C(X) = C(Y)$. $\square$

Theorem 2.12 is obvious if we replace $W \in \Lambda_c$ by $W \in D_\infty$ since D_∞ is extensional. The theorem says that the extensionality holds in Λ_c modulo $\underset{D_\infty}{=}$.

§3. Generalized λ-Expressions

As in the previous section, let $\mathbb{C} = \Delta \to V \cup \{\omega\}$. The C-function is the map $C: \Lambda \to \mathbb{C}$ where $C(\Lambda)$ is a proper subset of $\mathbb{C}$. We may ask how $C(\Lambda)$ can be characterized as a subset of $\mathbb{C}$. The following conditions determine the hierarchy of some interesting subclasses of $\mathbb{C}$.

Given $c \in \mathbb{C}$:

Condition 1: If $c(\delta) = z \in V$, then either z is in F or $z = t_{\delta'}$ for $\delta' \in \Delta$ such that $\delta' \leq \delta$ or $\delta' = \delta \circ m$ for some $m \in \mathbb{N}$ (i.e. if a variable is bound, it must be so in an outer context).

Condition 2: If $c(\delta) = \omega$ for some $\delta \in \Delta$, $c(\delta') = \omega$ for all $\delta' \in \Delta$ with $\delta < \delta'$ (i.e. once a subexpression turns out to be bottom, any of its descendants must be bottom, too).

Condition 3: If $c(\delta) \neq \omega$, there exists an integer k_δ^c, a positive integer N_δ^c such that

$$c(\delta \circ n) = t_{\delta \circ (n + k_\delta^c)} \quad \text{and} \quad c(\delta \circ n \circ \delta') = t_{\delta \circ n \circ \delta'} \quad \text{for all } \delta' \in \Delta - \{0\}$$

for all $n > N_\delta^c$ (i.e. c is 'finitely wide').

Condition 4: Let $Fr(c) = \{z \in F \mid c(\delta) = z \text{ for some } \delta \in \Delta\}$. Then $\#Fr(c) < \infty$ (i.e. the number of the distinct free variables which occur in $\{c(\delta) \mid \delta \in \Delta\}$ is finite).

Condition 5: There are partially computable functions $\phi_c\colon \Delta \to \mathbb{N}$ and $\psi_c\colon \Delta \to V$ such that

$$\left.\begin{aligned} \phi_c(\delta) &= N_\delta^c \\ \psi_c(\delta) &= z \end{aligned}\right\} \quad \text{if } c(\delta) = z \in V$$

$\phi_c(\delta)$ and $\psi_c(\delta)$ are undefined if $c(\delta) = \omega$ (i.e. $\{c(\delta) \mid c(\delta) \neq \omega, \delta \in \Delta\}$ is a recursively enumerable object and the width N_δ^c in Condition 3 is also partially computable).

3.1 Theorem. For $c \in \mathbb{C}$, $c \in C(\Lambda)$ and if only if c satisfies Conditions 1-5.

Proof. If $c \in C(\Lambda)$, it easy to see that c satisfies Conditions 1-5 by the definition of C. Suppose that $c \in \mathbb{C}$ satisfies Conditions 1-5. We give effective codings of $\mathbb{Z}$, Δ and V into Λ.

$$\text{En:} \quad \begin{aligned} n \in \mathbb{Z} &\mapsto \bar{n} \in \Lambda \\ \delta \in \Delta &\mapsto \tilde{\delta} \in \Lambda \\ t_\delta \in T &\mapsto \hat{t}_\delta \in \Lambda \\ f_i \in F &\mapsto \hat{f}_i = f_i \in \Lambda \end{aligned} \quad .$$

We can assume that $En(\mathbb{Z})$, $En(\Delta)$, $En(T)$ and $En(F)$ (=F) are mutually disjoint.

In the rest of the proof, the existence of π_c, P, Θ_c, f, g and $N \in \Lambda$ is assumed due to the fact that all partially recursive functions are λ-definable and we do not present their actual constructions.

Let Δ_c be the subset of Δ consisting of all δ such that $\phi_c(\delta)$ is defined. Obviously Δ_c is recursively enumerable. We define $\pi_c \in \Lambda$ by:

$$\pi_c\tilde{\delta} \xrightarrow{CNV} \begin{cases} \lambda\text{-expression without a head normal form} & \text{if } \delta \notin \Delta_c \\ \lambda x.x & \text{if } \delta \in \Delta_c \,. \end{cases}$$

A partially computable function $M_c\colon \Delta \to \mathbb{N}$ is defined by:

$$M_c(\delta) = \begin{cases} \text{undefined} & \text{if } c(\delta) = \omega \\ k^c_{\tilde{\delta}} + \phi_c(\delta) & \text{if } c(\delta) \neq \omega \end{cases}$$

$P \in \Lambda$ is defined by:

$$P\bar{i}\tilde{\delta} \xrightarrow{CNV} \widetilde{\delta\circ i} \quad \text{for each } i \in \mathbb{N} \text{ and } \delta \in \Delta \quad .$$

Θ_c is a λ-expression that is defined recursively in the following way:

$$\Theta_c\tilde{\delta}e \xrightarrow{CNV} \pi_c\tilde{\delta}(f\tilde{\delta}\bar{0}\overline{M_c(\delta)}\,\overline{\phi_c(\delta)}\,\widehat{c(\delta)}e)$$

where $f \in \Lambda$ is defined by:

$$f\tilde{\delta}\bar{i}\bar{m}\bar{n}\hat{z}e \xrightarrow{CNV} \begin{cases} g\tilde{\delta}\bar{0}\bar{n}\hat{z}e & \text{if } i = m \\ \lambda s.f\tilde{\delta}\overline{i+1}\bar{m}\bar{n}\hat{z}(N\tilde{\delta}\overline{i+1}e) & \text{otherwise} \end{cases}$$

where

$$g\tilde{\delta}\bar{j}\bar{n}\hat{z}e \xrightarrow{CNV} \begin{cases} e\hat{z} & \text{if } j = n \\ g\tilde{\delta}\overline{j+1}\bar{n}\hat{z}e(\Theta_c(P\overline{n-j}\tilde{\delta})e) & \text{otherwise} \end{cases}$$

and

$$N\tilde{\delta}\bar{i}e\hat{z} \xrightarrow{CNV} \begin{cases} s & \text{if } z = t_{\delta\circ i} \\ e\hat{z} & \text{otherwise .} \end{cases}$$

Note that Condition 4 is one of the necessary conditions for the existence of Θ_c.

Now $C(\Theta_c\tilde{0}I) = c$ where $I = \lambda x.x$. □

Let $\mathbb{C}_{fin} = \{c \in \mathbb{C} \mid c \text{ satisfies Conditions 1 to 5}\}$ and $\mathbb{C}_{inf} = \{c \in \mathbb{C} \mid c \text{ satisfies Conditions 1 to 3}\}$. Then $C(\Lambda) = \mathbb{C}_{fin} \subsetneq \mathbb{C}_{inf} \subsetneq \mathbb{C}$. Theorem 2.8 can be stated as $\mathbb{C}_{fin} \approx \Lambda/\underset{D_\infty}{=}$.

Our next stage is to define infinite λ-expressions, Λ^∞, which correspond to $\mathbb{C}_{inf}$. But before defining Λ^∞, we consider how a textually infinite program can be realized. Reynolds [6] presents the following programming environment: A person is programming in front of a terminal. He builds up his program in such a way that some of the integral parts (e.g. inside of begin$\cdots$end block, a procedure body or simply a statement) are left unspecified. He lets the system run this program. When it turns out that the system needs the specification of an undefined part to continue execution, the programmer is requested to fill it with a code which can have several undefined parts, too. The programmer meets this request probably considering the outcome of the execution he has obtained so far. This process can continue infinitely. Since a person with free will takes part in this process, it can become a non-recursively enumerable object.

One may ask how a λ-like expression can be infinite. Probably there are three ways:

(1) Infinite application: $\cdots(\cdots((X_1X_2)X_3)X_4)\cdots)X_n)\cdots$

(2) Infinite abstraction: $\lambda r_1 r_2 r_3 \cdots r_n \cdots . X$

(3) Infinite depth: $X_1(X_2(X_3\cdots(X_n(\cdots))\cdots))$

Let $S^{(n)} \equiv$ begin $S_1;S_2;\ldots;S_n$ end be an Algol-like program. We translate S_i into a λ-expression s_i for $i = 1,2,\ldots,n$. Using the technique of continuation, the translation of $S^{(n)}$ will be: $s^{(n)} = \lambda v.s_1(s_2(\cdots(s_n I))\cdots)(v)$ where $I = \lambda x.x$.

Letting $n \to \infty$, we have an infinitely deep λ-expression:
$s = \lambda v.s_1(s_2(s_3(\cdots)\cdots))(v)$ which will probably be the limit of
$A^n(s) = \lambda v.s_1(s_2(\cdots(s_n\Omega)\cdots))(v)$, i.e. $s = \bigcup_{n=1}^{\infty} A^n(s)$. We formalize this idea on Λ and have the following definition of Λ^∞.

3.2 Definition.

1) $\Lambda_\Box$ is the set defined by:

a) $v \in U$ alone is in $\Lambda_\Box$

b) $\Box$ alone is in $\Lambda_\Box$

c) If X, Y are in $\Lambda_\Box$, then X (Y) is in $\Lambda_\Box$.

d) If X is in $\Lambda_\Box$, then $\lambda v.X$ is in $\Lambda_\Box$ for $v \in U$.

2) Given X in $\Lambda_\Box$, $X^* \in \Lambda$ is the λ-expression derived from X by replacing each $\Box$ in $\Lambda_\Box$ by Ω.

3) Let X, Y be in $\Lambda_\Box$. We say that X is a specification of Y if either X = Y or X derives from Y by replacing some of $\Box$ in Y by elements in $\Lambda_\Box$.

We denote this relationship by X <u>spec</u> Y.

4) Λ^∞ is the set of all sequences $(X_1,X_2,X_3,\ldots,X_n,\ldots)$ where, for each $i \in \mathbb{N}$, $X_i \in \Lambda_\Box$ and X_{i+1} <u>spec</u> X_i.

5) Given $\zeta = (X_1,X_2,\ldots,X_n,\ldots) \in \Lambda^\infty$ and $\xi = (Y_1,Y_2,\ldots,Y_n,\ldots) \in \Lambda^\infty$, we define the application $\zeta(\xi) \in \Lambda^\infty$ by $\zeta(\xi) = (X_1(Y_1),X_2(Y_2),\ldots,X_n(Y_n),\ldots)$.

3.3 <u>Definition</u>. We define $\mathbb{V}_\infty\colon \Lambda^\infty \to (EN \to D_\infty)$ as follows: Given $\xi = (X_1,X_2,\ldots,X_n,\ldots) \in \Lambda^\infty$ and an environment $\rho \in EN$

$$\mathbb{V}_\infty[\![\xi]\!]\,\rho = \bigcup_{i=1}^{\infty} \mathbb{V}[\![X_i^*]\!]\,\rho \quad .$$

We say $\xi \underset{D_\infty}{\subseteq} \zeta$ for $\xi, \zeta \in \Lambda^\infty$ if $\mathbb{V}_\infty[\![\xi]\!]\,\rho \subseteq \mathbb{V}_\infty[\![\zeta]\!]\,\rho$ for all ρ.

3.4 <u>Definition</u>. $C_\infty\colon \Lambda^\infty \to \mathbb{C}$ is defined by: Let $\xi = (X_1,X_2,\ldots,X_n,\ldots) \in \Lambda^\infty$,

$$C_\infty(\xi) = \begin{cases} \omega & \text{if } C(X_i^*) = \omega \text{ for all } i \\ z & \text{if } C(X_i^*) = z \text{ for some } i \;. \end{cases}$$

C_∞ is well defined since

$$C(X_1^*) \le C(X_2^*) \le \cdots \le C(X_i^*) \le \cdots \;.$$

Any $X \in \Lambda$ can be embedded into Λ^∞ by

$$\iota\colon X \mapsto (X,X,X,\ldots) \;.$$

Obviously $\mathbb{V}_\infty[\![\iota(X)]\!]\,\rho = \mathbb{V}[\![X]\!]\,\rho$ and $C_\infty(\iota(X)) = C(X)$ for all $X \in \Lambda$. Let Λ_c^∞ be $\{\xi \in \Lambda^\infty \mid \xi = (X_1,X_2,\ldots,X_n,\ldots)$ where X_n contains no free variables for each $n \in \mathbb{N}\}$. Obviously $\iota|_{\Lambda_c}$ gives the inclusion: $\iota|_{\Lambda_c}\colon \Lambda_c \to \Lambda_c^\infty$.

3.5 Theorem.

1) For all $\zeta, \xi \in \Lambda^\infty$ and $\rho \in EN$, $\mathbb{V}_\infty [\![\zeta(\xi)]\!] \rho = (\mathbb{V}_\infty [\![\zeta]\!] \rho)(\mathbb{V}_\infty [\![\xi]\!] \rho)$

2) $C_\infty(\Lambda^\infty) = \mathbb{C}_{inf}$ and $C_\infty|_\Lambda = C$

3) For all $\xi, \zeta \in \Lambda^\infty$, $\xi \underset{D_\infty}{\subseteq} \zeta$ if and only if $C_\infty(\xi) \leq C_\infty(\zeta)$.

Proof. Refer to [4]. □

By this theorem, we have another correspondence $\mathbb{C}_{inf} \approx \Lambda^\infty / \underset{D_\infty}{=}$.

A real number is defined as the limit of a non-decreasing sequence of the rational numbers. A similar situation exists between Λ and Λ^∞. Thus we may as well call Λ^∞ as generalized λ-expressions.

3.6 Theorem. $\{\mathbb{V}_\infty [\![\xi]\!] \mid \xi \in \Lambda_C^\infty\} \subseteq D_\infty$ is the set of every element of D_∞ that is the limit of a directed subset of $\{\mathbb{V} [\![X]\!] \mid X \in \Lambda_C\} \subseteq D_\infty$. □

3.7 Corollary. The cardinality of D_∞ is strictly larger than denumerable.

Proof. By Theorem 3.5, each two elements ξ, ζ are mapped into two different elements in D_∞ if $C_\infty(\xi) \neq C_\infty(\zeta)$. Since $C(\Lambda_C^\infty)$ has the cardinality which is strictly larger than denumerable, we conclude the corollary. □

Now, we study some of the properties of Λ and Λ^∞ in D_∞. We refer to [4] for the proof of each proposition. Hereafter, we denote both C and C_∞ by C.

3.8 Definition. A subset $\mathcal{D}$ of Λ (Λ^∞) is said to be directed if, for any finite subset F of $\mathcal{D}$, there exists ξ in $\mathcal{D}$ such that $\zeta \underset{D_\infty}{\subseteq} \xi$ for all $\zeta \in F$.

3.9 Proposition. Given any $X \in \Lambda$ (Λ^∞) such that $X \underset{D_\infty}{\neq} \bot$, there exists a directed set $\mathcal{D} \subseteq \Lambda$ (Λ^∞) such that

$$X \underset{D_\infty}{=} \cup \mathcal{D}$$

and

$$Y \underset{D_\infty}{\subsetneq} X \text{ for all } Y \in \mathcal{D}. \quad \square$$

3.10 <u>Proposition</u>. Let $\mathcal{D}$ be a directed set of Λ (Λ^∞). We define $c_\mathcal{D} \in \mathbb{C}$ by

$$c_\mathcal{D}(\delta) = \begin{cases} \omega & \text{if } C(Y,\delta) = \omega \text{ for all } Y \in \mathcal{D} \\ z & \text{if } C(Y,\delta) = z \text{ for some } Y \in \mathcal{D}. \end{cases}$$

If $C(X) = c_\mathcal{D}$ for a given $X \in \Lambda$ (Λ^∞), then $X \underset{D_\infty}{=} \cup\mathcal{D}$. □

3.11 <u>Proposition</u>. For $X, Y \in \Lambda$ (Λ^∞), if $X \underset{D_\infty}{\subsetneq} Y$, there exists $Z \in \Lambda$ (Λ^∞) such that

$$X \underset{D_\infty}{\subsetneq} Z \underset{D_\infty}{\subsetneq} Y \quad . \quad \square$$

By the results we have reached, the behavior of each member of Λ or Λ^∞ can be completely determined by its map of C or C_∞. It can be said that $\mathbb{C}_{fin}$ and $\mathbb{C}_{inf}$ are the lattice structures which are inherently associated with Λ and Λ^∞ respectively, independent of D_∞. So we study $\mathbb{C}_{fin}$ and $\mathbb{C}_{inf}$ as partially ordered sets themselves to gain some insight into Λ and Λ^∞.

3.12 <u>Proposition</u>. $\mathbb{C}_{fin}$ and $\mathbb{C}_{inf}$ are partially ordered sets by $\leq$. □

3.13 <u>Proposition</u>. $\mathbb{C}_{fin}$ is directed-complete, i.e. any directed subset of $\mathbb{C}_{inf}$ has the least upper bound. □

The following theorem asserts that the lattice topology of Λ induced by D_∞ is equivalent to the lattice topology of $\mathbb{C}_{inf}$.

3.14 <u>Theorem</u>. For $X \in \Lambda$ (Λ^∞) and a directed set $\mathcal{D} \subseteq \Lambda$ (Λ^∞), $X \underset{D_\infty}{=} \cup\mathcal{D}$ if and only if $C(X) = \cup\{C(Y) \mid Y \in \mathcal{D}\}$. □

3.15 <u>Proposition</u>. $\mathbb{C}_{fin}$ and $\mathbb{C}_{inf}$ are lower semi-lattice, i.e. for all $a, b \in \mathbb{C}_{fin}$ ($\mathbb{C}_{inf}$), there exists $a \cap b \in \mathbb{C}_{fin}$ ($\mathbb{C}_{inf}$) where $c = a \cap b$ is defined inductively by:

Let $m_\delta = \max(N^a_\delta, N^b_\delta) + 1$ for N^a_δ, N^b_δ in Condition 3 of $\mathbb{C}_{inf}$.

1) $$c(0) = \begin{cases} a(0) & \text{if } a(0) = b(0) \neq \omega \text{ and } a(m_0) = b(m_0) \\ \omega & \text{otherwise} \end{cases}$$

2) Let $\delta = \delta' \circ i$:

i) If $c(\delta') = \omega$, then $c(\delta) = \omega$.

ii) If $c(\delta') \neq \omega$, then

$$c(\delta) = \begin{cases} a(\delta) & \text{if } a(\delta) = b(\delta) \neq \omega \text{ and } a(\delta \circ m_\delta) = b(\delta \circ m_\delta) \\ \omega & \text{otherwise} . \end{cases} \quad \square$$

3.16 Corollary. Given any $X, Y \in \Lambda$, there exists $Z \in \Lambda$ such that $C(X) \cap C(Y) = C(Z)$. $\square$

3.17 Definition. 1) For $a, b \in \mathbb{C}_{inf}$, we say a and b are compatible if there is no $\delta \in \Delta$ such that $a(\delta) \neq \omega$, $b(\delta) \neq \omega$ and $a(\delta) \neq b(\delta)$.

2) For $S \subseteq \mathbb{C}_{inf}$, S is said to be compatible if any two elements of S are compatible.

3.18 Proposition. Adding T (top) to $\mathbb{C}_{inf}$, $\mathbb{C}_{inf} \cup \{T\}$ is a complete lattice by the following definition of $\cup$:

For $a, b \in \mathbb{C}_{inf} \cup \{T\}$,

1) If $a = T$, $b = T$ or a and b are not compatible, then $a \cup b = T$.

2) Otherwise define $a \cup b \in \mathbb{C}_{inf}$ by:

$$(a \cup b)(\delta) = \begin{cases} \omega & \text{if } a(\delta) = b(\delta) = \omega \\ v & \text{if } a(\delta) = v \text{ or } b(\delta) = v. \end{cases} \quad \square$$

Unfortunately, $\cap$ and $\cup$ as defined above does not reflect the reality with Λ in D_∞ (except the case of Theorem 3.14).

3.19 Counterexample.[†] Let $X = \lambda xyz.x\Omega z$, $Y = \lambda xyz.xy\Omega$ and $Z = \lambda xyz.x\Omega\Omega$. $C(Z) = C(X) \cap C(Y)$, but, for $A = \lambda ab.a \cup \lambda ab.b$, $ZAII \xrightarrow{\beta} \Omega$, $XAII \xrightarrow{\beta} I$ and $YAII \xrightarrow{\beta} I$. So $Z \underset{D_\infty}{\subsetneq} X \cap Y$.

3.20 Counterexample. 1) Let $Z = \lambda xyz.xyz$, $X = \lambda xyz.x\Omega z$ and $Y = \lambda xyz.xy\Omega$. It is easy to see that X and Y are compatible and $C(Z) = C(X) \cup C(Y)$. However since, for $A = \lambda ab.ba$, $XAII \xrightarrow{\beta} \Omega$, $YAII \xrightarrow{\beta} \Omega$ and $ZAII \xrightarrow{\beta} I$, $X \cup Y \underset{D_\infty}{\subsetneq} Z$.

[†] The continuity of D_∞ is assumed, i.e. for $a, b, c \in D_\infty$, $a(c) \cap b(c) = (a \cap b)(c)$. Refer to [9].

2) Obviously $I = \lambda x.x$ and $\lambda x.xx$ are not compatible.

However $(I \cup \lambda x.xx)(\lambda x.xx) \underset{D_\infty}{=} I \cup \Omega \underset{D_\infty}{=} I$ so $I \cup \lambda x.xx \underset{D_\infty}{\neq} T$. So

$C(I) \cup C(\lambda x.xx) \underset{\mathbb{C}_{inf}}{=} T$ is artificially too strong.

We conclude from 3.20-1 that Proposition 3.10 and Corollary 3.21 are false if we remove the condition of directedness.

§4. An Axiomatization of the Extensional Model Theory of the λ-Calculus

The following diagram illustrates the relation among Λ, Λ^∞, $\mathbb{C}_{fin}$, $\mathbb{C}_{inf}$ and D_∞.

4.1 Diagram

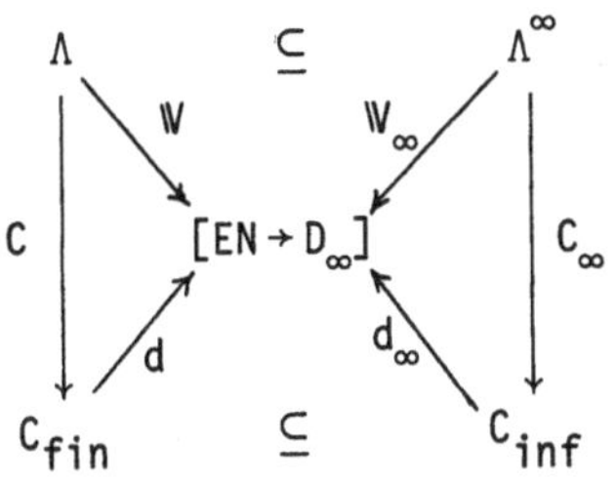

where d and d_∞ are defined as follows: For $c \in \mathbb{C}_{inf}$, $\rho \in EN$, $d_\infty[\![c]\!]\rho = \mathbb{W}_\infty[\![\xi]\!]\rho$ for $\xi \in \Lambda^\infty$ such that $\xi = C_\infty^{-1}(c)$. $d = d_\infty|_{\mathbb{C}_{fin}}$. d and d_∞ are well defined by Theorems 2.10 and 3.5.

4.2 Corollary. 1) C and C_∞ are surjective and $C = C|_\Lambda$.

2) d and d_∞ are injective and $d = d_\infty|_{\mathbb{C}_{fin}}$.

3) $d_\infty : \mathbb{C}_{inf} \to [EN \to D_\infty]$ is a monotonic and continuous function.

4) Diagram 3.22 is commutative.

Proof. 1) is deduced from Theorems 3.1 and 3.5 while 2) and 3) are from Theorems 2.10, 3.5 and 3.14 and 4) is from the definitions. □

Next, we state how $d_\infty[\![c]\!]$ is actually synthesized from $c \in \mathbb{C}_{inf}$.

4.3 Definition. An infinite subset, T, of Δ is said to be a Δ-tree if

1) $0 \in \Delta$

2) If $\delta \in T$, then there exists $N \in \mathbb{N}$ such that $\delta\circ 1, \delta\circ 2, \ldots, \delta\circ N \in T$ and $\delta\circ k \notin T$ for all $k > N$.

For a Δ-tree, T, and $\delta \in T$, $\underline{\gamma_T(\delta)}$ is N in (2), i.e. $\gamma_T(\delta) = \#\{\delta' \mid \delta' \in T$ and $\delta' = \delta\circ m$ for some $m \in \mathbb{N}\}$.

4.4 Definition. Given $c \in \mathbb{C}_{inf}$ and a Δ-tree T, we say T is admissible to c if, for all $\delta \in T$, $\gamma_T(\delta) \geqq N^c_\delta$ where N^c_δ is as in Condition 3 of $\mathbb{C}_{inf}$.

For each $\delta \in \Delta$, $|\delta| \in \mathbb{N}\cup\{0\}$, length of δ, is 0 if $\delta = 0$ and n if $\delta = (i_1, i_2, \ldots, i_n)$.

4.5 Definition. Given $c \in \mathbb{C}_{inf}$ and a Δ-tree T which is admissible to c, we define $A^n_p(c,T) \in \Lambda_\Box$ in the following way: $A^n_p(c,T) = A^0(c,T,n)$ where $A^\delta(c,T,n) \in \Lambda_\Box$ is inductively defined for each $\delta \in T$ as:

1) If $|\delta| < n$

i) $A^\delta(c,T,n) = \Omega$ if $c(\delta) = \omega$.

ii) $A^\delta(c,T,n) = \lambda t_{\delta\circ 1} t_{\delta\circ 2} \cdots t_{\delta\circ\gamma_T(\delta)+k^c_\delta} . z A^{\delta\circ 1}(c,T,n) A^{\delta\circ 2}(c,T,n) \cdots$
$\cdots A^{\delta\circ\gamma_T(\delta)}(c,T,n)$ if $c(\delta) = z$ where k^c_δ is as in Condition 3 of $\mathbb{C}_{inf}$.

2) If $|\delta| = n$, then $A^\delta(c,T,n) = \Box$.

Obviously, $A^{n+1}_p(c,T) = \text{spec}(A^n_p(c,T))$ for $n \in \mathbb{N}$ and $\xi_c = (A^1_p(c,T), A^2_p(c,T), \ldots) \in \Lambda^\infty$.

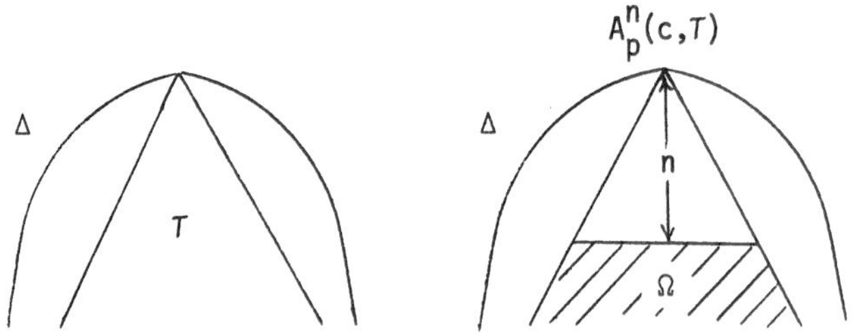

The following lemma is the key to prove Theorems 2.9, 3.5 and Proposition 3.10.

4.6 Lemma. Given $X \in \Lambda$ and a Δ-tree T, if T is admissible to $C(X)$, then

$$X \underset{D_\infty}{=} \bigcup_{n=1}^{\infty} (A^n_p(C(X),T))^*$$

where $*: \Lambda_{\Box} \to \Lambda$ is as in Definition 3.2.

Proof. Refer to [4]. □

Lemma 4.6 gives the justification to the following definition of d_∞.

4.7 Definition. $d_\infty: \mathbb{C}_{inf} \to (EN \to D_\infty)$ is the following map: $d_\infty[\![c]\!]\rho = \mathbb{V}_\infty[\![\xi_c]\!]\rho$ where $\xi_c = (A_p^1(c,T), A_p^2(c,T), \ldots, A_p^n(c,T), \ldots) \in \Lambda^\infty$ for some Δ-tree T which is admissible to c.

Lastly, we raise a question as to what properties of D_∞ are essentially necessary to develop the theory in this paper. The following is the answer.

4.8 Theorem. If a domain D satisfies the following Axioms 1-3, then Diagram 4.1 and Corollary 4.2 remain valid when "D_∞" is replaced by "D".

Axiom 1. D is a directed-complete partially ordered set with the least element $\perp = \sqcap D$ and $D \neq \{\perp\}$.

Axiom 2. There exists the following pair of maps (Φ, Ψ) which are bijective and continuous such that $\Phi \circ \Psi = 1_{[D \to D]}$ and $\Psi \circ \Phi = 1_D$.

$$D \underset{\Psi}{\overset{\Phi}{\rightleftarrows}} [D \to D]$$

Axiom 3. For $EN = [U \to D]$, we define the semantic function $\mathbb{V}: \Lambda \to [EN \to D]$ in Wadsworth's manner. Then

a) For each $X \in \Lambda$, if X has no head normal form, then $\mathbb{V}[\![X]\!]\rho = \perp$ for all $\rho \in EN$.

b) Lemma 4.6 holds for D, i.e. for $X \in \Lambda$ and a Δ-tree T. If T is admissible to C(X) then $X \underset{D}{=} \bigcup_{n=1}^{\infty} (A_p^n(C(X),T))^*$ where $*: \Lambda_{\Box} \to \Lambda$ is as in Definition 3.2.

Proof. See [4]. □

Since Axiom 3-b is the most complex, one might want to replace it by a simpler condition such as D's continuity. However, it is probably not possible since, for example, 3-b does not hold on Park's Pathological D_∞ [5].

Also note that if D satisfies Axioms 1-3, then all the results obtained on Λ vs. D_∞ due to Wadsworth [12,13] are valid on D such as $I \underset{D}{\equiv} J$ or $X \underset{D}{\equiv} \cup A(X)$ ($A(X)$ is the set of all reduced approximants of X).

Acknowledgments

The author is deeply indebted to Dr. James Morris for his support in all respects during this work. Special acknowledgment is due to Dr. Christopher Wadsworth for his many very helpful suggestions to this work.

References

[1] Barendregt, H.P., Some extensional term models for combinatory logics and λ-calculi, Thesis, Utrecht (1971).

[2] Böhm, C., Alcune proprieta della forme β-η-normali del λ-κ-calcolo, Publicazioni dell'Istituto per le Applicazioni Del Calcolo, No. 696, Rome (1968).

[3] Morris, J. and Nakajima, R., Mechanical characterization of the partial order in lattice model, D_∞, of the λ-calculus, Technical Report No. 18, Department of Computer Science, University of California at Berkeley (1973).

[4] Nakajima, R., Ph.D. Thesis, University of California at Berkeley (to appear).

[5] Park, D., The Y-combinator in Scott's λ-calculus models, Symposium on Theory of Programming, University of Warwick (1970).

[6] Plotkin, C.D., The λ-calculus is ω-incomplete, SAI-RM-2, School of Artificial Intelligence, University of Edinburgh (1973).

[7] Reynolds, J., Lattice theoretic approach to theory of computation, Unpublished lecture notes, Syracuse University (1971).

[8] Scott, D., Outline of a mathematical theory of computation, Oxford Monograph PRG-2, Oxford University (1970).

[9] Scott, D., Continuous lattices, Oxford Monograph PRG-7, Oxford University (1972).

[10] Scott, D., Lattice theory, data types and semantics, Formal Semantics of Programming Languages, Courant Computer Science Symposium 2 (1970), 65-106.

[11] Scott, D., The lattice of flow diagrams, Semantics of Algorithmic Languages, Springer Lecture Notes in Mathematics, Vol. 188 (1971), 311-366.

[12] Wadsworth, C.P., The relation between λ-expressions and their denotations in Scott's models for the λ-calculus, SIAM Journal of Computing (to appear).

[13] Wadsworth, C.P., Approximate reductions and λ-calculus models, SIAM Journal of Computing (to appear).

[14] Wadsworth, C.P., A general form of a theorem of Böhm and its application to Scott's model for the λ-calculus (to appear).

A SURVEY OF SOME USEFUL PARTIAL ORDER RELATIONS ON TERMS OF THE LAMBDA CALCULUS

J.M.E.Hyland (Christ Church, Oxford).

§0 Introduction. The equality in models for the λ-calculus gives rise to equality relations on terms of the λ-calculus, where by an equality relation we mean an equivalence relation preserved under context substitution. We focus attention on equality relations as often these are given syntactically and so prior to any model. Of course from a given equality relation one can always define a model (the model of terms factored out by the relation) which gives rise to it.

The most interesting purely semantic models for the λ-calculus, the continuous lattices of Dana Scott, are equipped with a partial order. This gives rise to what we call a partial order relation (p.o.r.) on terms of the λ-calculus, that is a pre-partial-order (i.e. transitive relation) preserved under context substitution. To any p.o.r. there corresponds the equality relation obtained by setting two terms equal iff each is less than or equal to the other. So the p.o.r. induces an ordinary partial order on the equivalence classes.

We take the view (arising out of the theses of Barendregt and Wadsworth) that terms with no head normal form (i.e. terms whose closure is unsolvable) have no computational value and so may sensibly be set equal. Thus we say that a p.o.r. is sensible iff it extends that p.o.r. obtained from β-equality by setting all terms with no head normal form equal, and less than any term; this latter p.o.r. is thus the minimal sensible p.o.r.

Our aim in this paper is to map out some of the main landmarks in the territory of sensible p.o.r.'s. To this end we make use of the $\lambda\Omega$-calculus as described in Wadsworth (1971). This arises by adding a constant Ω to the pure λ-calculus. Ω will be a minimal element in all our p.o.r.'s; that is to say Ω canonically represents the terms without head normal form. Thus the addition of Ω adds nothing to the expressive power of the λ-calculus as Ω can always be replaced by $(\lambda x.xx)(\lambda x.xx)$.

An equality relation is <u>consistent</u> iff it does not set all terms equal; a p.o.r. is <u>consistent</u> iff its induced equality relation is so. Barendregt (1971) shows that the minimal sensible p.o.r. is consistent. Our paper contains many consistent sensible p.o.r.'s, and thereby many alternative proofs of Barendregt's result; the interest of his analysis is that it shows directly the computational irrelevance of terms with no head normal form.

§1 <u>Head normal forms</u>. We define which terms of the λ-calculus are <u>head normal forms</u> (h.n.f.'s) as follows:

(a) all variables are h.n.f.'s;

(b) if $X_1, \ldots, X_k$ are terms, and x is a variable, then $xX_1 \ldots X_k$ is an h.n.f.;

(c) if P is an h.n.f. then so is $\lambda x.P$.

A term M <u>has</u> <u>h.n.f.</u> iff there is an h.n.f. N with $M =_\beta N$. Otherwise M has no h.n.f.. An h.n.f. has the form,

$$\lambda x_1 \ldots x_i . zX_1 \ldots X_j,$$

and z is the <u>head variable</u>. A non-h.n.f. has the form,

$$\lambda x_1 \ldots . x_i . (\lambda y . P) X_1 \ldots . X_j ;$$

the <u>head redex</u> is $(\lambda y.P)X_1$, and the (possibly infinite) reduction of a term, obtained by always reducing the head redex if any, is the <u>head reduction</u> of that term. By the Standardization Theorem, a term has h.n.f. iff its head reduction terminates; hence the set of terms with no h.n.f. has strong closure properties (Wadsworth (1971)). A term has h.n.f. iff its closure is solvable in the sense of Barendregt (1971).

Let $\lambda x_1 \ldots . x_m . z X_1 \ldots . X_i$ and $\lambda y_1 \ldots . y_n . w Y_1 \ldots . Y_j$ be two h.n.f.'s. By α-conversion we may take x_r to be y_r for $r \leq \min(m,n)$, and so we assume the two terms are,

(1) $\lambda x_1 \ldots . x_m . z X_1 \ldots . X_i$ and $\lambda x_1 \ldots . x_n . w Y_1 \ldots . Y_j$.

The two h.n.f.'s are

(i) <u>similar</u> iff (when arranged as in (1)) $m = n$, $i = j$ and z is w, and (ii) <u>inseparable</u> iff (when arranged as in (1)) $(m-i) = (n-j)$ and z is w.

<u>Proposition 1.1</u>. Let M be any term and let M β-reduce (respectively $\beta\eta$-reduce) to M_1 and to M_2 both h.n.f.'s. Then M_1 and M_2 are similar (respectively inseparable).

Proof: Immediate by the Church-Rosser Theorem.

The rest of this section presents a technical analysis of the theorem of Böhm (1968), by way of some lemmas which will be important later. Proofs are omitted as the methods are fairly well known, and details appear in Hyland (1975).

Lemma 1.2. (a) Suppose M, N have h.n.f.'s which are not inseparable; then there is a context $C[\ ]$ such that $C[M] =_\beta x$,

$C[N] =_\beta y$,

where x and y are distinct variables.

(b) Suppose M has no h.n.f. while N has an h.n.f.; then there is a context $C[\ ]$ such that $C[M]$ has no h.n.f.,

$C[N] =_\beta y$, for some variable y.

Proof: See Hyland (1975).

Now we define for $k \geqslant 1$, (a) the terms M and N have the same k-normal form (henceforth written $M =_k N$), and (b) the set of k-pairs of the pair (M,N). The definition is by induction on k as follows:

Case k = 1. $M =_1 N$ iff either both M and N have no h.n.f. or both M and N have h.n.f.'s, and the h.n.f.'s to which M and N reduce are inseparable. (Proposition 1.1 shows that this last requirement is unambiguous). In the first case, there are no 1-pairs of (M,N). It remains to consider the second case. We may assume that M and N reduce to the h.n.f.'s of (1) above (to fix things just consider β-reduction) where (m-i) = (n-j) and z is w. Suppose without loss of generality that $n \leqslant m$, and consider,

$$Mx_1 \ldots x_m =_\beta zX_1 \ldots X_i,$$

$$Nx_1 \ldots x_m =_\beta wY_1 \ldots Y_j x_{n+1} \ldots x_m, \text{ which is } zY_1 \ldots Y_i, \text{ say.}$$

Then the 1-pairs of (M,N) are the pairs (X_r, Y_r) for $1 \leqslant r \leqslant i$.

Induction step. $M =_{k+1} N$ iff $M =_1 N$ and for any 1-pairs (X,Y) of (M,N) we have $X =_k Y$. The (k+1)-pairs of (M,N) are the k-pairs of the 1-pairs of (M,N).

<u>Lemma 1.3</u>. Given terms M and N, with (X,Y) k-pairs of (M,N), there is a context C[ ] and substitutions (R/x,....) such that,

$C[M] =_\beta X(R/x,....)$ a substitution instance of X, and

$C[N] =_\beta Y(R/x,....)$ the same substitution instance of Y.

The terms R substituted are of the form $\lambda x_1 .... x_h . x_h x_1 ... x_{h-1}$, for h sufficiently large.

Proof: See Hyland (1975).

<u>Remark</u>. The substitutions of (1.3) have the following trivial effect on the similarity type (respectively inseparability type) of X and Y. X and Y β-reduce (respectively $\beta\eta$-reduce) to similar (respectively inseparable) h.n.f.'s iff X(R/x,....) and Y(R/x,....) do so.

<u>Corollary 1.4</u>. (Böhm). If terms M and N have distinct $\beta\eta$-normal forms then there is a context C[ ] such that $C[M] =_\beta x$,

$C[N] =_\beta y$,

where x and y are distinct variables.

Proof: By (1.2), (1.3) and the observation that if M and N have distinct $\beta\eta$-normal forms, then there is some k-pair (X,Y) of (M,N) such that X and Y have h.n.f.'s which are not inseparable.

§2 <u>Ω-approximants</u>. We recall that we have introduced a constant Ω into our language to represent the terms with no h.n.f.. The closure properties of the set of terms with no h.n.f. make it sensible to introduce <u>Ω-reductions</u> as follows. Terms of the forms ΩM and $\lambda x.\Omega$ are <u>Ω-redexes</u> and both <u>Ω-reduce</u> to Ω. A term M is in <u>$\beta\Omega$-normal form</u> iff it contains no β-redexes and no Ω-redexes; it is in <u>$\beta\eta\Omega$-normal form</u> iff it also contains no η-redexes.

Attempts to present arbitrary λ-terms as limits of normal forms which approximate them, give rise to the notion of an Ω-approximant. We shall need two such notions (depending on whether or not we are taking η-reduction into account). For a given term M, we define its sets of approximants $\omega(M)$ and $\omega\eta(M)$ as follows:

$\omega(M) = \{L \mid L$ is a $\beta\Omega$-normal form obtained from some N, where $N =_{\beta} M$, by replacing subterms of N by $\Omega\}$;

$\omega\eta(M) = \{L \mid L$ is a $\beta\eta\Omega$-normal form obtained from some N, where $N =_{\beta\eta} M$, by replacing subterms of N by $\Omega\}$.

<u>Proposition 2.1</u>. (a) $C[M]$ β-reduces (respectively $\beta\eta$-reduces) to the β-normal (respectively $\beta\eta$-normal) form N iff for some $L \in \omega(M)$ (respectively $L \in \omega\eta(M)$) $C[L]$ does so.

(b) $C[M]$ β-reduces (respectively $\beta\eta$-reduces) to a h.n.f. of a given similarity type (inseparability type) iff for some $L \in \omega(M)$ (respectively $L \in \omega\eta(M)$) $C[L]$ does so.

Proof: Wadsworth (1971) proves one of the cases in detail by a method which easily extends to the others.

<u>Lemma 2.2</u>. If the $\beta\Omega$-normal form L is not in $\omega(N)$, then for some (X,Y) k-pairs of (L,N) we have,

(i) X β-reduces to a h.n.f. X',

(ii) if Y has h.n.f. then Y β-reduces to a h.n.f. which is not similar to X'.

Proof: The lemma is easily proved for all N by induction on the structure of L.

Theorem 2.3. $\omega(M) \subset \omega(N)$ iff whenever $C[M]$ β-reduces to the h.n.f. M' then $C[N]$ β-reduces to a similar h.n.f.

Proof: That L.H.S. implies R.H.S. is immediate by a couple of applications of (2.1).

Suppose not L.H.S.. Then there is $L \varepsilon \omega(M)$, L not in $\omega(N)$. Now by (2.2) take k-pairs (X,Y) of (L,N) satisfying (i) and (ii) above. By (1.3) there is a context $C[\ ]$ such that $C[L]$ and $C[N]$ β-reduce to substitution instances of X and Y. By the remark following (1.3) we can conclude that $C[L]$ has h.n.f., but $C[L]$ and $C[N]$ do not β-reduce to similar h.n.f.'s. Hence by applying (2.1) we have not R.H.S.. This completes the proof of the theorem.

Corollary 2.4. (Independant result of Levy and of Welch) $\omega(M) \subset \omega(N)$ does define a (consistent) p.o.r. on λ-terms.

Proof: The relation on the R.H.S. of (2.3) is clearly preserved under context substitution.

Remark. The relation of (2.3) properly extends the minimal sensible p.o.r. as (for example) it sets all the members of the usual sequence Y_0, Y_1,... of fixed point operators, equal.

Lemma 2.5. If the $\beta\eta\Omega$-normal form L is not in $\omega\eta(N)$, then for some (X,Y) k-pairs of (L,N), we have,

(i) X $\beta\eta$-reduces to the variable x,

(ii) Y does not $\beta\eta$-reduce to x.

Proof: The lemma is easily proved for all N by induction on the structure of L.

<u>Theorem 2.6</u>. $\omega\eta(M) \subset \omega\eta(N)$ iff whenever $C[M]$ $\beta\eta$-reduces to the $\beta\eta$-normal form M' then $C[N]$ $\beta\eta$-reduces to M'.

Proof: That L.H.S. implies R.H.S. is immediate by a couple of applications of (2.1).

Suppose not L.H.S.. Then there is $L \in \omega\eta(M)$, L not in $\omega\eta(N)$. Things are not so simple now as they were in the proof of (2.3), so we dispose of the easy case first. Suppose there exist k-pairs (X,Y) of (L,N) such that X has h.n.f. but if Y has h.n.f. then it is not inseparable from that of X. Then not R.H.S. follows easily from (1.2), (1.3) and the remark following (1.3).

So henceforth assume that for all k-pairs (X,Y) of (L,N), if X has h.n.f. then Y has h.n.f. inseparable from that of X.

Now by (2.5) take k-pairs (X,Y) of (L,N) satisfying (i) and (ii) of (2.5).

Then $X =_{\beta\eta} x$,

and $Y =_{\beta\eta} \lambda y_1 \ldots y_k . x Y_1 \ldots Y_k$,

and it follows from our assumption anove that Y has no normal form. Consider the substitution instances X' and Y' of X and Y determined by (1.3). It suffices to show that Y' has no normal form. (This does not follow from the general nature of the substitutions, but from the special form of Y). Note that even if in the substitution instances X' and Y', some R has been substituted for the variable x, there must be (k+1)-pairs (X_i,Y_i) say satisfying (i) and (ii) of (2.5), where X_i is a variable y_i say and nothing is substituted for y_i by the appropriate context determined by (1.3). So we can assume that nothing is substituted for x in X and Y. But then for all r-pairs (A,B) of (X,Y) nothing has been substituted for the head variable of B. By considering normal reductions, since Y has no normal form, neither has Y'. The proof is now completed as for (2.3).

§3 Scott's models. In this section we outline the main results of Hyland (1975). We are concerned with the values of λ-terms in continuous lattice models for the λ-calculus. D denotes some (arbitrary) continuous lattice isomorphic to its function space, which is constructed from a continuous lattice D_0 and the initial maps,

$\phi_0 : D_0 \longrightarrow D_1$, defined by $\phi_0(d_0) = \lambda x.d_0$, and

$\psi_0 : D_1 \longrightarrow D_0$, defined by $\psi_0(d_1) = d_1(\perp)$.

$P\omega$ denotes the Graph Model described in Scott's "Data Types as Lattices". (The Scott Model D is fully described in Scott's "Continuous Lattices"). The value of a term M in these models will be denoted by $[\![M]\!]_D$ and $[\![M]\!]_{P\omega}$ respectively. $\sqsubseteq$ denotes the order relation and $\bigsqcup$ the sup operation in either lattice.

Proofs of all the results of this section appear in Hyland (1975), and we do not include them here. Furthermore, Wadsworth presented his considerable improvement on our original proof of (3.1)(a) and his own proof of (3.2)(a) at a conference in Orleans, 1972. So the basic ideas should be familiar.

Theorem 3.1. (a) $[\![M]\!]_D = \bigsqcup\{ [\![L]\!]_D | L \in \omega(M)\} = \bigsqcup\{ [\![L]\!]_D | L \in \omega\eta(M)\}$.

(b) $[\![M]\!]_{P\omega} = \bigsqcup\{ [\![L]\!]_{P\omega} | L \in \omega(M)\}$.

Next we make some definitions which extend those of §1. We introduce relations $<^s_k$ and $<^g_k$ for $k \geqslant 1$, by induction on k. The superscripts s and g are to indicate that the relations are important for the Scott and Graph Models respectively.

Definition. $M <^s_1 N$ iff whenever M has h.n.f. then $M =_1 N$. Then by induction, $M <^s_{k+1} N$ iff $M <^s_1 N$ and for any 1-pairs (X,Y) of (M,N) we have $X <^s_k Y$.

The h.n.f. $\lambda x_1 \ldots x_m . zX_1 \ldots X_i$ is <u>more functional than</u> the h.n.f. $\lambda y_1 \ldots y_n . wY_1 \ldots Y_j$ iff the two h.n.f.'s are inseparable and $m \geq n$.

<u>Definition</u>. $M <_1^g N$ iff whenever M has h.n.f., then N β-reduces to a h.n.f. which is more functional than that to which M β-reduces. Then by induction, $M <_{k+1}^g N$ iff $M <_1^g N$ and for any 1-pairs (X,Y) of (M,N) we have $X <_k^s Y$.

<u>Theorem 3.2</u>. (a) The following are equivalent:

(i) $[\![M]\!]_D \sqsubseteq [\![N]\!]_D$;

(ii) for all $k \geq 1$, $M <_k^s N$;

(iii) whenever $C[M]$ has h.n.f. then so has $C[N]$.

(b) The following are equivalent:

(i) $[\![M]\!]_{P\omega} \sqsubseteq [\![N]\!]_{P\omega}$;

(ii) for all $k \geq 1$, $M <_k^g N$;

(iii) whenever $C[M]$ has h.n.f. then $C[N]$ β-reduces to a h.n.f. more functional than that to which $C[M]$ β-reduces.

The p.o.r. induced by the Scott Model D has a beautiful uniqueness property.

<u>Theorem 3.3</u>. The p.o.r. characterised in (3.2)(a) is the unique maximal consistent sensible p.o.r.

§4 <u>Concluding remarks</u>. We have in (2.3), (2.6) and (3.2) characterised four sensible p.o.r.'s. It is easy to see that they are all distinct (though note that the induced equality of (2.3) and (3.2)(b) is the same). The most significant feature of the results to my mind is this. Each of the four p.o.r.'s has a characterisation in terms of contexts, of a natural form: if a context acting on one term does such and such, then so does the context acting on the other. In other words, each p.o.r. is characterised in terms of its computational significance. This in my view should be a feature of any interesting sensible p.o.r. so at least we know what to look for should we search for more sensible p.o.r.'s.

Bibliography.

Barendregt, H.P., Some extensional term models for combinatory logics and λ-calculi, Utrecht 1971.

Böhm, C., Alcune proprieta della forme $\beta\eta$-normali del λK-calcolo, C.N.R. No.696, Rome 1968.

Hyland, J.M.E., A syntactic characterization of the equality in some models for the Lambda Calculus, to appear Bull. L.M.S. 1975.

Scott, D., Continuous Lattices, Springer L.N.M. No.274, 97-136.

Scott, D., Data Types as Lattices, to appear.

Note on the papers of J.M.E.Hyland and J.-J.Lévy.

(Corrections and clarifications.)

J.M.E.Hyland (Christ Church, Oxford).

Some confusion is apparent in the reference in my paper to that of J.-J.Lévy, and in the conclusion of his to my own and to that of Welch. Since this is largely the fault of a careless abstract of my paper which appeared before the conference, it seems right that I should set the matter straight.

The facts are as follows.

1) Lévy's relationship "$A(M) \subset A(N)$" is not the same as my "$\omega(M) \subset \omega(N)$". My relation is clearly identical with the semantics (E_∞, N) of Welch. Indeed Lévy's relation is neither a continuous semantics in Welch's sense, nor is it sensible in mine. But it is interesting enough for all that.

2) Lévy's relation can be characterized along the lines of (2.3) of my paper, thus: $A(M) \subset A(N)$ iff both (i) whenever $C[M]$ β-reduces to a h.n.f. then $C[N]$ β-reduces to a similar h.n.f., and (ii) whenever $C[M]$ β-reduces to a term with n λ's at the front , then so does $C[N]$.

Conversely, Lévy's work suffices to show that both "$\omega(M) \subset \omega(N)$" and "$\omega\eta(M) \subset \omega\eta(N)$" are substitutive.

3) It is the equality relation "$\omega(M) = \omega(N)$" which is identical with the equality relation " $M_{P\omega} \subset N_{P\omega}$". The corresponding p.o.r.'s differ. One can get a similar result for Lévy's relation by tampering with the coding in $P\omega$. It suffices to start the enumeration of finite sets with e_1, and conventionally interpret $m \in \tau(e_0)$ in the definition of lambda abstraction, as always true. The equality in the resulting model is the same as the equality in Lévy's model. Again the p.o.r.'s differ.

λ-TERMS AS TOTAL OR PARTIAL FUNCTIONS ON NORMAL FORMS

Corrado Böhm - Istituto Matematico G. Castelnuovo (Università di Roma)
Mariangiola Dezani-Ciancaglini - Istituto di Scienza dell'Informazione (Università di Torino)

Abstract. In this paper the set of λ-terms is split into $2\omega+1$ disjoint classes $\mathcal{N}_h$ $(-\omega\leq h\leq\omega)$. This classification takes into account the meaning of a λ-term F as function on normal forms, and more precisely:

$F\in\mathcal{N}_{-\omega}$ iff when successively applied to any number of normal forms it gives a λ-term without normal form

$F\in\mathcal{N}_{-h}$ $(0<h<\omega)$ iff when successively applied to h-1 arbitrary normal forms it gives a λ-term without normal form, but there exist h normal forms $X_1,\ldots,X_h$ such that $FX_1\ldots X_h$ possesses normal form

$F\in\mathcal{N}_h$ $(0\leq h<\omega)$ iff when successively applied to h arbitrary normal forms it gives a λ-term which possesses normal form, but there exist h+1 normal forms $X_1,\ldots,X_{h+1}$ such that $FX_1\ldots X_{h+1}$ possesses no normal form

$F\in\mathcal{N}_\omega$ iff when successively applied to any number of normal forms it gives a λ-term which possesses normal form.

This classification can be effectively determined only for λ-terms in normal form.

Introduction

It is well known that recursive functions, programs, structured data can be represented by λ-terms and λ-terms are presently used to give an abstract definition of the current state of a concrete computer during a computation. Our approach is in some sense opposite to the previous. We start considering λ-calculus as a programming language (i.e. every term acts as program and/or input data) and we assume that the ensuing computations evolve following the β-reduction rule. This enables us to identify the whole state of computation with a λ-term, in particular it follows that terminal states must correspond to normal forms.

Let us investigate what kind of information can be gained from the assumption that convertibility means equivalence of programs inside the λ-calculus. The usual notion of equivalence ($\simeq$) in programming theory is weaker than the convertibility (=) of λ-terms, in fact we have:

$$U \simeq V \overset{Df}{\longleftrightarrow} \forall f\,[\text{either } Uf=Vf \text{ if } Uf \text{ and } Vf \text{ possess normal forms or } Uf \text{ and } Vf \text{ possess no normal forms}]$$

From this definition it follows, for example, that the fixed-point combinators $\mathbf{Y}_0$ and $\mathbf{Y}_1$ (see [4] p. 154-155) are equivalent as programs, but it is proved that they are not interconvertible [1]. Then to make coincide the notions of equivalence of programs and convertibility of λ-terms, we must require that:

1) the extensionality principle is valid, i.e. $\forall f\ (Uf=Vf) \supset U=V$
2) every λ-term standing for the application of a program to an input data has normal form.

Obviously the extensionality principle holds if f ranges over the set $\mathcal{N}$ of λ-terms in normal form. Therefore $\mathcal{N}$ is the correct framework, if we limit ourselves to consider applications of normal forms assuring the existence of normal form. Hence we must be able to decide when, given a normal form U, for every $f \in \mathcal{N}$, Uf possesses normal form. This request is not in contradiction with the fact that the existence of a normal form for a λ-term is proved to be semi-decidable [3].

In this paper we give an effective computation of the "arity" of a normal form, i.e. we associate to each normal form the maximum integer h ($h \geq 0$) such that all λ-terms obtained applying that normal form successively to h arbitrary normal forms possess normal form too. Obviously h can be infinite (ω). This splittes the set $\mathcal{N}$ of normal forms into $\omega+1$ disjoint classes.

Two more related researches are done. First it is proved that η-reduction does not modify the classification of normal forms just given.

In another direction, also a classification of λ-terms not in normal form is given. This classification cannot be effective and it is done using Wadsworth's approximants [7]. To this aim we introduce negative arities for λ-terms. A λ-term F has arity -h when applied successively to (h-1) normal forms gives always a λ-term without normal form, but there exist h normal forms such that F applied successively to them is reducible to normal form. The arity $-\omega$ means that the successive applications of the λ-term to any number of normal forms give always a λ-term without normal form. It is clear that a λ-term has negative arity iff it possesses no normal form.

1. Key notions and their properties

We shall adopt the following conventions:

a) normal form means β-normal form, $\geq$ denotes α-β-reducibility, = denotes α-β-convertibility and $\equiv$ denotes identity of objects

b) the combinators are denoted by uppercase boldface characters

c) $<X_1,\dots,X_n>$ denotes $\lambda z(zX_1\dots X_n)$ where $X_i (1\leq i\leq n)$ are λ-terms and z shall not occur free in any of them (Church n-tuple).

For reasons of clarity, we use the tree representation of normal forms described extensively in [2] and next summarized.

Taking into account the following syntax for normal forms:

$$\mathcal{N} ::= \lambda v_1 \dots \lambda v_n (v \mathcal{N}_1 \dots \mathcal{N}_m) \qquad n,m \geq 0$$

$$v ::= x_1 | x_2 | \dots$$

we can produce the tree representing a normal form according to the one-one correspondence:

$$N \equiv \lambda x_{j_1} \dots \lambda x_{j_n} (x_j N'_1 \dots N'_m) \quad \longleftrightarrow$$

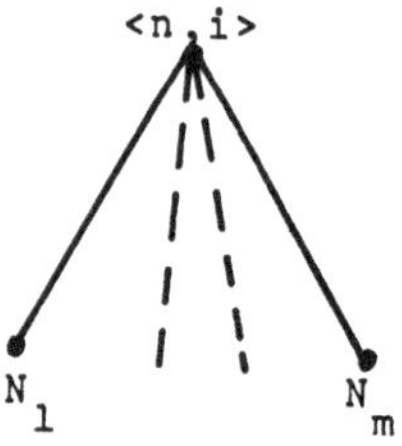

where $\quad i = \begin{cases} \ell & \text{if } \exists \ell \ (1\leq \ell \leq n) \ [j=j_\ell] \\ n+j & \text{otherwise} \end{cases}$

and $N_p \equiv \lambda x_{j_1} \dots \lambda x_{j_n} N'_p \ (1\leq p\leq m)$.

x_i is called the head variable of N, and N_p the p-th component of N $(1\leq p\leq m)$.

Example 1. The tree representing $N \equiv \lambda x_1 \lambda x_2 \lambda x_3 (x_2(x_1 \lambda x_4 \lambda x_5 (x_3 x_4)) \lambda x_4 \lambda x_5 (x_5 x_4)(x_4 x_2 x_1))$ is shown in fig. . This representation suggests a many one correspondence between λ-terms not in normal form and trees. If we replace in a λ-term F each subterm which is a β-redex by Ω, we obtain the "instant meaning" F^* of F according to [8]. F^* is representable by the unique tree in which some terminal nodes are labeled by a pair whose second component is Ω. As particular cases, the trees consisting of a single root labeled $<n,\Omega>(n\geq 0)$ correspond to all λ-terms which

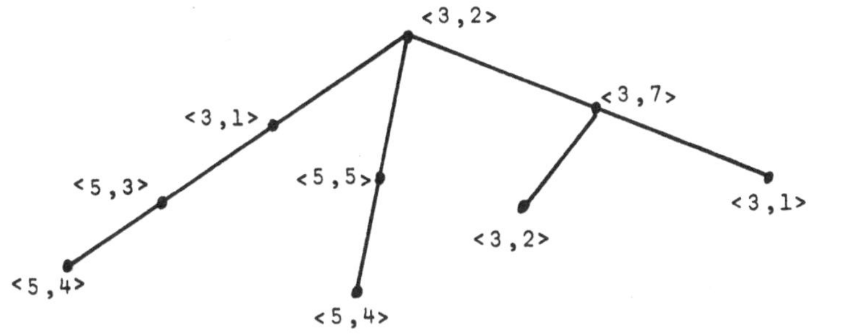

Fig. 1 - Tree representing the normal form of example 1.

are β-redexes abstracted relative to n variables.

We assume in this case the same definitions of head variable and components as in the case of normal forms. The two representations coincide for all λ-terms in normal form. According to [7], we shall say that a λ-term is in <u>head normal form</u> iff it has the following shape : $\lambda x_1 \ldots \lambda x_n (x_i X_1 \ldots X_m)$ where $m,n \geq 0$ and X_j ($1 \leq j \leq m$) is an arbitrary λ-term.

<u>Example 2.</u> The tree representing $F \equiv \lambda x_1 \lambda x_2 \lambda x_3 \; (x_2(x_1 \; \lambda x_4(\lambda x_5(x_3 x_4)x_1)) (\lambda x_4 \lambda x_5 (x_5 x_4) x_2)(x_4 x_2 x_1))$ whose instant meaning is $F^* \equiv \lambda x_1 \lambda x_2 \lambda x_3 (x_2 (x_1 \lambda x_4 \Omega)\Omega \; (x_4 x_2 x_1))$ is shown in fig.2.

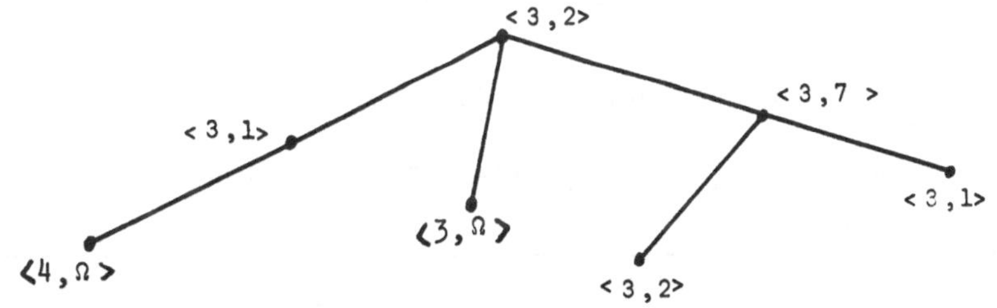

Fig. 2 - Tree representing the λ-term of example 2.

From now an, we shall assume that a bound variable represented by nodes whose second component is i is called x_i. This is always feasible by α-reductions.

We are now able to introduce our key notion, i.e. that of h-replaceable variable ($h \geq 1$). An h-replaceable variable will be proved to be replaceable by an arbitrary normal form when the λ-term, in which it occurs, is applied successively to h suitable normal forms. The notion of h-replaceability inserts a degree of boundness and a degree of freedom for the variables. To give an easier understanding of these facts and to visualize the proof we will use the tree representation of λ-terms.

Definition 1. In a tree representing an arbitrary λ-term whose root has label $<n,i>$, the h-replaceable nodes are:

- the nodes whose label is $<p,q>$ with $q \leq h \leq n$
- if a node with a label $<p,q>$ is h-replaceable and it possesses as son a node with label $<r,s>$,then all nodes descendant from the last one itself included and whose label is $<u,v>$ with $p<v\leq r$ are h-replaceable.

All nodes not satisfying the preceding recursive definition, for a given h, are said non-h-replaceable.

Let us note that as a consequence of the definition:

- no node is 0-replaceable
- in a λ-term with n initial abstractions there are no nodes h-replaceable for h>n and if h<n then a node h-replaceable is always (h+1)-replaceable.

Definition 1'. A node is non-replaceable iff it is non-h-replaceable for every $h \geq 1$.

Example 3. Fig. 3 shows the tree corresponding to the λ-term $\lambda x_1 \lambda x_2 \lambda x_3 (x_1 x_4 \lambda x_4 \lambda x_5 (x_4 (\lambda x_6 x_5 x_2)) (x_2 \lambda x_4 \lambda x_5 x_5) x_3)$ where the h-replaceable nodes are surrounded by h circles.

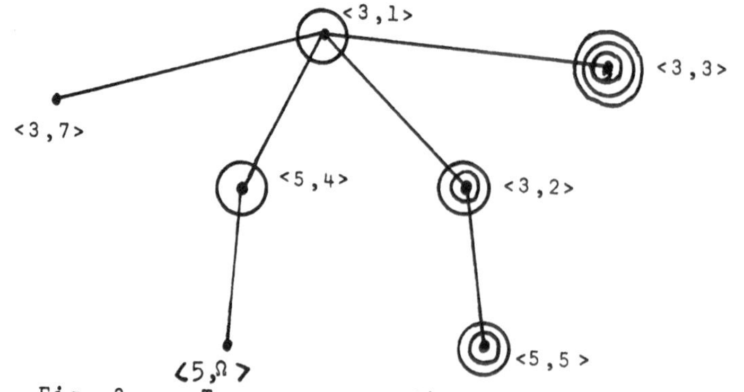

Fig. 3 - Tree corresponding to the λ-term of example 3

Since the tree representation of λ-terms is not invariant by β-reduction, two β-convertible λ-terms may have different h-replaceable nodes.
Let us note that in the tree representation of a λ-term every node corresponds to an occurrence of a variable, but the contrary is not true. In fact, occurrences of variables inside a β-redex correspond to no node in the tree.
Clearly, from definition 1 it follows that two nodes corresponding to two different occurrences of the same variable are both h-replaceable or both non-h-replaceable. Then we say that a <u>variable</u> is <u>h-replaceable</u> (<u>non-h-replaceable</u>) iff all nodes corresponding to its occurrences are h-replaceable (non-h-replaceable).

The following two lemmas characterize the replaceability and the non-replaceability of a given variable by an arbitrary normal form.
<u>Lemma 1</u>. Given a λ-term F having an h-replaceable variable and given one arbitrary normal form Y, we can always find h normal forms $X_1,\ldots,X_h$ such that inside $FX_1\ldots X_h$ the current variable is replaced by Y.
<u>Proof</u>. Let be $<n,i>$ the label of the root of the tree representing F. If a given variable is h-replaceable in F, it follows from definition 1 that in the corresponding tree there is at least one path from a node of label $<p,q>$ with $q\leq h\leq n$ to that node where the current variable is bound. Let be t the number of h-replaceable variables occurring in this path. We perform the desired proof by induction on this number t, that is the number of applications of the recursive definition rule of h-replaceability.
If t=0 the current variable is bound by the root of the tree and therefore its occurrences are represented by nodes whose labels $<p,q>$ satisfy $q\leq h\leq n$. Then it is sufficient to choose $X_q\equiv Y$.
Given a positive integer w, let us assume that we have done the proof for t<w. If we wish to replace an h-replaceable variable for which is t=w by Y, it follows from definition 1 that this variable is bound in a node (with label $<r,s>$) whose father is an h-replaceable node (with label $<p,q>$). The variable x_q is then h-replaceable and its t<w, and therefore by inductive hypothesis it can be replaced by an arbitrary normal form Y'. Then, if v is the label of the current variable and $<r,s>$ is the j-th son of $<p,q>$ between m sons, it is sufficient to

choose:

$$Y' \equiv \lambda x_1 \ldots \lambda x_m (x_j \psi_{p+1} \ldots \psi_{v-1} Y)$$

where $\psi_\ell (p+1 \leq \ell \leq v-1)$ are arbitrary normal forms, to obtain the desired result.

□

Let us note that the proof of this lemma is constructive, i.e. it exhibits $X_1, \ldots, X_h$.

<u>Lemma 2</u>. Given a λ-term F, do not exist h normal forms $X_1, \ldots, X_h$ such that a variable non-h-replaceable in F is replaced in F $X_1 \ldots X_h$.

<u>Proof</u>. If the current variable of F is free, this lemma is obvious. If the current variable of F is bound and non-h-replaceable, this means that, in the tree representing F, it is bound in a node son of a node non-h-replaceable. We prove this case by induction on the level t of the node bounding the current variable.

If t=0, this means that the current variable x_q is bound in the root of the tree representing F and q>h. Then this variable cannot obviously be replaced when F is applied to h arbitrary normal forms. Given a positive integere w, let us assume that a variable non-h-replaceable whose bounding node is at a level t<w cannot be replaced. We prove that also a variable non-h-replaceable whose bounding node is at a level t=w cannot be replaced. By definition 1, the bounding node of a variable satisfying these conditions is son of a non-h-replaceable variable, whose bounding node is then at a level t<w. This latter variable cannot be replaced by inductive hypothesis, and this means in particular that the current variable cannot be replaced.

□

Now we define a mapping from a set of subterms of a first λ-term into the power set of subterms of a second λ-term obtained by applying the first one to h normal forms. The idea behind this mapping is to use a non-h-replaceable variable as detector of the transformation supported by the subtermshaving this variable as head variable. To interpret correctly this definition it is necessary to consider as new and distinguished (possibly marking it with some auxiliary label) each occurrence of the same non-h-replaceable variable in the first λ-term.

Since this mapping and its properties are used only in the classification of the set $\mathcal{N}$ of λ-terms in normal form, we agree that the first λ-term be in normal form.

Definition 2. Let be $N \in \mathcal{N}$ and M a subterm of N. We call "image set" of M in $N\, X_1 \ldots X_h$ for $X_1, \ldots, X_h \in \mathcal{N}$ the set $\mathcal{M}$ of subterms of $N\, X_1 \ldots X_h$ so defined:

- if the head variable x_i of M is non-h-replaceable, then exactly the subterms of $N\, X_1 \ldots X_h$ whose head variables are x_i belong to $\mathcal{M}$
- if the head variable of M is h-replaceable, but M is the n-th component of a term whose head variable x_ℓ is non-h-replaceable, then exactly the subterms of $N\, X_1 \ldots X_h$ which are the n-th components of terms whose head variables are x_ℓ belong to $\mathcal{M}$
- otherwise, $\mathcal{M}$ is undefined.

To avoid that $\mathcal{M}$ will contain λ-terms which differ only by one initial abstraction, we agree to cross out from $\mathcal{M}$ each λ-term which can be obtained by another λ-term belonging to $\mathcal{M}$ by abstraction relative to a given variable.

As a particular case, only $N\, X_1 \ldots X_h$ belongs to the image set of N in $N\, X_1 \ldots X_h$.

Example 4. If $N \equiv \lambda x_1 \lambda x_2 \lambda x_3 (x_1 x_4 \lambda x_4 \lambda x_5 x_4 (x_2 \lambda x_4 \lambda x_5 x_5) x_3)$, $M_1 \equiv x_4$, $M_2 \equiv \lambda x_4 \lambda x_5 x_4$, $M_3 \equiv x_2 \lambda x_4 \lambda x_5 x_5$, $M_4 \equiv x_3$, $X_1 \equiv \lambda y_1 \lambda y_2 \lambda y_3 \lambda y_4 (y_1 y_2 y_1 (y_1 y_3)(y_3 y_2))$, then the image sets $\mathcal{M}_1, \mathcal{M}_2, \mathcal{M}_3, \mathcal{M}_4$ respectively of M_1, M_2, M_3, M_4 in $N\, X_1 \geq \lambda x_2 \lambda x_3 (x_4 \lambda x_4 \lambda x_5 x_4 x_4 (x_4 (x_2 \lambda x_4 \lambda x_5 x_5))(x_2 \lambda x_4 \lambda x_5 x_5 \;\; \lambda x_4 \lambda x_5 x_4)))$ are: $\mathcal{M}_1 = \{x_4, x_4 (x_2 \lambda x_4 \lambda x_5 x_5), N X_1\}$, $\mathcal{M}_2$ = undefined, $\mathcal{M}_3 = \{x_2 \lambda x_4 \lambda x_5 x_5 \lambda x_4 \lambda x_5 x_5, x_2 \lambda x_4 \lambda x_5 x_5\}$, $\mathcal{M}_4 = \Phi$ (1).

The following two lemmas assure that some subformulas of a normal form N, being in suitable conditions, have as images in $N\, X_1 \ldots X_h$ sets of λ-terms in head normal form or reducible to head normal forms.

Lemma 3. Let be $N \in \mathcal{N}$ and M a subterm of N whose head variable is non-h-replaceable. Then for h arbitrary normal forms $X_1, \ldots, X_h$ each term

---

(1) Φ denotes the empty set.

belonging to the image set of M in N $X_1 \dots X_h$ is in head normal form.
Proof. By lemma 2 the head variable of M cannot be replaced in N $X_1 \dots X_h$, and then each λ-term belonging to the image set of M is in head normal form.

□

Lemma 4. Let be $N \in \mathcal{N}$ and M a subterm of N whose head variable x_p is h-replaceable, but such that:

- M is component of a term whose head variable is non-h-replaceable
- each component of M has as head variable a non-h-replaceable variable.

Then for h arbitrary normal forms $X_1, \dots, X_h$ such that the term which will replace x_p in N $X_1 \dots X_h$ is an head normal form, each term belonging to the image set of M in N $X_1 \dots X_h$ is reducible to an head normal form.

Proof. Let $M \equiv \lambda x_1 \dots \lambda x_s (x_p M_1 \dots M_r)$ and let $Y \equiv \lambda y_1 \dots \lambda y_v (y_q Y_1 \dots Y_u)$ be the head normal form which must replace x_p. We consider a term M' belonging to the image set of M. Let be M'_ℓ $(1 \leq \ell \leq r)$ the terms belonging respectively to the image sets of M_ℓ $(1 \leq \ell \leq r)$ in N $X_1 \dots X_h$ and moreover occurring as subterms of M'. Since all head variables of M_ℓ $(1 \leq \ell \leq r)$ are non-h-replaceable, by lemma 3 M'_ℓ $(1 \leq \ell \leq r)$ are head normal forms. The condition that M be component of a term whose head variable is non-h-replaceable assures that M' will be reducible to:

$$\lambda x_1 \dots \lambda x_s (\lambda y_1 \dots \lambda y_v (y_q Y_1 \dots Y_u) M'_1 \dots M'_r).$$

We distinguish the following three cases:

1) if $q > v$ or $q > r$ then the head variable y_q of Y cannot be replaced and therefore M' is reducible to an head normal form.

2) if $q \leq v \leq r$ then M' will be reducible to:

$$\lambda x_1 \dots \lambda x_s (M'_q \bar{Y}_1 \dots \bar{Y}_u \, M'_{v+1} \dots M'_r)$$

where $\bar{Y}_\ell \equiv Y_\ell [y_j / M'_j]$ $(1 \leq \ell \leq u)(1 \leq j \leq v)$.

M'_q is an head normal form whose head variable is non-h-replaceable by hypothesis and therefore M' is reducible to an head normal form.

3) if $q \leq r \leq v$ then M' will be reducible to:

$$\lambda x_1 \dots \lambda x_s \lambda y_{r+1} \dots \lambda y_v (M'_q \bar{Y}_1 \dots \bar{Y}_u)$$

where $\bar{Y}_\ell \equiv Y_\ell [y_j / M'_j]$ $(1 \leq \ell \leq u)$ $(1 \leq j \leq r)$.

Then M' is reducible to an head normal form by the same reason of case 2.

□

2. Classification of normal forms.

We give now the formal definition of the arity of a normal form N.

Definition 3. $N \in \mathcal{N}_h$ ($h \geq 0$) iff the following conditions are both satisfied:

- $\forall X_1,\dots,X_h \in \mathcal{N}$: $N\ X_1 \dots X_h \geq N' \in \mathcal{N}$
- $\exists X_1,\dots,X_{h+1} \in \mathcal{N}$: $N\ X_1 \dots X_{h+1}$ possesses no normal form.

Definition 4. $N \in \mathcal{N}_\omega$ iff the following conditions are both satisfied:

- $\forall h\ (h \geq 0)\ \forall X_1,\dots,X_h \in \mathcal{N}$: $N\ X_1 \dots X_h \geq N' \in \mathcal{N}$.

The classification of a normal form N according to its arity is given by means of the following two theorems:

Theorem 1. A normal form N with n initial abstractions belongs to $\mathcal{N}_h$ ($0 \leq h \leq n-1$) iff in the corresponding tree:

1) there exist no two nodes both h-replaceable and joint be one arc
2) there exist at least two nodes (h+1)-replaceable and joint by one arc.

Theorem 2. A normal form N with n initial abstractions belongs to $\mathcal{N}_n \cup \mathcal{N}_\omega$ iff in the corresponding tree there exist no two nodes both n-replaceable and joint by one arc. More specifically $N \in \mathcal{N}_n$ or $N \in \mathcal{N}_\omega$ according to whether the head variable of N is bound or free.

Proof of theorem 1.

"If" part. We must prove that conditions 1 and 2 imply:

a) $\forall X_1,\dots,X_h \in \mathcal{N}$: $N\ X_1 \dots X_h \geq N' \in \mathcal{N}$

b) $\exists X_1,\dots,X_{h+1} \in \mathcal{N}$: $N\ X_1 \dots X_{h+1}$ possesses no normal form.

By lemmas 3 and 4 it follows that if N satisfies condition 1, then for h arbitrary normal forms $X_1,\dots,X_h$, each subterm in $N\ X_1 \dots X_h$ is reducible to an head normal form. Then $N\ X_1 \dots X_h$ is reducible to a normal form. If N satisfies condition 2, this means that there exist a subterm M of N such that, in the tree representation of N, the root of M is $\langle m,i \rangle$, this root possesses f sons, the j-th son is called $\langle s, \ell \rangle$, it possesses in its turn r sons and moreover the variables x_i and x_ℓ are both (h+1)-replaceable. This situation is sketched in fig.4.

We will distinguish the only 5 possible cases:

i) $m < \ell$

ii) $i = \ell$

iii) $i \neq \ell$, $\ell \leq m$, i and ℓ are independently (h+1)-replaceable.

iv) $i < \ell \leq m$, ℓ is (h+1)-replaceable thanks to some other occurrence of i

v) $\ell \leq m$, $i > \ell$, i is (h+1)-replaceable thanks to some other occurrence of ℓ.

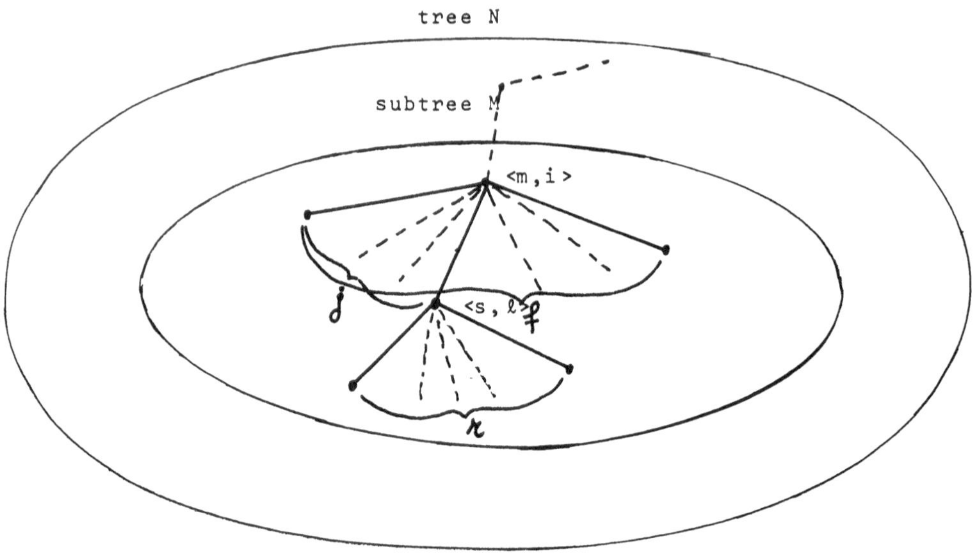

Fig.4. Sketch of a tree satisfying condition 2 of theorem 1, if the variables x_i and x_ℓ are both (h+1)-replaceable.

In all these cases, the proof is done by exhibiting the normal forms which must replace x_i and/or x_ℓ in M in order to obtain a subterm M' not possessing normal form. The corresponding normal forms $X_1,\ldots,X_{h+1}$ may be obtained, for a given N, using the construction of lemma 1.
We must always take in consideration in the tree representation of N the path joining the root of N with the root of M, to avoid that M' will be erased. Let be p and q respectively the numbers of occurrences of the variables x_i and x_ℓ in this path, and let this path selects the u_t-th (v_t-th) son in the t-th occurrence of the variable $x_i(x_\ell)$ for $1\leq t\leq p$ (for $1\leq t\leq q$). At last, we will denote by u(v) the maximum number of sons of the variable $x_i(x_\ell)$ in this path. In what follows, indexed ϕ and ψ shall denote arbitrary normal forms.
<u>Case i</u>. Since $m<\ell$, the variable x_ℓ cannot occur in the path joining the root with the subtree representing M, i.e., q=0. Called $\nabla\equiv\langle\psi_{m+1},\ldots$ $\ldots,\psi_{\ell-1},\mathbf{K}^r\mathbf{I},\psi_{\ell+1},\ldots,\psi_s\rangle$, it is sufficient to replace the variable x_i

by the normal form of:

$X\equiv\lambda z_1\ldots\lambda z_u<z_{u_1},\ldots,z_{u_p},\nabla z_j(\mathbf{W}(\nabla z_j))(\mathbf{W}(\nabla z_j))>$.

The first p components of this (p+1)-tuple assure that the replacement of x_i by X does not erase M'. The last component gives rise to a formula without normal form, as it can be verified by β-reductions (see Appendix I).

<u>Case ii</u>. If $i=\ell$, then $p=q$, $u_t=v_t$ $(1\leq t\leq p)$, $u=v$.

Called $w=\max.[u,r+1]$ and $\Gamma\equiv\lambda z_1\ldots\lambda z_w<z_{r+1}z_{r+1},z_j\psi_{m+1}\ldots\psi_s(z_j\psi_{m+1}\ldots\ldots\psi_s)>$, it is sufficient to replace x_i by the normal form of:

$$X\equiv\lambda z_1\ldots\lambda z_w<z_{u_1},\ldots,z_{u_p},\Gamma z_1\ldots z_w>.$$

The aims of the components of this (p+1)-tuple are the same as in case i.

<u>Case iii</u>. In this case it is sufficient to replace variables x_i and x_ℓ respectively by the normal forms of:

$$X\equiv\lambda z_1\ldots\lambda z_u<z_{u_1},\ldots,z_{u_p},\Omega\, z_j(\mathbf{W}(\Omega z_j))(\mathbf{W}(\Omega\, z_j))>$$

and

$$Y\equiv\lambda z_1\ldots\lambda z_v<z_{v_1},\ldots,z_{v_q},\mathbf{I}>$$

where $\Omega\equiv<\phi_{m+1},\ldots,\phi_s,\psi_{r+1},\ldots,\psi_v,\mathbf{U}^{(q+1)}_{q+1}>$ (1).

<u>Case iv</u>. By hypothesis, the node bounding the variable x_ℓ occurs in a subtree whose root is son of a node with label <w,i> (obviously all these nodes occur in the path joining the root of N with the root of M). Let such a subtree be the t-th of k. This means that we can **find** a normal form Z such that replacing the variable x_i by:

$$\lambda z_1\ldots\lambda z_k(z_t\psi_{w+1}\ldots\psi_{\ell-1}(ZV)),$$

the variable x_ℓ is replaced by an arbitrary normal form V.

The expression of Z will depend on the path joining the node <w,i> with the node bounding the variable x_ℓ and may be built using the technique of lemma 1.

In this case it is sufficient to replace the variable x_i by the normal form of:

$$X\equiv\lambda z_1\ldots\lambda z_u<z_{u_1},\ldots,z_{u_p},z_t\psi_{w+1}\ldots\psi_{\ell-1}(ZY),\ \Omega\, z_j(\mathbf{W}(\Omega\, z_j))(\mathbf{W}(\Omega z_j))>$$

---

(1) The normal combinators $\mathbf{U}^{(n)}_i$ $(1\leq i\leq n)$ are defined by:

$$\mathbf{U}^{(n)}_i=\lambda z_1\ldots\lambda z_n z_i.$$

where Y and Ω are defined as in case iii.

Case v. In this case x_i and x_ℓ play the inverse rôles as in case 4, i.e. we can find by lemma 1 a normal form $\bar{Z}$ such that replacing the variable x_ℓ by:

$$\lambda z_1 \ldots \lambda z_{\bar{k}}(z_{\bar{t}}\psi_{\bar{w}+1}\ldots\psi_{i-1}(\bar{Z}V))$$

the variable x_i will be replaced by an arbitrary normal form V. The meanings of $\bar{k},\bar{t},\bar{w}$ can be obtained from those of k,t,w as defined in case iv replacing systematically x_i by x_ℓ and viceversa. Then it is sufficient to replace x_ℓ by the normal form of:

$$Y \equiv \lambda z_1 \ldots \lambda z_v <z_{v_1},\ldots,z_{v_q},z_{\bar{t}}\psi_{\bar{w}+1}\ldots\psi_{i-1}(\bar{Z}X),\mathbf{I}>$$

where X is defined as in case iii with $\bar{\Omega}$ instead of Ω and

$$\bar{\Omega} \equiv <\phi_{m+1},\ldots,\phi_s,\psi_{r+1},\ldots,\psi_v,\mathbf{U}^{(q+2)}_{q+2}>.$$

Appendix I includes also the verifications of cases ii, iii, iv and v.

"Only-if" part. If $N \in \mathcal{N}_h$ this means that:

a) $\forall X_1,\ldots,X_h \in \mathcal{N}$: $N X_1 \ldots X_h \geq N' \in \mathcal{N}$

b) $\exists X_1,\ldots,X_{h+1} \in \mathcal{N}$: $N X_1 \ldots X_{h+1}$ possesses no normal form.

Condition a implies that there exist no two nodes both h-replaceable and joint by one arc, because otherwise with the technique of the "if" part of this theorem we could find h normal forms $X_1,\ldots,X_h$ such that $N X_1 \ldots X_h$ possesses no normal form.

Condition b implies that there esist two nodes both (h+1)-replaceable and joint by one arc, because otherwise by lemmas 3 and 4 for h+1 arbitrary normal forms $X_1,\ldots,X_{h+1}$, each subterm in $N X_1 \ldots X_{h+1}$ is reducible to an head normal form, and therefore $N X_1 \ldots X_{h+1}$ is reducible to normal form too.

□

Proof of theorem 2.

For $h \leq n$, $N X_1 \ldots X_h \geq N' \in \mathcal{N}$ because there are no two nodes both n-replaceable and joint by one arc.

If the head variable x_i of N is bound, then $N \in \mathcal{N}_n$, because, if m is the number of components of N, it is sufficient to choose $X_i = \mathbf{K}^m(\mathbf{WI})$, $X_{n+1} = \mathbf{WI}$ to obtain $N X_1 \ldots X_{n+1} \geq \mathbf{WI}(\mathbf{WI})$. Otherwise, if the head variable of N is free, for $h \geq n$ $N X_1 \ldots X_h$ is reducible to a normal form without initial abstractions, and therefore $N \in \mathcal{N}_\omega$.

□

From theorems 1 and 2 it follows immediately that:

<u>Corollary 1</u>. If N is a normal form with n initial abstractions, then $N\in\mathcal{N}_h(0\leq h\leq n)$ or $N\in\mathcal{N}_\omega$.

The following corollary enables the evaluation of the arity of a normal form N from:

- the head variable of N
- the arities of the components of N
- the head variables of the components of N
- the emptyness or the non-emptyness of the set of k-replaceable nodes joint by one arc in the components of N, for all k greater than the number of initial abstractions of N.

More precisely, we have:

<u>Corollary 2</u>. If $N=\lambda x_1\ldots\lambda x_n(x_i(N_1x_1\ldots x_n)\ldots(N_mx_1\ldots x_n)),N_j=\lambda x_1\ldots\lambda x_{n_j}(x_{i_j}N_1^{(j)}\ldots N_{m_j}^{(j)})\in\mathcal{N}_{h_j}$ and $\mathcal{R}_j^{(n)}$ is the set of k-replaceable node pairs joint by one arc in the tree representation of N_j for $k>n$ (1) $(1\leq j\leq m)$, then $N\in\mathcal{N}_h$ where:

a) if $i>n$ and $\forall j(1\leq j\leq m)\,[h_j\geq n]$ then $h=\omega$

b) if $i>n$ and $\exists j(1\leq j\leq m)\,[h_j<n]$ then $h=\min.[h_1,\ldots,h_m]$

c) if $i\leq n$ and $\forall j(1\leq j\leq m)\,[h_j=\omega]$ then $h=n$

d) if $i\leq n$, $\exists j(1\leq j\leq m)\,[h_j<n]$ and $\forall j(1\leq j\leq m)\,[(i_j>n_j)\wedge(\mathcal{R}_j^{(n)}=\Phi)]$ then $h=\min.[h_1,\ldots,h_m]$

e) if $i\leq n$, $\exists j(1\leq j\leq m)[i_j\leq n]$ and $\forall j(1\leq j\leq m)[\mathcal{R}_j^{(n)}=\Phi]$ then $h=\min.[h_1,\ldots,h_m,\max.[i-1,\min.[i_1-1,\ldots,i_m-1]]]$

f) if $i\leq n$, $\exists j(1\leq j\leq m)\,[(n\leq i_j\leq n_j)\vee(\mathcal{R}_j^{(n)}\neq\Phi)]$ then $h=\min.[i-1,h_1,\ldots,h_m]$.

<u>Proof</u>. We split the proof according to the mutually exclusive cases a,...,f.

<u>Case a</u>. For $t>n$, $N\,X_1\ldots X_t\geq x_i(N_1X_1\ldots X_n)\ldots(N_mX_1\ldots X_n)X_{n+1}\ldots X_t$ possesses normal form, since, by hypothesis, each its subterm possesses normal form.

<u>Case b</u>. $i>n$ means that the head variable of N is free, and therefore we must take into account only the arcs joining k-replaceable nodes inside the components of N.

<u>Case c</u>. By hypothesis every component of N applied to any number of normal forms still reduces to normal form, i.e. as a special case $\forall X_1,\ldots,X_n\in\mathcal{N}$: $X_i(N_1X_1\ldots X_n)\ldots(N_mX_1\ldots X_n)$ possesses normal form. But if we choose $X_i=K^m(WI),X_{n+1}=WI$ then $N\,X_1\ldots X_{n+1}\geq WI(WI)$, i.e. a formula

---

(1) This is possible because by construction $n_j\geq n$ $(1\leq j\leq m)$.

without normal form.

To prove cases d, e and f let us note that $i \leq n$ means that the head variable of N is i-replaceable. The classification of the variables inside N_j will be modified or not, when N_j $(1 \leq j \leq m)$ becomes a component of N, according to the following rules immediate consequences of definition 1:

1) the k-replaceable variables for $k \leq n$ remain k-replaceable
2) the k-replaceable variables for $k > n$ become i-replaceable.

Case d. $\forall j(1 \leq j \leq m) i_j > n_j$ means that the head variables of the components are free; $\forall j(1 \leq j \leq m) \mathfrak{R}_j^{(n)} = \Phi$ means that no pair of i-replaceable nodes is created, and therefore the head variable of N may be replaced. We must take into account only the arcs joining k-replaceable nodes inside the components of N.

Case e. Also in this case no pair of i-replaceable nodes is created. We must take into account not only the arcs joining k-replaceable nodes inside the components of N, but also the arcs joining the node $\langle n,i \rangle$ with the node $\langle n_j, i_j \rangle$ such that $i_j \leq n$. The expression $\max.[i-1, \min.[i_1-1, \ldots, i_m-1]]$ avoids exactly that x_i and x_{i_j} are both replaced.

Case f. In this case at last a pair of i-replaceable nodes joint by one arc is created, i.e. either a pair x_i, x_{i_j} for some value of j $(1 \leq j \leq m)$ or a pair inside a component of N. In both cases the head variable of N cannot be replaced.

□

At last, we will prove that η-reduction does not modify our classification of normal forms. The following lemma assures that if $N \in \mathcal{N}_h (\mathcal{N}_\omega)$ and it is an η-redex, then also its η-contractum $N' \in \mathcal{N}_h (\mathcal{N}_\omega)$.

Lemma 5. If $N \equiv \lambda x_1 \ldots \lambda x_n (x_i N_1 \ldots N_m) \in \mathcal{N}_h (0 \leq h \leq \omega)$ and it is η-reducible to N' relative to the variable x_n, then also $N' \in \mathcal{N}_h$

Proof. We distinguish the following four cases:

a) $N \in \mathcal{N}_\omega$
b) $N \in \mathcal{N}_n$
c) $N \in \mathcal{N}_{n-1}$
d) $N \in \mathcal{N}_h (h < n-1)$

which are the only possible cases by corollary 1.

case a. If $N\in\mathcal{N}_\omega$, then in its tree representation there are no two nodes both n-replaceable and joint by one arc and moreover $i\geq n$. Its η-contractum N' is:

$$N'\equiv\lambda x_1\ldots\lambda x_{n-1}(x_i N_1\ldots N_{m-1}).$$

No pair of (n-1)-replaceable nodes joint by are arc can be created in the tree representation of N' and $i-1\geq n-1$. Therefore $N'\in\mathcal{N}_\omega$.

case b. This case cannot occur. In fact, if $N\in\mathcal{N}_n$ theorem 2 implies that $i\leq n$. On the other hand, if N is an η-redex $N_m\equiv x_n$. Therefore the arc from the root corresponding to x_i to N_m joins two n-replaceable nodes, and then $N\notin\mathcal{N}_n$.

case c. If $N\in\mathcal{N}_{n-1}$, this means that there exist at least one arc joining two n-replaceable nodes. If N is an η-redex, the variable x_n occurs only in $N_m\equiv x_n$. It follows that the current pair of n-replaceable variables is x_i, x_n, i.e. $i\leq n-1$. $N'\equiv\lambda x_1\ldots\lambda x_{n-1}(x_i N_1\ldots N_{m-1})$ cannot belong to $\mathcal{N}_\omega$ because $i\leq n-1$, nor to $\mathcal{N}_h$ for $h<n-1$ because no pair of (n-1)-replaceable nodes joint by one arc can be created, and therefore $N'\in\mathcal{N}_{n-1}$.

case d. If $N\in\mathcal{N}_h$ for $h<n-1$ this means that there exist two (h+1)-replaceable nodes joint by one arc. $N_m\equiv x_n$ for the hypothesis of η-reducibility and therefore this pair of nodes survives in N'. It follows that $N'\in\mathcal{N}_h$.

□

We conclude with the following theorem about the invariance of our classification respect to the η-reduction:

Theorem 3. If a β-normal form $N\in\mathcal{N}_h$ $(0\leq h\leq\omega)$ and N is η-reducible to N', then also $N'\in\mathcal{N}_h$.

Proof. N is η-reducible to N' means that $N'\equiv N[M/M']$ where M is an η-redex and M' its η-contractum. By lemma 5, if $M\in\mathcal{N}_h$ then also $M'\in\mathcal{N}_h$ $(0\leq h\leq\omega)$. Moreover M and M' have the same head variable and the same sets $\mathcal{R}^{(n)}$ $(n\geq 0)$ as defined in corollary 2. Corollary 2 proves that the arity, the head variables and the sets $\mathcal{R}^{(n)}$ $(n\geq 0)$ are the unique informations on each component useful in determining the arity of a whole λ-term, independently from the specific expression of the components themselves. This means in particular that the replacement of M by M' does not modify the arity of N, i.e. that N' has the same arity as N.

□

3. Classification of λ-terms not in normal form.

We give now the formal definition of negative arities.

Definition 5. $F \in \mathcal{N}_{-h}$ $(h \geq 1)$ iff:

- $\forall X_1 \ldots X_{h-1} \in \mathcal{N}$: $F\, X_1 \ldots X_{h-1}$ possesses no normal form
- $\exists X_1 \ldots X_h \in \mathcal{N}$: $F\, X_1 \ldots X_h \geq F' \in \mathcal{N}$.

Definition 6. $F \in \mathcal{N}_{-\omega}$ iff:

- $\forall h(h \geq 0)$ $\forall X_1, \ldots, X_h$: $F\, X_1 \ldots X_h$ possesses no normal form.

It is clear that a λ-term has a negative arity iff it possesses no normal form.

A non-effective method to classify λ-terms not in normal form is based on the instant meaning and its tree representation as defined in section 1. We are able to classify the instant meanings of λ-terms according to the following definitions:

Definition 7. An instant meaning F^* with n initial abstractions belongs to $\mathcal{N}_{-h}$ $(1 \leq h \leq n)$ iff in the corresponding tree:

- there exist at least one node with label $\langle p, \Omega \rangle (p \geq n)$ descendant exclusively from non-(h-1)-replaceable nodes.
- all nodes with label $\langle p, \Omega \rangle (p \geq n)$ are descendant from at least one h-replaceable node.

Example 5. $F^* \equiv \lambda x_1 \lambda x_2 \lambda x_3 (x_2 (x_1 \overset{\lambda x_4}{\Omega}) \Omega (x_4 x_2 x_1)) \in \mathcal{N}_{-2}$, because the variable x_2 is 2-replaceable. The tree corresponding to F^* is shown in fig. 2.

Definition 8. An instant meaning $F^* \in \mathcal{N}_{-\omega}$ iff in the corresponding tree there exist at least one node with label $\langle p, \Omega \rangle (p \geq n)$ descendant exclusively from non-replaceable nodes.

Example 6. $F^* \equiv \lambda x_1 \lambda x_2 \lambda x_3 (x_4 (x_1 \lambda x_4 \Omega) \Omega (x_4 x_2 x_1)) \in \mathcal{N}_{-\omega}$,because the variable x_4 is free. The tree corresponding to F^* is shown in fig.5.

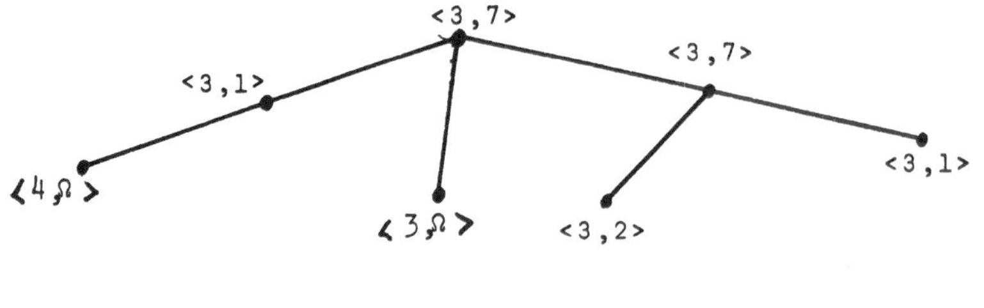

Fig. 5. Tree representing the instant meaning of example 6.

Obviously if the instant meaning of a λ-term contains no Ω, this λ-term is in normal form and the classification given in section 2 will apply. To classify λ-terms not in normal form, we must consider a complete reduction algorithm, i.e. an algorithm which gives in some order the (possibly infinite) set of all λ-terms obtained by reductions from a given λ-term. Given a λ-term F, we consider then the countable sequence of all λ-terms to which F is reducible (here possibly lies the non effective argument):

$$F \equiv F_0, F_1, \ldots$$

and the corresponding sequence of instant meanings:

$$F_0^*, F_1^*, \ldots \ .$$

If $F_i^* \in \mathcal{N}_{k(i)}$ we say that the arity of F is $k = \max_{i \geq 0}.[k(i)]$.

We will prove that the arities so obtained satisfy definitions 5 and 6. First we show that the arities of the instant meanings are non-decreasing relative to the relation of reducibility and that every sequence of arities of instant meanings corresponding to the same λ-term has only a finite number of different values.

Lemma 6. If F is reducible to $\bar{F}$,F^* and $\bar{F}^*$ are respectively the corresponding instant meanings, then the arity of F^* is less or equal to the arity of $\bar{F}^*$.

Proof. It is clearly sufficient to prove that, if $\bar{F}$ is obtained from F by a single β-reduction, the arity of F^* is less or equal to the arity of $\bar{F}^*$. It is well known that in this case $\bar{F}$ is obtained from F by replacing a β-redex by the corresponding β-contractum. A β-contractum may be, in its turn, a β-redex or an head normal form which may contain some β-redexes. In the first case, (i.e. $\bar{F}$ is obtained from F by replacing a β-redex by a β-redex) the instant meanings F^* and $\bar{F}^*$ coincide and therefore the arity remains unchanged. In the second case (i.e. $\bar{F}$ is obtained from F by replacing a β-redex by an head normal form) the tree representing $\bar{F}^*$ is obtained from the tree representing F^* by replacing a terminal node labeled $\langle p,\Omega\rangle$ ($p \geq 0$) by a tree whose root represents the head variable of the head normal form and which may have some terminal nodes labeled $\langle p,\Omega\rangle$ ($p \geq 0$). From definitions 7 and 8 it follows that the arity of F^* must be less or equal to the arity of $\bar{F}^*$.

□

Lemma 7. The set of the arities of the instant meanings corresponding to a λ-term F has only a finite number of different values.

Proof. If F is a β-redex and all λ-terms to which F is reducible are β-redexes, then for every $i \geq 0$ $k(i)=-\omega$.

Otherwise it can occur that in the infinite sequence of λ-terms to which F is reducible according to a complete reduction algorithm there exists an integer j such that for $0 \leq i \leq j-1$ F_i is a β-redex and F_j is an head normal form. This means that for $i>j$ each F_i will be a β-redex or an head normal form with the same number n of initial abstractions as F_j.

Then by definitions 7 and 8 the arities of the corresponding instant meanings may assume only the values $-\omega$, $-n$, $-n+1, \ldots, -1, 0, 1 \ldots, n$, ω.

□

Now we can prove the correctness of our (non effective) method of calculation of arities of λ-terms not in normal form with respect to definitions 5 and 6.

Theorem 4. If $k(0), k(1), \ldots$ is the countable sequence of arities of the instant meanings corresponding to a λ-term F, and $k = \max_{i \geq 0}[k(i)]$ then $F \in \mathcal{N}_k$ and viceversa.

Proof. The case $k \geq 0$ corresponding to the existence of the normal form inside the sequence of instant meanings of F was treated in section 2.

"If" part. Let be j an integer such that $k(j) = k = \max_{i \geq 0}[k(i)]$.

If $k=-h$ $(0<h<\omega)$ definition 7 applies to F_j^* and therefore for $h-1$ arbitrary normal forms $X_1, \ldots, X_{h-1}$, $F X_1 \ldots X_{h-1}$ possesses no normal form, because it survives always at least one β-redex which corresponds to a node $\langle p, \Omega \rangle$ $(p \geq 0)$ satisfying the first clause of definition 7. But here it is possible to construct h normal forms $X_1, \ldots, X_h$ such that $F X_1 \ldots X_h$ possesses normal form. In fact the second clause of definition 7 assures that it is sufficient to replace all interested h-replaceable variables by suitable normal forms, in such a way that in the so obtained λ-term all β-redexes disappear, i.e. the so obtained λ-term be reducible to normal form. The possibility of such a replacement is assured by lemma 1 and conflicts cannot arise, since:

a) if the same h-replaceable variable must delete in one occurrence the sons $p_1,\dots,p_s$ between u and in another occurrence the sons $q_1,\dots,q_r$ between v it is sufficient to replace this variable by the normal form which deletes the sons $p_1,\dots,p_s$, $q_1,\dots,q_r$ between max.[u, v] sons.

b) it is never necessary to replace two variables x_p and x_q such that x_q is h-replaceable thanks to an occurrence of x_p. In this case in fact it is sufficient to replace x_p by a normal form which deletes the subtree bounding the variable x_q to obtain that all sons of x_q are deleted.

If at the contrary $k=-\omega$ applying definition 8 one obtains that for every h and for h arbitrary normal forms $X_1,\dots,X_h$, $F\ X_1\dots X_h$ possesses no normal form, since it survives always a β-redex without normal form, i.e. a β-redex which corresponds to a node $\langle p,\Omega\rangle (p\geq 0)$ satisfying definition 8.

"Only-if" part. If $F\in\mathcal{N}_k$ for $k=-h$ $(0<h<\omega)$ from the first clause of definition 5 it follows that for h-1 arbitrary normal forms $X_1,\dots,X_{h-1}$, at least one β-redex without normal form always survives in $F\ X_1\dots X_{h-1}$, i.e. this β-redex cannot be deleted by replacing (h-1)-replaceable variables by suitable normal forms. This means that in the tree representation all instant meanings corresponding to F satisfy the first clause of definition 7.

The second clause of definition 5 implies that there exist h normal form $X_1,\dots,X_h$ such that in $F\ X_1\dots X_h$ no β-redex without normal form survives, i.e. all β-redexes without normal form can be deleted by replacing h-replaceable variables by suitable normal forms. This means that there exists an integer j such that F^*_j satisfies the second clause of definition 7.

For $k=-\omega$ by definition 6 it follows that for every $h\geq 0$, for h arbitrary normal forms $X_1,\dots,X_h$, $F\ X_1\dots X_h$ possesses no normal form, i.e. in F there is at least one β-redex which can never be deleted. This means that in the tree representation all instant meanings corresponding to F satisfy definition 8.

□

The cause to consider a complete reduction algorithm, i.e. an algorithm which gives the whole sequence of λ-terms to which a given λ-term is reducible, is to obtain a classification invariant by β-convertibility. This choice will be further discussed in the conclusion. The invariance by β-convertibility is intuitively clear if we think that the sequences of λ-terms obtained from two β-convertible λ-terms may differ only for λ-terms whose arities are less or equal to the maximum one. This invariance is implicitly proved in theorem 4, since definitions 5 and 6 are obviously invariant by β-convertibility. Appendix II gives a direct proof of this fact.

Conclusion.

Given a λ-term F, we could construct the sequence of all λ-terms to which F is reducible according to a determined reduction algorithm: $F=F_0, F_1, \ldots$, the corresponding sequence of instant meanings $F_0^*, F_1^*, \ldots$, and then define $k = \lim_{i\to\infty} k(i)$.

The choice of a reduction algorithm which reaches always the normal form (if it exists) assures that negative arities are not assigned to λ-terms possessing a normal form. This is not true for example for the leftmost--innermost reduction algorithm [6]; in fact using this algorithm the λ-term $\mathbf{YO} \in \mathcal{N}_{-\omega}$, what is absurdum because $\mathbf{YO} \geq \mathbf{I} \in \mathcal{N}_1$. This difficulty is avoided for example by Church's reduction algorithm. But unfortunately using this algorithm we may assign different arities to β-convertible λ-terms not possessing normal forms. For example, we found that $\lambda x(a(x(\mathbf{WI}(\mathbf{WI}))(\mathbf{I}a)) \in \mathcal{N}_{-\omega}$ and $\lambda x(a(x(\mathbf{WI}(\mathbf{WI}))a) \in \mathcal{N}_{-1}$, because in the first λ-term the infinite reducibility of $\mathbf{WI}(\mathbf{WI})$ prevents the reduction $\mathbf{I}a \geq a$. It is possible that the invariance with respect to β-convertibility could be achieved by means of a reduction algorithm which is in some sense parallel, as for example Knuth's reduction algorithm [5].

In any case the present partition of the set of λ-terms in $2\omega+1$ classes may be considered as a simple kind of type introduction, in which every λ-term possesses an unique type. This partition is visualized in fig. 6, where as examples of λ-terms of arity h are chosen λ-terms, which in the usual type theory (see [4] p.265) do not possess type. Moreover this classification can be effectively determined for λ-terms in normal form.

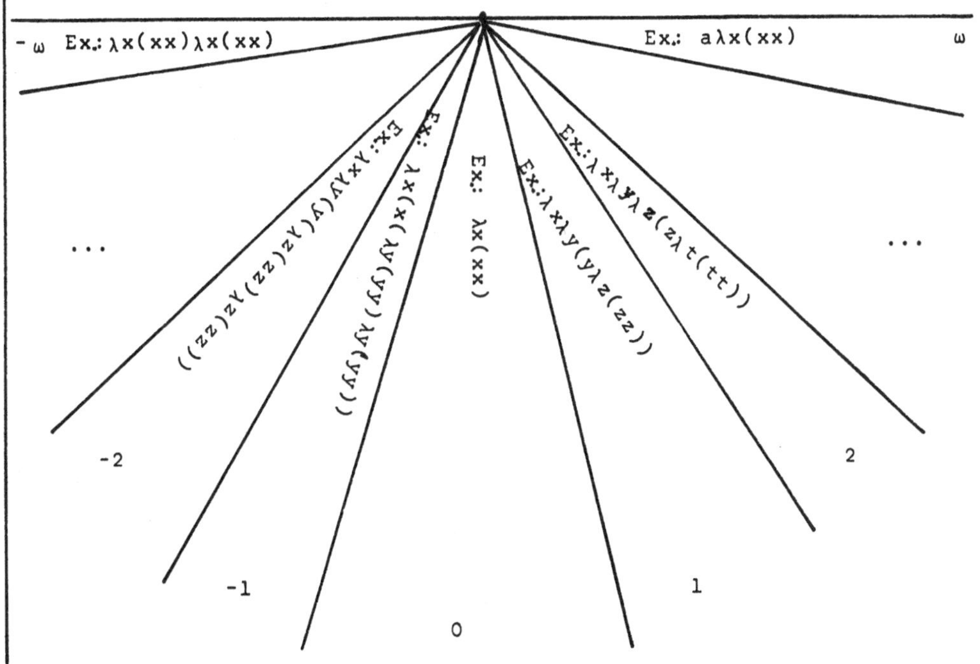

Fig. 6 - Classification of the set of λ-terms with some examples of λ-terms of arity h.

Appendix I.

Next follow the β-reductions to verify the "if" part of the proof of theorem 1, which is split into five cases. Let be $M=\lambda x_1\ldots\lambda x_m(x_i(A_1x_1\ldots\ldots x_m)\ldots(A_jx_1\ldots x_m)\ldots(A_fx_1\ldots x_m))$ and $A_j=\lambda x_1\ldots\lambda x_s(x_\ell(D_1x_1\ldots x_s)\ldots\ldots(D_rx_1\ldots x_s))$. We denote by indexed χ the normal forms replacing the first i-1 or ℓ-1 bound variables of M.

case i.

$$M\chi_1\ldots\chi_{i-1}X\geq\lambda x_{i+1}\ldots\lambda x_m\lambda z_{f+1}\ldots\lambda z_u<\zeta_{u_1},\ldots,\zeta_{u_p},(\nabla\zeta_j(W(\nabla\zeta_j))(W(\nabla\zeta_j))>$$

where $\zeta_t=\begin{cases}A_t\chi_1\ldots\chi_{i-1}Xx_{i+1}\ldots x_m & \text{if } t\leq f\\ z_t & \text{otherwise}\end{cases}$ for $1\leq t\leq u$.

To be short, we evaluate separately:

$\nabla\zeta_j \geq \zeta_j\psi_{m+1}\cdots\psi_{\ell-1}(\mathbf{K}^r\mathbf{I})\psi_{\ell+1}\cdots\psi_s \geq \mathbf{K}^r\mathbf{I}\partial_1\cdots\partial_r \geq \mathbf{I}$

where $\partial_t = D_t\chi_1\cdots\chi_{i-1}Xx_{i+1}\cdots x_m\psi_{m+1}\cdots\psi_{\ell-1}(\mathbf{K}^r\mathbf{I})\psi_{\ell+1}\cdots\psi_s$ $(1\leq t\leq r)$.

Therefore the last component of the (p+1)-tuple is convertible to $\mathbf{WI}(\mathbf{WI})$.

<u>case ii</u>.

$M\chi_1\cdots\chi_{i-1}X \geq \lambda x_{i+1}\cdots\lambda x_m\,\lambda z_{f+1}\cdots\lambda z_w <\zeta_{u_1},\ldots,\zeta_{u_p},\Gamma\zeta_1\cdots\zeta_w>$

where $\zeta_t\ (1\leq t\leq w)$ are defined as in case i.

We need to consider only the last component of the preceding (p+1)-tuple:

$\Gamma\zeta_1\cdots\zeta_w \geq <\zeta_{r+1}\zeta_{r+1},\zeta_j\psi_{m+1}\cdots\psi_s(\zeta_j\psi_{m+1}\cdots\psi_s)>$.

Again we have to consider only the second component of the previous pair:

$\underline{\zeta_j\psi_{m+1}\cdots\psi_s(\zeta_j\psi_{m+1}\cdots\psi_s)} \geq X\gamma_1\cdots\gamma_r(\zeta_j\psi_{m+1}\cdots\psi_s)$

where $\gamma_t = D_t\chi_1\cdots\chi_{i-1}Xx_{i+1}\cdots x_m\psi_{m+1}\cdots\psi_s$ $(1\leq t\leq r)$

$X\gamma_1\cdots\gamma_r(\zeta_j\psi_{m+1}\cdots\psi_s) \geq \lambda z_{r+2}\cdots\lambda z_w<\eta_{u_1},\ldots,\eta_{u_p},\Gamma\gamma_1\cdots\gamma_r(\zeta_j\psi_{m+1}\cdots\psi_s)z_{r+2}\cdots$
$\cdots z_w>$

$$\text{where } \eta_t = \begin{cases} \gamma_t & \text{if } t\leq r \\ \zeta_j\psi_{m+1}\cdots\psi_s & \text{if } t=r+1 \\ z_t & \text{otherwise} \end{cases} \qquad \text{for } 1\leq t\leq w.$$

Finally once more we consider the last component of the preceding (p+1)--tuple:

$\Gamma\gamma_1\cdots\gamma_r(\zeta_j\psi_{m+1}\cdots\psi_s)z_{r+2}\cdots z_u \geq <\underline{\zeta_j\psi_{m+1}\cdots\psi_s(\zeta_j\psi_{m+1}\cdots\psi_s)},\eta_j\psi_{m+1}\cdots$
$\cdots\psi_s(\eta_j\psi_{m+1}\cdots\psi_s)>$

The initial λ-term is then without normal form, since its underlined subterm produces itself.

<u>case iii</u>.

We assume $i<\ell$, but the proof in the case $i>\ell$ can be obtained simply by changing i and ℓ and the relative positions of X and Y.

$M\chi_1\cdots\chi_{i-1}X\chi_{i+1}\cdots\chi_{\ell-1}Y \geq \lambda x_{\ell+1}\cdots\lambda x_m\lambda z_{f+1}\cdots\lambda z_u<\varepsilon_{u_1},\ldots,\varepsilon_{u_p},\Omega\varepsilon_j(\mathbf{W}(\Omega\varepsilon_j))$
$(\mathbf{W}(\Omega\varepsilon_j))>$

$$\text{where } \varepsilon_t = \begin{cases} A_t\chi_1\cdots\chi_{i-1}X\chi_{i+1}\cdots\chi_{\ell-1}Yx_{\ell+1}\cdots x_m & \text{if } t\leq f \\ z_t & \text{otherwise} \end{cases} \qquad \text{for } 1\leq t\leq u$$

We evaluate analogously to case i:

$\Omega\varepsilon_j \geq Y\beta_1\cdots\beta_r\psi_{r+1}\cdots\psi_v\mathbf{U}_{q+1}^{(q+1)}$

where $\beta_t = D_t\chi_1\cdots\chi_{i-1}X\chi_{i+1}\cdots\chi_{\ell-1}Yx_{\ell+1}\cdots x_m\phi_{m+1}\cdots\phi_s$ $(1\leq t\leq r)$

$\Omega\varepsilon_j \geq \langle \alpha_{v_1}, \ldots, \alpha_{v_q}, \mathbf{I} \rangle \, \mathbf{U}^{(q+1)}_{q+1} \geq \mathbf{I}$

where $\alpha_t = \begin{cases} \beta_t & \text{if } t \leq r \\ \psi_t & \text{otherwise} \end{cases}$ for $1 \leq t \leq v$.

Therefore the last component of the preceding (p+1)-tuple is reducible to **WI(WI)**.

<u>case iv.</u>

In this case we have to consider the subterm M inside the (p+1)-th component of X, where by construction the variable x_ℓ is replaced by Y:

$MX_1 \ldots X_{i-1} XX_{i+1} \ldots X_{\ell-1} Y \geq \lambda x_{\ell+1} \ldots \lambda x_m \lambda z_{f+1} \ldots \lambda z_u \langle \varepsilon_{u_1}, \ldots, \varepsilon_{u_p}, \varepsilon_t \psi_{w+1} \ldots$
$\ldots \psi_{\ell-1}(ZY), \Omega\varepsilon_j (\mathbf{W}(\Omega\varepsilon_j))(\mathbf{W}(\Omega\varepsilon_j)) \rangle$

where ε_e $(1 \leq e \leq u)$ are defined as in case iii.

The last component of this (p+2)-tuple is the same as that one of case iii, except a different definition of X, and therefore it is reducible to **WI(WI)**.

<u>case v.</u>

In this case we consider the subterm M inside the (q+1)-th component of Y, where by construction the variable x_i is replaced by X:

$MX_1 \ldots X_{\ell-1} YX_{\ell+1} \ldots X_{i-1} X \geq \lambda x_{i+1} \ldots \lambda x_m \lambda z_{f+1} \ldots \lambda z_u \langle \pi_{u_1}, \ldots, \pi_{u_p}, \bar{\Omega}\pi_j (\mathbf{W}$
$(\bar{\Omega}\pi_j))(\mathbf{W}(\bar{\Omega}\pi_j)) \rangle$

where $\pi_e = \begin{cases} A_e X_1 \ldots X_{\ell-1} YX_{\ell+1} \ldots X_{i-1} Xx_{i+1} \ldots x_m & \text{if } e \leq f \\ z_e & \text{otherwise} \end{cases}$ for $1 \leq e \leq u$.

Then it is sufficient to show that $\bar{\Omega}\pi_j \geq \mathbf{I}$.

$\bar{\Omega}\pi_j \geq Y \sigma_1 \ldots \sigma_r \psi_{r+1} \ldots \psi_v \mathbf{U}^{(q+2)}_{q+2}$

where $\sigma_e = D_e X_1 \ldots X_{\ell-1} YX_{\ell+1} \ldots X_{i-1} Xx_{i+1} \ldots x_m \phi_{m+1} \ldots \phi_s$ $(1 \leq e \leq r)$

$\bar{\Omega}\pi_j \geq \langle \mu_{v_1}, \ldots, \mu_{v_q}, \mu_t \psi_{w+1} \ldots \psi_{i-1}(\bar{Z}X), \mathbf{I} \rangle \mathbf{U}^{(q+2)}_{q+2} \geq \mathbf{I}$

where $\mu_e = \begin{cases} \sigma_e & \text{if } e \leq r \\ \psi_e & \text{otherwise} \end{cases}$ for $1 \leq e \leq v$.

<u>Appendix II.</u>

<u>Theorem 5</u>. Two β-convertible λ-terms have the same arity.

Proof. Let be F and G two β-convertible λ-terms, whose arities are respectively h and k. Let be $\mathfrak{F}$ and $\mathcal{G}$ respectively the sequences of λ-terms to which F and G are β-reducible. We can always find two integers j and ℓ such that:

$F_j \in \mathfrak{F}$ and the arity of its instant meaning F_j^* is h

$G_\ell \in \mathcal{G}$ and the arity of its instant meaning G_ℓ^* is k

The λ-terms F_j and G_ℓ are β-convertible, then by Church-Rosser's theorem there exists a λ-term M to which F_j and G_ℓ are both β-reducible. The arity of M^* (instant meaning of M) by lemma 6 must be greater or equal to the arity of F_j^* ; but since $M \in \mathfrak{F}$ and h is by definition the maximum arity of all instant meanings corresponding to λ-terms belonging to $\mathfrak{F}$, then the arity of M^* is just h. By the symmetric argument the arity of M^* is just k, and then $h=k$.

□

An alternative proof could be done by using a result of Welch [8] ; he shows that the "maximum" instant meaning (in the order derived from making Ω "less than" anything else) of two β-convertible λ-terms is the same. Trivially the maximum instant meaning possesses the maximum arity.

References.

[1] Böhm,C., The CUCH as a Formal and Description Language, Formal Language Description Languages for Computer Programming, ed.T.B. Steel, North-Holland, Amsterdam, (1966), 179-197.

[2] Böhm,C., M.Dezani-Ciancaglini, Combinatorial Problems Combinator Equations and Normal Forms, Automata Languages and Programming, ed. J. Loeckx, Lecture Notes in Computer Science, 14, Springer -Verlag, (1974), 185-199.

[3] Church,A., J.Barkley Rosser, Some properties of λ-conversion, Trans.Amer. Math Soc., 39, (1936), 472-482.

[4] Curry, H.B., J.R.Hindley, J.P.Seldin, Combinatory Logic, 2, North--Holland, Amsterdam, (1972).

[5] Knuth, D.E., Examples of Formal Semantics, Symposium on Semantics of Algorithmic Languages, 188, ed. E. Engeler, Springer-Verlag,

Berlin, (1970), 212-235.

[6] Simone, C., Un modello di λ-calcolo fortemente equivalente agli schemi ricorsivi, Calcolo, 1, (1974), 111-125.

[7] Wadsworth, C.P., Semantics and Pragmatics of λ-calculus, Ph.D. Thesis, Oxford, (1971).

[8] Welch, P.H., Continuous Semantics and Inside-Outside Reductions, these Proceedings.

CONTINUOUS SEMANTICS AND INSIDE-OUT REDUCTIONS.

P.H.Welch,
Computer Laboratory,
Cornwallis Building,
The University, CANTERBURY,
KENT - CT2 7NF, ENGLAND.

0: Introduction:-

A "natural" semantics of the λ-K-β-calculus can be achieved in the following way. First, define an "instant" semantics which gives $\tilde{\varepsilon}$ from a λ-expression ε. All that happens here is that we take the meaning of the expression to be its own symbols, except for those sub-parts that are β-redexes. These parts are still "unevaluated" and, since we do not wish to do any work in our instant semantics, we just replace them with some symbol, Ω say, which means "unknown" or "undefined" (this is Wadsworth's direct approximant - [9]). Then, the natural meaning of an expression is just the "union" of the instant meanings of all the expressions to which it reduces - i.e. :-

$$N[\![\varepsilon]\!] := \cup\{\tilde{\varepsilon}' \mid \varepsilon \xrightarrow{\beta} \varepsilon'\}.$$

This "union" is a limiting process with respect to a natural ordering on the direct approximants derived from making Ω "less than" anything else - for, clearly, it should make no difference to the meaning if we were to leave out some of the "smaller" elements. This ordering induces an ordering on λ-expressions themselves, which we shall write as $\tilde{\sqsubseteq}$ and can read as "directly approximates" :-

$$\varepsilon \;\tilde{\sqsubseteq}\; \delta \text{ iff } \tilde{\varepsilon} \text{ "less than" } \tilde{\delta}.$$

We will characterise $\tilde{\sqsubseteq}$ formally in the next section (1.2), but a few properties can be seen already. Certainly, if $\varepsilon \xrightarrow{\beta} \delta$, then $\varepsilon \;\tilde{\sqsubseteq}\; \delta$ and so, by a simple appeal to the Church-Rosser theorem, we see that the set $\{\varepsilon' \mid \varepsilon \xrightarrow{\beta} \varepsilon'\}$ is directed under $\tilde{\sqsubseteq}$ and that the natural semantics models β-reduction - i.e. :-

$$\varepsilon \xrightarrow{\beta} \delta \Rightarrow N[\![\varepsilon]\!] = N[\![\delta]\!].$$

The trouble with N as a semantics is that it is non-trivial to prove that it has the following property : if we have two expressions, ε and δ, such that $N[\![\varepsilon]\!] = N[\![\delta]\!]$, and then embed them in turn as sub-expressions of a larger expression, the semantics of the overall expression should be independent of whether ε or δ was used (e.g. $\mathcal{N}[\![\varepsilon]\!] = \mathcal{N}[\![\delta]\!] \Rightarrow \mathcal{N}[\![\varepsilon(\gamma)]\!] = \mathcal{N}[\![\delta(\gamma)]\!]$). This is a very desirable property for $\mathcal{N}$ to have and the semantics would not be very good without it.

This approach is similar to that studied independently by Lévy - [2] and [3].

1:Basic Concepts:-

1.0:DEF:-

We define our terms with the following context-free grammars. We do not worry about redundant brackets as we are only concerned with the structure and not the "syntactic sugar".

I	::=	a|b|c|..................	(Variables - countable)
EXP	::=	I|λI.EXP|(EXP)(EXP)	(λ-K-expressions)
NF	::=	λI.NF|HD	(β-normal forms)
HD	::=	I|(HD)(NF)	
HNF	::=	λI.HNF|HEAD	(Head Normal Forms)
HEAD	::=	I|(HEAD)(EXP)	- Wadsworth [9]
NOH	::=	(λI.EXP)(EXP)|λI.NOH|(NOH)(EXP)	(Not in HNF)
SOL	:=	$\{\varepsilon \in \text{EXP} \mid \varepsilon \xrightarrow{\beta} \varepsilon' \in \text{HNF}\}$	(Solvable exps.)
INSOL	:=	EXP\SOL	(Unsolvable exps.) - Wadsworth [9] & Barendregt [0]

An ordered pair, <F,$\mathcal{F}$>, is a SEMANTICS of the λ-calculus if F is a set and $\mathcal{F} : \text{EXP} \to \text{F}$.

A semantics, <F,$\mathcal{F}$>, is a β-MODEL if :-

$$(\varepsilon \xrightarrow{\beta} \delta) \Rightarrow (\mathcal{F}[\![\varepsilon]\!] = \mathcal{F}[\![\delta]\!]).$$

A semantics, <F,$\mathcal{F}$>, is SUBSTITUTIVE if :-

$$(\mathcal{F}[\![\varepsilon]\!] = \mathcal{F}[\![\varepsilon']\!])_\wedge(\mathcal{F}[\![\delta]\!] = \mathcal{F}[\![\delta']\!]) \Rightarrow (\mathcal{F}[\![\lambda x.\varepsilon]\!] = \mathcal{F}[\![\lambda x.\varepsilon']\!])_\wedge(\mathcal{F}[\![\varepsilon(\delta)]\!] = \mathcal{F}[\![\varepsilon'(\delta')]\!]).$$

A semantics, <F,$\mathcal{F}$>, is NORMAL if :-

$$(\varepsilon \xrightarrow{\beta} \nu \in \text{NF})_\wedge(\delta \text{ has no normal form}) \Rightarrow (\mathcal{F}[\![\varepsilon]\!] \neq \mathcal{F}[\![\delta]\!]).$$

A semantics, $\langle F, \mathcal{F}\rangle$, is <u>SOLVABLE</u> if :-

$(\varepsilon \in SOL)_\wedge(\delta \in INSOL) \Rightarrow (\mathcal{F}[\![\varepsilon]\!] \neq \mathcal{F}[\![\delta]\!])$.

We order semantics by inclusion of the induced equivalences :-

$(\langle F,\mathcal{F}\rangle \leq \langle G,\mathcal{G}\rangle)$ iff $(\mathcal{F}[\![\varepsilon]\!] = \mathcal{F}[\![\delta]\!]) \Rightarrow (\mathcal{G}[\![\varepsilon]\!] = \mathcal{G}[\![\delta]\!])$.

<u>1.1:LEMMA:-</u>

(i) $NF \subset HNF$. (ii) $EXP = HNF \cup NOH$. (iii) $\emptyset = HNF \cap NOH$.

(iv) $(\varepsilon \in HNF)_\wedge(\varepsilon \xrightarrow{\beta} \varepsilon') \Rightarrow (\varepsilon' \in HNF)$.

(v) $(\varepsilon \in INSOL)_\wedge(\varepsilon \xrightarrow{\beta} \varepsilon') \Rightarrow (\varepsilon' \in INSOL)$.

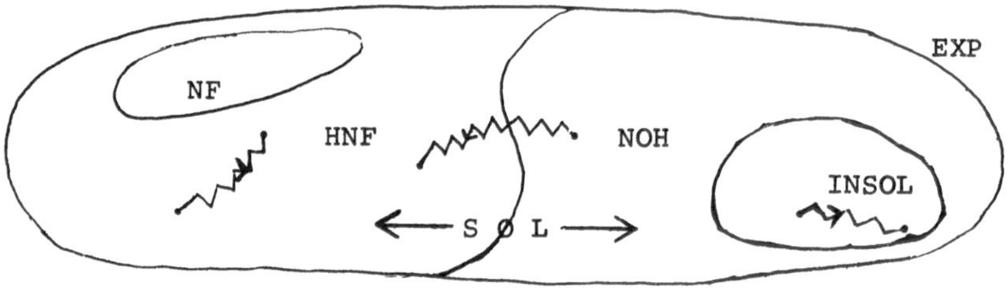

(vi) $\langle\{*\},const\rangle$, where $const[\![\varepsilon]\!] := *$, is the maximal semantics. It is a substitutive β-model.

(vii) $\langle EXP,id\rangle$, where $id[\![\varepsilon]\!] := \varepsilon$, is the minimal semantics. It is substitutive, normal and solvable.

<u>Proof:-</u>

(i),...,(iv) -By structural inductions over the grammars.

(v) -Clear, by definition.

(vi) and (vii) -Obvious.

ǂ

<u>1.2:DEF:-</u>

We formalise the notion of directly approximates :-

$\varepsilon \mathrel{\tilde{\sqsubseteq}} \delta$ **iff** <u>either</u> $\varepsilon \in NOH$

<u>or</u> $\varepsilon = x = \delta$

<u>or</u> $\varepsilon \xrightarrow{\alpha} \lambda x.\varepsilon' \in HNF$, $\delta \xrightarrow{\alpha} \lambda x.\delta'$, $\varepsilon' \mathrel{\tilde{\sqsubseteq}} \delta'$

<u>or</u> $\varepsilon = \omega(\gamma) \in HEAD$, $\delta = \omega'(\gamma')$, $\omega \mathrel{\tilde{\sqsubseteq}} \omega'$, $\gamma \mathrel{\tilde{\sqsubseteq}} \gamma'$.

<u>1.3:LEMMA:-</u>

(i) $(\varepsilon \xrightarrow{\beta} \varepsilon') \Rightarrow (\varepsilon \mathrel{\tilde{\sqsubseteq}} \varepsilon')$.

(ii) $(\varepsilon \in NF)_\wedge(\varepsilon \mathrel{\tilde{\sqsubseteq}} \delta) \Rightarrow (\varepsilon \xrightarrow{\alpha} \delta)$.

<u>Proof:-</u>

-By structural inductions on $\varepsilon \in EXP$.

ǂ

2:Continuous Semantics:-

2.0:REMARK:-

We assume the reader is familiar with the concepts and notation of the Scott lattice-theoretic approach to semantics - [5],[7],[9].

Our motivation for considering this problem comes from the following theorem Wadsworth proved about some of Scott's models of the λ-calculus - [9] and [10]. Let D_∞ be the Scott-lattice such that either $D_\infty \cong [D_\infty \to D_\infty]$ or $D_\infty \cong A + [D_\infty \to D_\infty]$, where A is some "atomic" lattice. Let ENV be the set of environments - the set of maps : (Variables $\to D_\infty$). Then, we have two substitutive β-models, $<[ENV \to D_\infty],\mathcal{D}>$. As in section 0, define an "instant" semantics, $<[ENV \to D_\infty],\tilde{\mathcal{D}}>$, in precisely the same way as $\mathcal{D}$ except that, in the combination case :-

$$\tilde{\mathcal{D}}[\![\varepsilon(\delta)]\!](\rho) := \begin{cases} \bot, \text{ if } \varepsilon \in \lambda I.EXP. \\ (\tilde{\mathcal{D}}[\![\varepsilon]\!](\rho))(\tilde{\mathcal{D}}[\![\delta]\!](\rho)), \text{ if not.} \end{cases}$$

Again, we do no work with $\tilde{\mathcal{D}}$ when looking at β-redexes. Now, Wadsworth proved in [10] :-

$$\mathcal{D}[\![\varepsilon]\!] = \bigsqcup\{\tilde{\mathcal{D}}[\![\varepsilon']\!] \mid \varepsilon \xrightarrow{\beta} \varepsilon'\}.$$

We consider this theorem to be so essential to decent semantics (for instance, it is not true about Park's pathological version of Scott's model in which the **Y** combinator goes all wrong - see 7.2(iv)) that we make it into a definition. This property reflects a kind of "continuity" of the semantic function if you believe in the natural interpretation of λ-expressions :-

$$\mathcal{D}[\![\sqcup\{\tilde{\varepsilon}' \mid \varepsilon \xrightarrow{\beta} \varepsilon'\}]\!] = \bigsqcup\{\tilde{\mathcal{D}}[\![\varepsilon']\!] \mid \varepsilon \xrightarrow{\beta} \varepsilon'\}.$$

2.1:DEF:-

Let $<\mathbf{F},F>$ be a semantics where **F** is a directedly complete semi-lattice. Then, $\tilde{F}$ is a <u>**WELL-BEHAVED APPROXIMATE**</u> to F if :-

(i) $\tilde{F}$ is monotone w.r.t. $\overset{\beta}{\sqsubseteq}$

and (ii) $F[\![\varepsilon]\!] = \bigsqcup\{\tilde{F}[\![\varepsilon']\!] \mid \varepsilon \xrightarrow{\beta} \varepsilon'\}$.

Then, $<\mathbf{F},F>$ is a <u>CONTINUOUS</u> semantics if it has a well-behaved approximate.

2.2:LEMMA:-

Let $<\mathbf{F},F,\tilde{F}>$ be a continuous semantics. Then,

(i) $\tilde{F} \subseteq F$,

(ii) $\{\tilde{F}[\![\varepsilon']\!] \mid \varepsilon \xrightarrow{\beta} \varepsilon'\}$ is directed,

(iii) $<\mathbf{F},F>$ is a β-model,

(iv) If $\langle G, \mathcal{G}\rangle$ is continuously derivable from $\langle F, \mathcal{F}\rangle$ - i.e. there is a continuous map $f \in [F \to G]$ such that $\mathcal{G} = f \circ \mathcal{F}$ - then $\langle G, \mathcal{G}\rangle$ is continuous with $\tilde{\mathcal{G}} := f \circ \tilde{\mathcal{F}}$,

(v) $(\forall \varepsilon, \delta \in \mathrm{INSOL})(\mathcal{F}[\![\varepsilon]\!] = \mathcal{F}[\![\delta]\!])$
and $(\forall \varepsilon \in \mathrm{INSOL})(\forall \gamma \in \mathrm{EXP})(\mathcal{F}[\![\varepsilon]\!] \sqsubseteq \mathcal{F}[\![\gamma]\!])$,

(vi) $\langle [\mathrm{ENV} \to D_\infty], \mathcal{D}\rangle$, with either D_∞ as described in 2.0, is continuous,

(vii) The "natural" semantics is continuous,

(viii) Unfortunately, continuity does not imply substitutivity.

Proof:-

(i) -Trivial.

(ii) and (iii) -By the Church-Rosser theorem.

(iv) -By part (ii) and the properties of continuous maps.

(v) $-(\forall \varepsilon \in \mathrm{NOH})(\varepsilon \mathrel{\tilde{\sqsubseteq}} \Delta\Delta$ and $\Delta\Delta \mathrel{\tilde{\sqsubseteq}} \varepsilon)$, where $\Delta = \lambda x.xx$.

-So, $(\forall \varepsilon \in \mathrm{NOH})(\tilde{\mathcal{F}}[\![\varepsilon]\!] = \tilde{\mathcal{F}}[\![\Delta\Delta]\!])$.

-Hence, $(\forall \varepsilon \in \mathrm{INSOL})(\mathcal{F}[\![\varepsilon]\!] = \bigsqcup\{\tilde{\mathcal{F}}[\![\Delta\Delta]\!]\} = \tilde{\mathcal{F}}[\![\Delta\Delta]\!])$.

-Finally, $(\forall \gamma \in \mathrm{EXP})(\Delta\Delta \mathrel{\tilde{\sqsubseteq}} \gamma)$.

-Thus, $(\forall \varepsilon \in \mathrm{INSOL})(\forall \gamma \in \mathrm{EXP})(\mathcal{F}[\![\varepsilon]\!] = \tilde{\mathcal{F}}[\![\Delta\Delta]\!] \sqsubseteq \tilde{\mathcal{F}}[\![\gamma]\!] \sqsubseteq \mathcal{F}[\![\gamma]\!])$.

(vi) -This is what Wadsworth proved in [10].

(vii) -More or less by definition : we have not yet rigorously specified $\mathcal{N}$ - see 3.7.

(viii) -Let $\sim$ be the following equivalence relation on EXP :-

$\varepsilon \sim \delta$ **iff** <u>either</u> ε and δ have the same normal form
<u>or</u> neither ε or δ have normal form.

-Let $\langle \mathrm{EXP}/\sim, [\ ]\rangle$ be the semantics obtained by taking equivalence classes - i.e. $[\ ][\![\varepsilon]\!] := [\varepsilon]$.

-Clearly, we have a β-model. However, it is not substitutive, for while $Y \sim \Delta\Delta$ (where $Y = \lambda f.(\lambda y.f(yy))(\lambda y.f(yy))$, the "fixed-point" combinator), $Y(\lambda a.b) \not\sim (\Delta\Delta)(\lambda a.b)$.

-Now, $\mathrm{EXP}/\sim$ becomes a directedly complete "simple atomic" semi-lattice by making $[\Delta\Delta] \sqsubseteq [\varepsilon]$, for all $\varepsilon \in \mathrm{EXP}$, and leaving the rest incomparable.

-Define : $\tilde{\mathcal{F}}[\![\varepsilon]\!] := \begin{cases} [\varepsilon], & \text{if } \varepsilon \in \mathrm{NF}, \\ [\Delta\Delta], & \text{if not.} \end{cases}$

-Clearly, $\tilde{\mathcal{F}}$ is a well-behaved approximate to $[\ ]$, and so the semantics is continuous.

⫳

2.3:REMARK:-

We see that continuity imposes some good properties on semantics. In particular, it insists upon β-modelship and satisfies our intuition that says that unsolvable expressions should be lumped together as rubbish. We note that the maximal semantics is clearly continuous and that the "natural" semantics is probably going to be the minimal continuous one. We think that the minimal continuous semantics should be interesting and we hope very much that it is substitutive.

3:The Semantics $\langle E_\infty, \mathcal{E}\rangle$:-

3.0:REMARK:-

Instead of trying to prove the substitutivity of $\mathcal{N}$ directly, we will define, in a very constructive way, a semantic function, $\mathcal{E}$, to objects that resemble $\mathcal{N}[\![\mathrm{EXP}]\!]$. This semantics will be substitutive trivially (as Scott's models are) and the trick will be to show it is the same as $\mathcal{N}$. In the hope of gaining insight, we impose a strict discipline on the approximate objects, $\widetilde{\mathrm{EXP}}$, by building them up as a direct limit of a sequence of lattices whose elements are syntactic objects - in fact, normal forms + $\perp$'s : the larger the lattice in the sequence, the longer the "approximate normal form", $\tilde{\varepsilon}$, that can be represented. This technique is, of course, inspired by that used by Scott in "The Lattice of Flow Diagrams" - [8]. We are guided by the analogies :-

"never-terminating" ↔ "unsolvable"

"loop-free" ↔ "normal form".

3.1:CONSTRUCTION OF E_∞:-

Let I be a countable set of variable names. Let I′ be the simple atomic lattice obtained by adding a top and bottom element. This is the obvious place to start. Now, the flow-diagram lattice constructed a sequence of lattices to accomodate loop-free diagrams, so we will concentrate on normal forms. Following the grammar for **NF** in 1.0, we define :-

$$\left\{\begin{array}{l} E_0 := I' \\ A_0 := I' \end{array}\right\} \text{ and } \left\{\begin{array}{l} E_{i+1} := \lambda I.E_i + A_i \\ A_{i+1} := I' + A_i(E_{i+1}) \end{array}\right\}$$

Now, $\lambda I.E_i := (I \times E_i)/\alpha_i$, where α_i is a relation intended to equivalence the "α-convertible" elements of $(I \times E_i)$: we write $\lambda x.\varepsilon_i$ for the equivalence class $[\langle x,\varepsilon_i\rangle]$ and so, for instance, we want $\lambda x.x(\bot) = \lambda y.y(\bot)$. To define α_i, we have to have (by induction) a notion of a variable being <u>not free in</u> an element of E_i and a <u>change of bound variable operator</u>, $[x/y] \in [E_i \to E_i]$. Then :-

$$(\lambda x.\varepsilon_i = \lambda y.\delta_i) \text{ iff } \begin{array}{l}(\text{there exists } z \in I)\\ (z \text{ is not free in } \varepsilon_i,\delta_i)\\ ([z/x]\varepsilon_i = [z/y]\delta_i).\end{array}$$

Of course, these notions will have to exist for A_i and we will have to define them on E_{i+1} and A_{i+1} as we carry through the induction.

Also, $A_i(E_{i+1}) := (A_i \times E_{i+1})/\sim$, where $\sim$ is the equivalence relation so that $\langle\bot,\varepsilon_{i+1}\rangle \sim \langle\bot,\varepsilon'_{i+1}\rangle$ and $\langle\top,\varepsilon_{i+1}\rangle \sim \langle\top,\varepsilon'_{i+1}\rangle$. Thus, we build in at this stage our intuition that under/over-defined functions give under/over-defined answers. We write $\alpha_i(\varepsilon_{i+1})$ for the equivalence class $[\langle\alpha_i,\varepsilon_{i+1}\rangle]$.

Finally, the + is the old-fashioned lattice sum,  rather than,

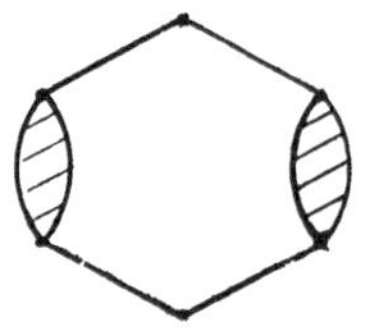

. This does not matter since all the lattices we construct have isolated $\top$'s and so their continuity is preserved.

In fact, all the constructs, $\{E_n,A_n \mid n \geq 0\}$, are countable lattices with "finite depth" - i.e. no infinitely long chains exist in them. As such, they are like finite lattices in that they are automatically complete and continuous with all their elements isolated and functions over them are continuous iff monotone.

Now, two lattices, L and M, form a <u>retraction pair</u>, $L \lhd M$, if there exist $f \in [L \to M]$ and $g \in [M \to L]$ such that $g\circ f(l) = l$ and $f\circ g(m) \sqsubseteq m$. We make $E_0 \lhd E_1$ and $A_0 \lhd A_1$ as follows :-

$$\phi_{1,0} : E_1 \longrightarrow E_0 \qquad \left\{\begin{matrix}\lambda x.\varepsilon_0 \neq \top\\ \alpha_0\end{matrix}\right\} \mapsto \left\{\begin{matrix}\bot\\ \alpha_0\end{matrix}\right\} \qquad \text{and} \qquad \phi_{0,1} : E_0 \subset E_1 \qquad \varepsilon_0 \mapsto \varepsilon_0$$

$$\theta_{1,0} : A_1 \longrightarrow A_0 \qquad \text{and} \qquad \theta_{0,1} : A_0 \subset A_1$$

$$\begin{Bmatrix} x \\ \alpha_0(\varepsilon_1) \neq \top \end{Bmatrix} \longmapsto \begin{Bmatrix} x \\ \bot \end{Bmatrix} \qquad\qquad \varepsilon_0 \longrightarrow \varepsilon_0$$

Then, we proceed by induction to make $E_i \lhd E_{i+1}$ and $A_i \lhd A_{i+1}$ for all $i \geq 1$:-

$$\phi_{i+1,i} : E_{i+1} \longrightarrow E_i$$

$$\begin{Bmatrix} \lambda x.\varepsilon_i \\ \alpha_i \end{Bmatrix} \longmapsto \begin{Bmatrix} \lambda x.\phi_{i,i-1}(\varepsilon_i) \\ \theta_{i,i-1}(\alpha_i) \end{Bmatrix}$$

$$\theta_{i+1,i} : A_{i+1} \longrightarrow A_i$$

$$\begin{Bmatrix} x \\ \alpha_i(\varepsilon_{i+1}) \end{Bmatrix} \longmapsto \begin{Bmatrix} x \\ \theta_{i,i-1}(\alpha_i)(\phi_{i+1,i}(\varepsilon_{i+1})) \end{Bmatrix}$$

Similarly, we define $\phi_{i,i+1}$ and $\theta_{i,i+1}$. These maps are all well-defined, monotonic and, hence, continuous. Then,

$$E_\infty := \text{Inverse limit of } \langle E_{i+1}, \phi_{i+1,i} \rangle_{i=0}^{\infty}$$

$$A_\infty := \text{Inverse limit of } \langle A_{i+1}, \theta_{i+1,i} \rangle_{i=0}^{\infty}$$

$$\lambda I.E_\infty := \text{Inverse limis of } \langle \lambda I.E_i, \phi_{i+1,i} \text{ (restricted)} \rangle_{i=0}^{\infty}$$

$$A_\infty(E_\infty) := \text{Inverse limit of } \langle A_i(E_{i+1}), \theta_{i+1,i} \text{ (restricted)} \rangle_{i=0}^{\infty}$$

We have :-

$$\begin{Bmatrix} E_\infty = \lambda I.E_\infty + A_\infty \\ A_\infty = I' + A_\infty(E_\infty) \end{Bmatrix}$$

(which is just another notation for the β-normal form grammar.)

For a rigorous exposition of the above and the results stated in the rest of this paper, see [11].

3.2: CONSTRUCTION OF *E* :-

First we must define a suitable application function for E_∞ - i.e. $Ap \in [E_\infty \times E_\infty \to E_\infty]$. Then, defining $\lambda x.\langle \varepsilon_0, \varepsilon_1, \varepsilon_2, \ldots\ldots \rangle$ to be $\langle \bot, \lambda x.\varepsilon_0, \lambda x.\varepsilon_1, \lambda x.\varepsilon_2, \ldots\ldots \rangle$, we can define our semantics, $\langle E_\infty, \mathcal{E} \rangle$, by :-

$$\mathcal{E}[\![x]\!] := \langle x,x,x,\ldots\ldots\rangle$$
$$\mathcal{E}[\![\lambda x.\varepsilon]\!] := \lambda x.\mathcal{E}[\![\varepsilon]\!]$$
$$\mathcal{E}[\![\varepsilon(\delta)]\!] := Ap(\mathcal{E}[\![\varepsilon]\!],\mathcal{E}[\![\delta]\!])$$

If **Ap** has been correctly defined, we should have a β-model. Anyway, the semantics will clearly be substitutive.

We define coordinate applications, $Ap_i : E_i \times E_i \to E_{i+1}$, show them to be monotonic and, hence, continuous and define Ap as their limit :-

$$Ap(\langle\varepsilon_i\rangle_{i=0}^{\infty},\langle\varepsilon_i'\rangle_{i=0}^{\infty}) := \bigsqcup_{i=1}^{\infty}\phi_{i+1,\infty}\circ Ap_i(\varepsilon_i,\varepsilon_i').$$

Thus, Ap will be continuous.

Unfortunately, for arbitrary $\langle\varepsilon_i\rangle_{i=0}^{\infty}$, $\langle\varepsilon_i'\rangle_{i=0}^{\infty} \in E_\infty$, we only have :-

$$\phi_{i+2,i+1}\circ Ap_{i+1}(\varepsilon_{i+1},\varepsilon_{i+1}') \sqsupseteq Ap_i(\varepsilon_i,\varepsilon_i'),$$

and so we cannot avoid defining Ap as a limit. The mathematics of the model would be much simpler if the Ap_i's produced a sequence of elements that exactly fitted the inverse limit - but this is wishful thinking! The problem arises because of the presence of the retraction map, $\phi_{i,i+1}$, in the following definition of Ap_i - no matter how the indices are juggled, it seems impossible to escape such an occurence somewhere. Note that the same awkwardness appears when defining application over the Scott-lattices, D_∞ - [6].

As with the definition of β-reduction, which we are trying to emulate with the Ap_i's, we cannot avoid the use of substitution operators. The following seems to be about the simplest presentation :-

$$[\varepsilon_i/x]_I : I' \longrightarrow E_i$$
$$y \longmapsto \begin{Bmatrix} y, \text{ if } x \neq y. \\ \varepsilon_i, \text{ if } x = y. \end{Bmatrix}$$

$$[\varepsilon_1/x]_{E_0} := [\varepsilon_1/x]_I \in [E_0 \to E_1]$$

$$[\varepsilon_2/x]_{A_0} := [\varepsilon_2/x]_I \in [A_0 \to E_2]$$

$$[\varepsilon_i'/x]_{E_{i-1}} : E_{i-1} \longrightarrow E_i \qquad (i \geq 2)$$

$$\begin{Bmatrix} \lambda y.\varepsilon_{i-2} \\ \alpha_{i-2} \end{Bmatrix} \longmapsto \begin{Bmatrix} \lambda z.[\varepsilon_{i-1}'/x]_{E_{i-2}}[z/y]\varepsilon_{i-2} \\ [\varepsilon_i'/x]_{A_{i-2}}\alpha_{i-2} \end{Bmatrix},$$

where $z \neq x$ and z is not free in $\varepsilon_{i-1}', \varepsilon_{i-2}$.

$$[\varepsilon_i'/x]_{A_{i-2}} : A_{i-2} \longrightarrow E_i \qquad (i \geq 3)$$

$$\begin{Bmatrix} y \\ \alpha_{i-3}(\varepsilon_{i-2}) \end{Bmatrix} \longmapsto \begin{Bmatrix} [\varepsilon_i'/x]_I y \\ Ap_{i-1}\langle [\varepsilon_{i-1}'/x]_{A_{i-3}}\alpha_{i-3}, [\varepsilon_{i-1}'/x]_{E_{i-2}}\varepsilon_{i-2}\rangle \end{Bmatrix}$$

$$Ap_i : E_i \times E_i \longrightarrow E_{i+1} \qquad (i \geq 1)$$

$$\left\langle \begin{Bmatrix} \lambda x.\varepsilon_{i-1} \\ \alpha_{i-1} \end{Bmatrix}, \varepsilon_i' \right\rangle \longmapsto \begin{Bmatrix} \phi_{i,i+1}([\varepsilon_i'/x]_{E_{i-1}}\varepsilon_{i-1}) \\ \alpha_{i-1}(\varepsilon_i) \end{Bmatrix}$$

All the above maps are well-defined and continuous. The obvious way to prove the modelship of $\langle E_\infty, \mathcal{E}\rangle$ is to define the substitution operator :-

$$[\langle\varepsilon_i'\rangle_{i=0}^{\infty}/x]_E\ (\langle\varepsilon_i\rangle_{i=0}^{\infty}) := \bigsqcup_{i=1}^{\infty} \phi_{i,\infty} \circ [\varepsilon_i'/x]_{E_{i-1}}\varepsilon_{i-1},$$

and then prove the "substitution lemma" :-

$$\mathcal{E}[\![[\varepsilon/x]\delta]\!] = [\mathcal{E}[\![\varepsilon]\!]/x]_E\, \mathcal{E}[\![\delta]\!], \qquad (?)$$

for it is clear that :-

$$\mathcal{E}[\![(\lambda x.\delta)(\varepsilon)]\!] = [\mathcal{E}[\![\varepsilon]\!]/x]_E\, \mathcal{E}[\![\delta]\!].$$

However, this direct approach defeats us.

3.3:EXAMPLES:-

(i) $\mathcal{E}[\![x]\!] = \langle x,x,x,x,x,\ldots\ldots\rangle$

$\mathcal{E}[\![xy]\!] = \langle \bot,\bot,xy,xy,xy,\ldots\ldots\rangle$

$\mathcal{E}[\![\lambda x.y]\!] = \langle \bot,\lambda x.y,\lambda x.y,\lambda x.y,\ldots\ldots\rangle$

$\mathcal{E}[\![\lambda y.xy]\!] = \langle \bot,\bot,\bot,\lambda y.xy,\lambda y.xy,\ldots\ldots\rangle$

$\mathcal{E}[\![f(yy)]\!] = \langle \bot,\bot,f(\bot),f(yy),f(yy),\ldots\ldots\rangle$

$\mathcal{E}[\![\lambda y.f(yy)]\!] = \langle \bot,\bot,\bot,\lambda y.f(\bot),\lambda y.f(yy),\ldots\ldots\rangle$

$\mathcal{E}[\![b(\lambda c.dd)]\!] = \langle \bot,b(\bot),b(\bot),b(\bot),b(\lambda c.dd),\ldots\ldots\rangle$

$$\mathcal{E}[\![(\lambda y.xy)(b)]\!] = \langle\bot,\bot,xb,xb,\ldots\ldots\ldots\ldots\ldots\ldots\ldots\ldots\rangle$$
$$\mathcal{E}[\![\Delta\Delta]\!] = \langle\bot,\bot,\bot,\bot,\bot,\bot,\ldots\ldots\ldots\ldots\ldots\ldots\ldots\rangle$$
$$\mathcal{E}[\![x(\Delta\Delta)]\!] = \langle\bot,\bot,x(\bot),x(\bot),x(\bot),\ldots\ldots\ldots\ldots\ldots\rangle$$
$$\mathcal{E}[\![Y(f)]\!] = \langle\bot,\bot,f(\bot),f(f(\bot)),f(f(f(\bot))),\ldots\ldots\rangle.$$

(ii) There is a family of "fixed-point" combinators defined by $\{Y_0 := Y\ ;\ G := \lambda x.\lambda f.f(xf)\ ;\ Y_{i+1} := Y_i G\}$, so called because $Y_i(\varepsilon)$ β-cnv $\varepsilon(Y_i(\varepsilon))$. Now, these are not β-convertible amongst themselves, but :-

$$(\forall i \geq 0)(\mathcal{E}[\![Y_i]\!] = \mathcal{E}[\![Y]\!]).$$

(iii) If we "Curry" the application function :-

$$Ap_C(\varepsilon)(\delta) := Ap(\varepsilon,\delta),$$

then, for any $\varepsilon \in E_\infty$, $Ap_C(\varepsilon) \in [E_\infty \to E_\infty]$. As such, it has a minimal fixed point, $\mu Ap_C(\varepsilon)$, which we shall just write as $\mu\varepsilon$. Then,

$$\mathcal{E}[\![Y(\varepsilon)]\!] = \mu\mathcal{E}[\![\varepsilon]\!].$$

Proof:-

(i) and (ii) -By direct computation. The examples involving $\Delta\Delta$ and Y require their own individual induction arguments.

(iii) -If $\mathcal{E}[\![\varepsilon]\!] = \langle\varepsilon_i\rangle_{i=0}^{\infty}$, then we can reduce both sides of the equation to the form :-

$$\bigsqcup_{i=1}^{\infty} \phi_{i+1,\infty}\circ Ap_i(\varepsilon_i, Ap_{i-1}(\varepsilon_{i-1},\ldots,Ap_1(\varepsilon_1,\bot)\ldots)).$$

ǂ

3.4:REMARK:-

From the above examples, $\langle E_\infty,\mathcal{E}\rangle$ looks like a β-model with good properties concerning unsolvable elements and fixed-point combinators. Further, normal forms seem maximal and isolated and so the semantics should be normal.

3.5:DEF:-

We define an "approximate" semantics :-

$$\tilde{\mathcal{E}}[\![x]\!] := \langle x,x,x,\ldots\ldots\rangle$$
$$\tilde{\mathcal{E}}[\![\lambda x.\varepsilon]\!] := \lambda x.\tilde{\mathcal{E}}[\![\varepsilon]\!]$$
$$\tilde{\mathcal{E}}[\![\varepsilon(\delta)]\!] := \begin{cases}\bot, \text{ if } \varepsilon \in \lambda I.EXP. \\ Ap(\tilde{\mathcal{E}}[\![\varepsilon]\!],\tilde{\mathcal{E}}[\![\delta]\!]), \text{ if not.}\end{cases}$$

Again, $\tilde{\mathcal{E}}$ is something that can be "instantly" computed and it is very close to the "instant" semantics of section 0, as is shown by the following lemma.

3.6:LEMMA:-

(i) $(\varepsilon \in \mathrm{HNF}) \Longleftrightarrow (\tilde{E}[\![\varepsilon]\!] \neq \bot)$.

(ii) $(\varepsilon \overset{\sim}{\sqsubseteq} \delta) \Longleftrightarrow (\tilde{E}[\![\varepsilon]\!] \sqsubseteq \tilde{E}[\![\delta]\!])$.

Proof:-

-Structural inductions on ε.

ǂ

3.7:DEF:-

We can now define our "natural" semantics :-

$$N[\![\varepsilon]\!] := \bigsqcup\{\tilde{E}[\![\varepsilon']\!] \mid \varepsilon \xrightarrow{\beta} \varepsilon'\}.$$

3.8:LEMMA:-

(i) $\langle E_\infty, N, \tilde{E}\rangle$ is a continuous semantics.

(ii) $(\varepsilon \in \mathrm{SOL}) \Longleftrightarrow (N[\![\varepsilon]\!] \neq \bot)$ - i.e. $\langle E_\infty, N\rangle$ is solvable.

(iii) (ε has a normal form) => ($N[\![\varepsilon]\!]$ is isolated and maximal). Hence, $\langle E_\infty, N\rangle$ is normal.

(iv) $E \sqsubseteq N$. Hence,

$(\langle E_\infty, E\rangle$ is a β-model) $\Longleftrightarrow (E = N)$.

Proof:-

(i) -Clear, by 3.6(ii).

(ii) -Clear, by 3.6(i).

(iii) -$\langle E_\infty, N\rangle$ is a β-model, by part (i) above and 2.2(iii).

-Hence, if $\varepsilon \xrightarrow{\beta} \nu \in \mathrm{NF}$, then $N[\![\varepsilon]\!] = N[\![\nu]\!] = \tilde{E}[\![\nu]\!] \in E_n$, for some finite $n \geq 0$, and is therefore isolated.

-Now, if $N[\![\varepsilon]\!] \sqsubseteq N[\![\delta]\!]$, then $\tilde{E}[\![\nu]\!] \prec N[\![\delta]\!]$ and, so, $\tilde{E}[\![\nu]\!] \sqsubseteq \tilde{E}[\![\delta']\!]$, for some $\delta \xrightarrow{\beta} \delta'$.

-But, then, $\nu \xrightarrow{\alpha} \delta'$, by 3.6(ii) and 1.3(ii).

-Hence, ε β-<u>cnv</u> δ and $N[\![\varepsilon]\!]$ is maximal and $\langle E_\infty, N\rangle$ is normal.

(iv) -A straightforward, but tedious, induction on i to establish :-

$$(\varepsilon_i, \delta_i \,\underline{(\text{resp.})} \sqsubseteq \phi_{\infty,i}\circ\tilde{E}[\![\varepsilon]\!], \phi_{\infty,i}\circ\tilde{E}[\![\delta]\!]\,\underline{(\text{resp.})})$$
$$\Rightarrow (\varepsilon(\delta) \xrightarrow{\beta} \gamma) \wedge (Ap_i(\varepsilon_i, \delta_i) \sqsubseteq \phi_{\infty,i+1}\circ\tilde{E}[\![\gamma]\!]).$$

-Use this in a structural induction on ε to establish :-

$$E[\![\varepsilon]\!] \sqsubseteq N[\![\varepsilon]\!].$$

-Now, if $\langle E_\infty, E\rangle$ were a β-model, then :-

$$(\varepsilon \xrightarrow{\beta} \varepsilon') \Rightarrow (\tilde{E}[\![\varepsilon']\!] \sqsubseteq E[\![\varepsilon']\!] = E[\![\varepsilon]\!]),$$

and so $N[\![\varepsilon]\!] \sqsubseteq E[\![\varepsilon]\!]$. Thus, $E = N$.

-On the other hand, if $E = N$, then $\langle E_\infty, E\rangle$ is a β-model since $\langle E_\infty, N\rangle$ is one.

ǂ

4: I'th Reductions:-

4.0: REMARK:-

Failing sadly, for the moment, to prove $E = N$, we examine how the Ap_i's work. We define an "i'th application", Ap^i : EXP × EXP → EXP, that "models", so to speak, the "model" Ap_i. We just use superscripts instead of subscripts, the inclusions $\phi_{i,i+1}$ of the lattice definition vanish and we have expressions in NOH where there were $\perp$'s. Then, we use Ap^i to define a notion of "i'th reduction" from an expression in a way that is analogous to the definition of the semantic function E using Ap. Thus, we shall see that our problem in semantics gives rise to a problem concerning the "correctness" or "completeness" of certain reduction mechanisms which have not been previously studied (to our knowledge) and which may be of interest in their own right.

4.1: DEF:-

$$Ap^1(\varepsilon,\delta) := \varepsilon(\delta)$$
$$[\delta/x]^1\varepsilon := [\delta/x]\varepsilon$$

For $i \geq 2$,

$$Ap^i(\varepsilon,\delta) := \begin{Bmatrix} [\delta/x]^i\varepsilon', \text{ if } \varepsilon = \lambda x.\varepsilon' \in \text{EXP}. \\ \varepsilon(\delta), \text{ if not.} \end{Bmatrix}$$

and

$$[\delta/x]^i\varepsilon := \begin{Bmatrix} [\delta/x]\varepsilon, \text{ if } \varepsilon \in \text{I} \cup \text{NOH}. \\ \lambda z.[\delta/x]^{i-1}[z/y]\varepsilon', \text{ if } (*). \\ Ap^{i-1}([\delta/x]^{i-1}\omega,[\delta/x]^{i-1}\eta), \text{ if } (**). \end{Bmatrix}$$

where $(*) \equiv (\varepsilon = \lambda y.\varepsilon' \in \text{HNF}) \wedge (z \neq x$ and z is not free in $\delta,\varepsilon')$
and $(**) \equiv (\varepsilon = \omega(\eta) \in \text{HEAD})$.

4.2: LEMMA:-

(i) Ap^i is well-defined up to α-conversion.

(ii) $(\varepsilon \in \text{NOH}) \Rightarrow (Ap^i(\varepsilon,\delta) \in \text{NOH})$.

(iii) $[x/y]Ap^i(\varepsilon,\delta) \xrightarrow{\alpha} Ap^i([x/y]\varepsilon,[x/y]\delta)$.

(iv) $\varepsilon(\delta) \xrightarrow{\beta} Ap^i(\varepsilon,\delta)$.

(v) $(\varepsilon \xrightarrow{\beta} \varepsilon') \wedge (\delta \xrightarrow{\beta} \delta') \Rightarrow (Ap^i(\varepsilon,\delta) \xrightarrow{\beta} Ap^i(\varepsilon',\delta'))$.

(vi) $Ap^i(\varepsilon,\delta) \xrightarrow{\beta} Ap^{i+1}(\varepsilon,\delta)$.

(vii) $(\varepsilon \,\tilde{\sqsubseteq}\, \varepsilon') \wedge (\delta \,\tilde{\sqsubseteq}\, \delta') \Rightarrow (Ap^i(\varepsilon,\delta) \,\tilde{\sqsubseteq}\, Ap^i(\varepsilon',\delta'))$.

(viii) $\tilde{E}[\![Ap^i(\varepsilon,\delta)]\!] \sqsubseteq Ap(\tilde{E}[\![\varepsilon]\!],\tilde{E}[\![\delta]\!])$.

(ix) $(\varepsilon_i,\delta_i\,(\underline{\text{resp.}}) \sqsubseteq \phi_{\infty,i}\circ\tilde{E}[\![\varepsilon]\!],\phi_{\infty,i}\circ\tilde{E}[\![\delta]\!]\,(\underline{\text{resp.}}))$
$\Rightarrow (Ap_i(\varepsilon_i,\delta_i) \sqsubseteq \phi_{\infty,i+1}\circ\tilde{E}[\![Ap^i(\varepsilon,\delta)]\!])$.

Proof:-

-All straightforward inductions.

-N.B.1: if we did not insist on doing no work when evaluating $[\delta/x]^{i}\varepsilon$ for $\varepsilon \in NOH$, parts (v) and (vi) would not be right.

-N.B.2: part (viii) is isomorphic to the induction in 3.8(iv).

‡

4.3:DEF:-

$$i\langle x\rangle := x$$
$$i\langle \lambda x.\varepsilon\rangle := \lambda x.i\langle\varepsilon\rangle \qquad \text{(c/f definition of } E)$$
$$i\langle\varepsilon(\delta)\rangle := Ap^{i}(i\langle\varepsilon\rangle, i\langle\delta\rangle)$$

4.4:LEMMA:-

(i) $i\langle\varepsilon\rangle$ is well-defined up to α-conversion.

(ii) $[x/y]i\langle\varepsilon\rangle \xrightarrow{\alpha} i\langle[x/y]\varepsilon\rangle$.

(iii) $i\langle\varepsilon\rangle \xrightarrow{\beta} i+1\langle\varepsilon\rangle$.

Proof:-

-More inductions, using 4.2(i),...,(vi).

‡

4.5:EXAMPLES:-

(i) $1\langle\varepsilon\rangle = \varepsilon$ (ii) $(\varepsilon \in NF) \Rightarrow (i\langle\varepsilon\rangle \xrightarrow{\alpha} \varepsilon)$

(iii) $2\langle(\lambda y.xy)(b)\rangle = xb$ (iv) $[\delta/x]\varepsilon \xrightarrow{\beta} 2\langle(\lambda x.\varepsilon)(\delta)\rangle$

(v) $\varepsilon \xrightarrow{\beta} 2\langle I(\varepsilon)\rangle$ (vi) $\varepsilon \xrightarrow{\beta} 3\langle K(\varepsilon)(\delta)\rangle$ (vii) $i\langle\Delta\Delta\rangle \xrightarrow{\alpha} \Delta\Delta$

(viii) $i\langle Y\rangle = \lambda f.f^{i\div 2}(\lambda y.f(yy))(\lambda y.f(yy))$.

Proof:-

(i) -Structural induction on ε.

(ii),...,(viii) -By direct computation, using 4.2 and 4.4 (the last part requiring its own individual induction as in 3.3).

‡

4.6:THEOREM:-

(i) $\tilde{E}[\![i\langle\varepsilon\rangle]\!] \sqsubseteq E[\![\varepsilon]\!]$. Hence, $\bigsqcup_{i=1}^{\infty}\tilde{E}[\![i\langle\varepsilon\rangle]\!] \sqsubseteq E[\![\varepsilon]\!]$.

(ii) $E[\![\varepsilon]\!] \sqsubseteq \bigsqcup_{i=1}^{\infty}\tilde{E}[\![i\langle\varepsilon\rangle]\!]$. Hence, $E[\![\varepsilon]\!] = \bigsqcup_{i=1}^{\infty}\tilde{E}[\![i\langle\varepsilon\rangle]\!]$..

Proof:-

(i) -Structural induction on ε, using 4.2(viii).

(ii) -Structural induction on ε, using 4.2(ix).

‡

4.7:DEF:-

A reduction rule, R, is STRONGLY COMPLETE if :-

$$(\varepsilon \xrightarrow{\beta} \varepsilon') \Rightarrow (\varepsilon \xrightarrow{R} \delta) \wedge (\varepsilon' \xrightarrow{\beta} \delta).$$

It is WEAKLY COMPLETE if :-

$$(\varepsilon \xrightarrow{\beta} \varepsilon') \Rightarrow (\varepsilon \xrightarrow{R} \delta)_{\wedge}(\varepsilon' \tilde{\equiv} \delta).$$

4.8:LEMMA:-

(i) Standard reductions are strongly complete.

(ii) Normal reductions are not even weakly complete.

(iii) I'th reductions are not strongly complete.

(iv) R is weakly complete $\Leftrightarrow$ $N[\![\varepsilon]\!] = \bigsqcup\{\tilde{E}[\![\varepsilon']\!] \mid \varepsilon \xrightarrow{R} \varepsilon'\}$.

(v) $(E = N) \Leftrightarrow$ (i'th reductions are weakly complete).

Proof:-

(i) -By the second Church-Rosser theorem - see Curry [1] - any β-reduction can be made standard. Hence, the result with $\delta = \varepsilon'$.

(ii) -Let $\varepsilon = x(\Delta\Delta)(Iy)$. The only normal reductions are α-convertible to ε. But, $\varepsilon \xrightarrow{\beta} x(\Delta\Delta)(y) \not\tilde{\equiv} \varepsilon$. (This example is from [9])

(iii) -Let $T = \lambda x.xxx$. Write ε^n for $\varepsilon\varepsilon\varepsilon\ldots\ldots\varepsilon$ (n times).

-Now, $(\forall n \geq 2)(T^2 \xrightarrow{\beta} T^n)$, but $(\forall n \geq 3)(T^n \not\xrightarrow{\beta} T^2)$.

-Let $\varepsilon = (\lambda x.\Delta^2(xx))(T)$. Then, $(\forall n \geq 2)(\varepsilon \xrightarrow{\beta} \Delta^2 T^n)$.

-But, $(\forall i \geq 2)(i\langle\varepsilon\rangle \xrightarrow{\alpha} \Delta^2 T^2)$.

-Hence, $(\forall i \geq 1)(\forall n \geq 3)(\varepsilon \xrightarrow{\beta} \Delta^2 T^n \not\xrightarrow{\beta} i\langle\varepsilon\rangle)$.

(iv) $(\Rightarrow)$ -Clear, from the definition.

$(\Leftarrow)$ -Since $\tilde{E}$ maps only to isolated points.

(v) -By part (iv) above and 4.6(ii).

ǂ

5:Inside-Out Reductions:-

5.0:REMARK:-

Alas, it seems just as difficult to prove i'th reductions are weakly complete as it is to prove $\langle E_\infty, E\rangle$ is a β-model directly. Let us broaden our outlook.

How do i'th reductions work? To evaluate $i\langle\varepsilon(\delta)\rangle$, first we have to evaluate $i\langle\varepsilon\rangle$ and $\iota\langle\delta\rangle$ and then combine them with an Ap^i. The way $Ap^i(\lambda x.\varepsilon', \delta')$ is defined "protects" sub-redexes of ε' and δ' - i.e. does not contract their residuals - and only evaluates new ones that are created by $[\delta'/x]\varepsilon'$. This acts in almost the opposite way to standard reduction sequences and leads to the following generalisation.

5.1:DEF:-

Let $\varepsilon_0 \xrightarrow{R_1} \varepsilon_1 \xrightarrow{R_2} \varepsilon_2 \xrightarrow{R_3} \ldots\ldots \xrightarrow{R_n} \varepsilon_n$ be a β-reduction sequence where R_i is the redex contracted in going from ε_{i-1} to ε_i. Then, the sequence is INSIDE-OUT if, whenever $1 \leq i < j \leq n$, the redex R_j is not the residual of any sub-redex of R_i. We write :-

$$\varepsilon_0 \rightsquigarrow \varepsilon_n$$

5.2:EXAMPLES:-

The following reduction sequences are all inside-out :-

(i) $(\lambda x.xax)(\lambda y.y((\lambda z.z)y)) \rightarrow (\lambda x.xax)(\lambda y.yy)$
$\rightarrow (\lambda y.yy)(a)(\lambda y.yy) \rightarrow aa(\lambda y.yy)$.

(ii) $(\lambda f.f^3(a))((\lambda x.\lambda y.yx)b) \rightarrow (\lambda f.f^3(a))(\lambda y.yb)$
$\rightarrow (\lambda y.yb)((\lambda y.yb)((\lambda y.yb)(a)))$
$\rightarrow (\lambda y.yb)((\lambda y.yb)(ab)) \rightarrow (\lambda y.yb)(abb) \rightarrow abbb$.

(iii) $(\lambda x.y)(\Delta\Delta) \rightarrow y$.

(iv) $(\lambda y.\Delta(ya))(I) \rightarrow (\lambda y.(ya)(ya))(I) \rightarrow (Ia)(Ia)$
$\rightarrow (Ia)(a) \rightarrow aa$.

5.3:REMARK:-

Examples (i) and (ii) above were taken from Wadsworth - [9] They show inside-out reductions to be fairly efficient, being as good as Wadsworth's "normal graph reductions" in (ii) and better in (i). Example (iii) is to show that inside-out reductions do not fall into the same trap as the "call-by-value" mechanism, which would get stuck in this case. Example (iv) shows that inside-out reductions are not always the most efficient way, since it takes 4 steps while it is possible to do it in 3 (e.g. by "n.g.r.").

5.4:DEF:-

Let γ be a sub-expression of ε. Then, $\varepsilon \xrightarrow{\gamma} \delta$ means that $\varepsilon \xrightarrow{\beta} \delta$ but no residual of any sub-redex of γ is contracted.

5.5:DEF:-

Let R and S be two reduction rules. Then, R is STRONGLY COMPLETE RELATIVE to S if :-

$$(\varepsilon \xrightarrow{S} \varepsilon') \Rightarrow (\varepsilon \xrightarrow{R} \delta)_\wedge(\varepsilon' \xrightarrow{\beta} \delta).$$

R is WEAKLY COMPLETE RELATIVE to S if :-

$$(\varepsilon \xrightarrow{S} \varepsilon') \Rightarrow (\varepsilon \xrightarrow{R} \delta)_\wedge(\varepsilon' \lessapprox \delta).$$

Also, we define R and S to be STRONGLY or WEAKLY EQUIVALENT when we have the above relations, respectively, holding between them both ways.

5.6:LEMMA:-

(i) $(\varepsilon(\delta) \rightsquigarrow \gamma) \Leftrightarrow (\varepsilon \rightsquigarrow \varepsilon') \wedge (\delta \rightsquigarrow \delta')$
$\wedge (\varepsilon'(\delta') \overset{\varepsilon' \ \ \delta'}{\rightsquigarrow} \gamma)$.

(ii) $(\varepsilon \in NOH) \wedge (\varepsilon(\delta) \overset{\varepsilon}{\rightarrow} \gamma) \Rightarrow (\gamma \in NOH)$.

(iii) I'th reductions are not strongly complete relative to inside-out reductions.

(iv) $\varepsilon(\delta) \overset{\varepsilon \ \ \delta}{\rightsquigarrow} Ap^{i}(\varepsilon,\delta)$.

(v) $\varepsilon \rightsquigarrow i\langle\varepsilon\rangle$.

(vi) $(\varepsilon(\delta) \overset{\varepsilon \ \ \delta}{\rightsquigarrow} \eta) \Rightarrow (\eta \overset{\beta}{\equiv} Ap^{i}(\varepsilon,\delta)$, for some $i \geq 1)$.

(vii) I'th reductions are weakly equivalent to inside-out reductions.

(viii) (I'th reductions are weakly complete) <=> (inside-out reductions are weakly complete).

Proof:-

(i) -Clear, by the definition.
(ii) -By structural induction on $\varepsilon \in NOH$.
(iii) -Same counter-example as in 4.8(iii), since $\varepsilon \rightsquigarrow \Delta^2 T^n$.
(iv) -By induction on i.
(v) -By structural induction on $\varepsilon \in EXP$, using parts (i) and (iv).
(vi) -By induction on the length of the reduction sequence together with a structural induction on $\varepsilon \in EXP$.
-N.B.: a "strong" version of this (i.e. using $\overset{\beta}{\rightarrow}$ instead of $\overset{\beta}{\equiv}$) will fail since, otherwise, we would contradict part (iii).
(vii) -We prove that i'th reductions are weakly complete relative to inside-out reductions by a structural induction over EXP, using parts (i) and (vi). Hence, the result, using part (v).
(viii) -Obvious, by part (vii).

ǂ

5.7:PROPERTY A:-

Inside-out reductions are strongly complete.

5.8:PROPERTY B:-

Inside-out reductions are weakly complete.

6:Outline of Proof of Property B:-

6.0:REMARK:-

Before launching into this too deeply, the reader is referred to an alternative proof by J-J.Lévy - [2] or [3]. This is quite surprising, somewhat elegant, establishes the stronger property **A** and may be easier to understand than the "sledgehammer" approach which follows.

6.1:WEAK CHURCH-ROSSER THEOREM FOR INSIDE-OUT REDUCTIONS:-

$$(\varepsilon \rightsquigarrow \delta) \wedge (\varepsilon \rightsquigarrow \gamma) \Rightarrow (\varepsilon \rightsquigarrow \eta) \wedge (\delta, \gamma \;\tilde{\equiv}\; \eta).$$

Proof:-

-By 5.6(vii), $\delta \;\tilde{\equiv}\; i\langle\varepsilon\rangle$ and $\gamma \;\tilde{\equiv}\; j\langle\varepsilon\rangle$, for some $i, j \geq 1$.

-So, $\delta, \gamma \;\tilde{\equiv}\; k\langle\varepsilon\rangle$, where $k := \max(i,j)$, by 4.4(iii).

-But, $\varepsilon \rightsquigarrow k\langle\varepsilon\rangle$, by 5.6(v). Take $\eta := k\langle\varepsilon\rangle$.

ǂ

6.2:REMARK:-

If we could replace $\tilde{\equiv}$ by $\xrightarrow{\beta}$ in the above, we could forget results involving $\tilde{\equiv}$ in what follows and it would establish property **A**.

6.3:WEAK PARALLEL MOVES:-

$$(\varepsilon(\delta) \overset{\varepsilon \quad \delta \quad \leq n}{\rightsquigarrow} \eta) \wedge (\delta \;\tilde{\equiv}\; \delta') \wedge (\varepsilon \;\tilde{\equiv}\; \varepsilon')$$

$$\Rightarrow (\varepsilon'(\delta') \overset{\varepsilon' \quad \delta' \quad \leq n}{\rightsquigarrow} \eta') \wedge (\eta \;\tilde{\equiv}\; \eta').$$

Proof:-

-The "$\leq n$" in the above means that there are $\leq n$ β-reductions in the sequences.

-The result comes by induction on **n** during which a structural induction on $\varepsilon \in$ EXP is also required (similar to 5.6(vi)).

ǂ

6.4:DEF:-

Suppose we have a diagram of reduction paths with a unique source expression ε. Then, the diagram COMMUTES if, for any expression δ on the diagram and paths ① and ② such that :-

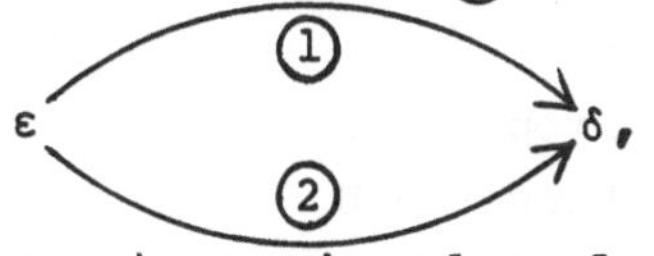

we have that whenever ρ is a sub-redex of ε, its residuals in δ

relative to either of the paths are the same : we write :-

$$(\forall \text{ sub-redexes } \rho \text{ of } \varepsilon)(\{\rho\}/① = \{\rho\}/②).$$

6.5:REMARK:-

We require this property in order to carry through the inductions on the diagrams that follow - in particular 6.13 (this was pointed out by J.R.Hindley - to whom thanks). The first example of the property is when the paths ① and ② in the above diagram are different complete relative reductions of some set, R_ε, of sub-redexes of ε : this is Curry's "strong property E" - [1]. This property tends to fail if ever we get "looping" in the diagram :-

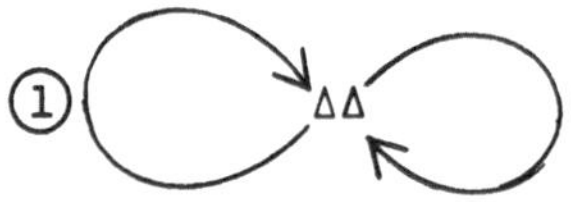

where ① is one β-reduction and ② is just α-conversions.

6.6:A SIMPLE PRESERVATION OF INSIDE-OUTNESS:-

Let $\varepsilon \longrightarrow \varepsilon'$ and $\delta \longrightarrow \delta'$. Then, we may construct the diagram :-

$$\begin{array}{ccc} (\lambda y.\varepsilon)(\delta) & \longrightarrow & (\lambda y.\varepsilon')(\delta') \\ \downarrow & & \downarrow \\ [\delta/y]\varepsilon & \longrightarrow & [\delta'/y]\varepsilon', \end{array}$$

such that it commutes.

Proof:-

-Trivial : trick is to do the reductions on δ first.

$\ddagger$

6.7:PROPERTY C:-

$$(\varepsilon \xrightarrow{1\beta} \delta \longrightarrow \eta) \Rightarrow (\varepsilon \longrightarrow \eta') \wedge (\eta \overset{\alpha}{\equiv} \eta').$$

6.8:PROPERTY D:-

$$([\delta/x]\varepsilon \longrightarrow \eta) \Rightarrow (\delta \longrightarrow \delta') \wedge (\varepsilon \longrightarrow \varepsilon')$$

$$\wedge ([\delta'/x]\varepsilon' \xrightarrow{\ \delta'\ \ \varepsilon'\ } \eta') \wedge (\eta \overset{\alpha}{\equiv} \eta').$$

6.9:PROPERTY E:-

$$([\delta/x]\omega(\varepsilon) \xrightarrow{\ \omega\ \ \varepsilon\ \ \delta\ } \eta) \Rightarrow (\omega(\varepsilon) \xrightarrow{\ \omega\ \ \varepsilon\ } \rho)$$

$$\wedge ([\delta/x]\rho \xrightarrow{\ \delta\ \ \rho\ } \eta') \wedge (\eta \overset{\alpha}{\equiv} \eta').$$

6.10:LEMMA:-

(i) Property B <=> Property C.

(ii) Property C <=> Property D.

(iii) Property D <= Property E.

Proof:-

(i) -Easy.

(ii) $(\Rightarrow)$ $-(\lambda x.\varepsilon)(\delta) \xrightarrow{1\beta} [\delta/x]\varepsilon \rightsquigarrow \eta$. Hence, the result, by property C and 5.6(i).

$(\Leftarrow)$ -By structural induction on $\varepsilon \in$ EXP, using 5.6(i) and 6.3.

(iii) -By structural induction on $\varepsilon \in$ EXP, using 5.6(i),6.1,6.3 and 6.6 : property E is just what is required to finish the induction in the case of combinations.

ǂ

6.11:STRONG PARALLEL MOVES:-

$(\varepsilon(\delta) \overset{\varepsilon\ \ \delta\ \ \leq n\ \ (1)}{\rightsquigarrow} \eta) \wedge (\varepsilon \overset{R_\varepsilon}{\rightsquigarrow} \varepsilon') \wedge (\delta \overset{R_\delta}{\rightsquigarrow} \delta')$

$\Rightarrow$ (we can construct the diagram :-

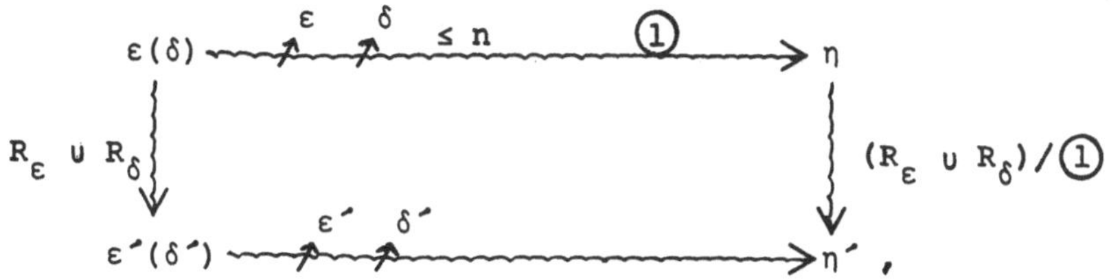

such that it commutes).

Proof:-

-By $\varepsilon \overset{R_\varepsilon}{\rightsquigarrow} \varepsilon'$, we mean a complete relative reduction of R_ε, which is some set of sub-redexes of ε.

-The proof is similar in outline to that of 6.3 and uses 6.6.

ǂ

6.12:COR:-

$(\varepsilon(\delta) \overset{\varepsilon\ \ \delta}{\rightsquigarrow} \eta) \wedge (\varepsilon \xrightarrow{\beta} \varepsilon') \wedge (\delta \xrightarrow{\beta} \delta')$ $\Rightarrow$ (we can construct the diagram :-

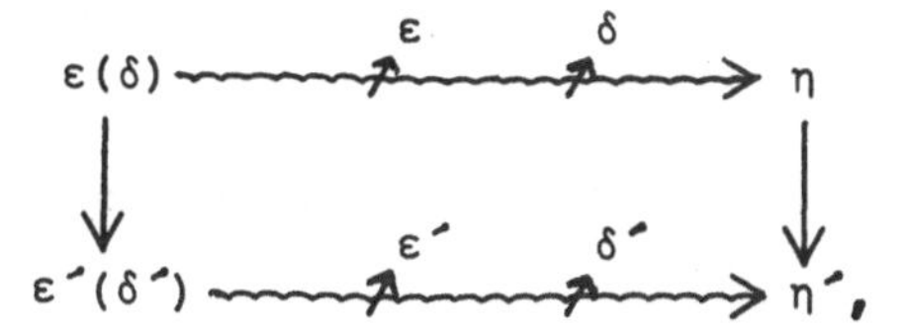

such that it commutes).

Proof:-

-Build up the diagram downwards, contracting one redex at a time.

ǂ

6.13:STRONG SERIAL MOVES:-

$([\delta/x]\varepsilon \xrightarrow{\overset{\delta}{\nearrow}\ \overset{\varepsilon}{\nearrow}\ \leq n\ (1)} \eta \xrightarrow{R_\eta\ (2)} \hat{\eta})$

$((1);(2) \nearrow \{\delta,\varepsilon\}) \Rightarrow$ (we can construct :-

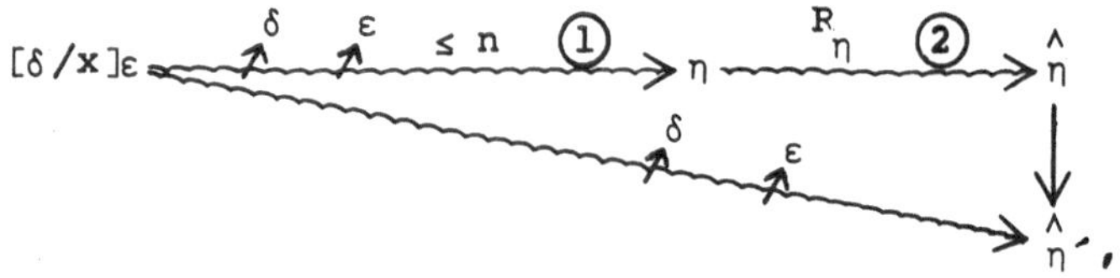

such that it commutes).

Proof:-

-By $(1);(2) \nearrow \{\delta,\varepsilon\}$, we mean that the concatenated sequences, (1) and (2), does not contract any residuals of sub-redexes of δ or ε.

-Again, the proof is similar in outline to 6.3 and uses 6.6, 6.11 and 6.12.

╪

6.14:STRONG SERIAL AND PARALLEL MOVES:-

Let $[\delta/x]\varepsilon \xrightarrow{\overset{\delta}{\nearrow}\ \overset{\varepsilon}{\nearrow}} \eta$ and $\eta \xrightarrow{R_\eta} \hat{\eta}$. Then, there exist sets of sub-redexes R_δ and R_ε, of δ and ε respectively, such that we can construct the following diagram :-

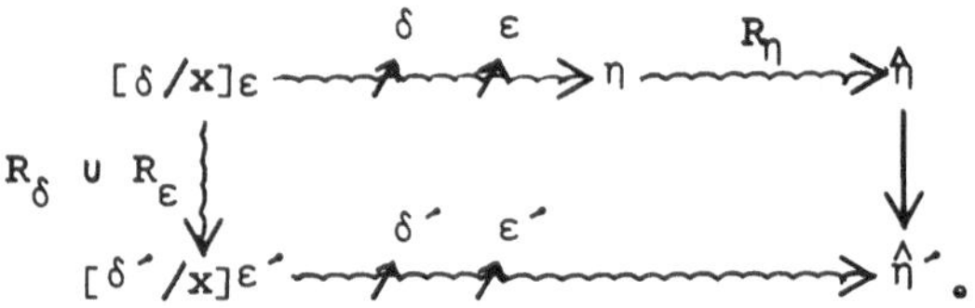

Proof:-

-Falls out from proof of 6.13.

╪

6.15:STRONG PROPERTY E:-

Let $[\delta/x]\omega(\varepsilon) \xrightarrow{\overset{\delta}{\nearrow}\ \overset{\omega}{\nearrow}\ \overset{\varepsilon}{\nearrow}} \eta$. Then, we can construct the following commuting diagram :-

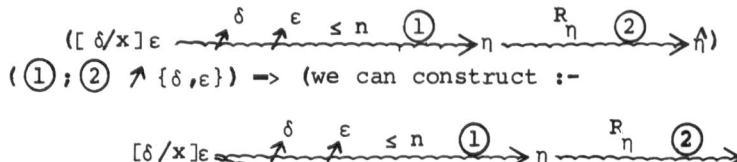

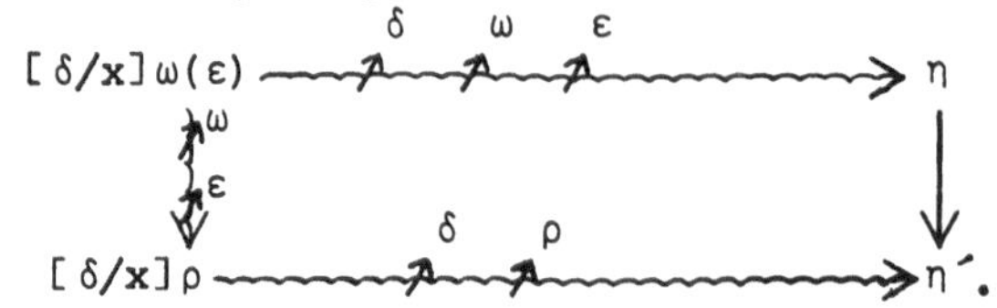

Proof:-

-By 5.6(i), $[\delta/x]\omega(\varepsilon) \xrightarrow{\ \delta\ \ \ \omega(\varepsilon)\ } \mu(\nu) \xrightarrow{\ \mu\ \ \ \nu\ \ \ (1)\ } \eta$.

-We can fill out the diagram by creeping along sequence (1), one redex at a time, using 6.14 in a suitable induction hypothesis.

ǂ

6.16:COR:-

Properties E, D, C and B.

Proof:-

-By 1.3(i) and 6.10.

ǂ

7:Consequences of Property B:-

7.0:THEOREM:-

(i) I'th reductions are weakly complete.

(ii) $\langle E_\infty, \mathcal{E}\rangle$, = $\langle E_\infty, \mathcal{N}\rangle$, is a normal solvable substitutive continuous β-model of the λ-calculus in which expressions with normal form are maximal and isolated and unsolvable expressions are $\perp$.

(iii) If $\varepsilon \in$ SOL, then $\varepsilon \rightsquigarrow \varepsilon' \in$ HNF. Also, $i\langle\varepsilon\rangle \in$ HNF, for some $i \geq 1$.

(iv) If $\varepsilon \xrightarrow{\beta} \nu \in$ NF, then $\varepsilon \rightsquigarrow \nu$. Also, $i\langle\varepsilon\rangle \xrightarrow{\alpha} \nu$, for some $i \geq 1$.

(v) In a continuous semantics $\langle F, \mathcal{F}, \tilde{\mathcal{F}}\rangle$, we need only look at the inside-out or i'th reductions to get the whole semantics :-

$$\mathcal{F}[\![\varepsilon]\!] = \bigsqcup\{\tilde{\mathcal{F}}[\![\varepsilon']\!] \mid \varepsilon \rightsquigarrow \varepsilon'\} = \bigsqcup\{\tilde{\mathcal{F}}[\![i\langle\varepsilon\rangle]\!] \mid i \geq 1\}.$$

Proof:-

-Trivial.

ǂ

7.1:THEOREM:-

Let $\langle F, \mathcal{F}, \tilde{\mathcal{F}}\rangle$ be a continuous semantics. Extend F to a complete lattice by adjoining a top element. Then, we can construct a map, $\bar{\mathcal{F}} \in [E_\infty \to F \cup \{\top\}]$, such that :-

(i) $\bar{\mathcal{F}} \circ \tilde{\mathcal{E}} = \tilde{\mathcal{F}}$

and (ii) $\bar{\mathcal{F}} \circ \mathcal{E} = \mathcal{F}$.

Proof:-

-The image of EXP under $\mathcal{E}$ lies within a directedly complete sub-semi-lattice, Low(E_∞), of E_∞.

-Low(E_∞) is the inverse limit of a sequence of semi-lattices, $\langle Low(E_i)\rangle_{i=0}^{\infty}$, where each Low($E_i$) $\subset E_i$.

-Because of the "syntactical" nature of E_∞, it is easy to define :-

$$syn_i : Low(E_i) \longrightarrow EXP,$$

that is monotonic in the sense that :-

$$(\varepsilon_i \sqsubseteq \varepsilon_i') \Rightarrow (syn_i(\varepsilon_i) \tilde{\sqsubseteq} syn_i(\varepsilon_i')).$$

-Also, for any $\varepsilon \in EXP$, by a simple structural induction, there is an $i \geq 0$ such that for all $j \geq i$:-

$$syn_j \circ \phi_{\infty,j} \circ \tilde{E}[\![\varepsilon]\!] \tilde{\equiv} \varepsilon \qquad (*)$$

where $\tilde{\equiv}$ means $\tilde{\sqsubseteq}$ and $\tilde{\sqsupseteq}$.

-Then, we define :-

$$\bar{F}_i : E_i \longrightarrow F \cup \{\top\}$$

$$\varepsilon_i \longmapsto \left\{\begin{array}{l}\top, \text{ if } \varepsilon_i \notin Low(E_i) \\ \tilde{F}\circ syn_i(\varepsilon_i), \text{ otherwise}\end{array}\right\}$$

and,

$$\bar{F} : E_\infty \longrightarrow F \cup \{\top\}$$

$$\langle \varepsilon_i \rangle_{i=0}^{\infty} \mapsto \bigsqcup\{\bar{F}_i(\varepsilon_i) \mid i \geq 0\}$$

-Now, $\bar{F}_i$ is well-defined and monotonic and, therefore, continuous.

-Hence, $\bar{F}$ is also.

-Then,

$$\begin{aligned}\bar{F}\circ\tilde{E}[\![\varepsilon]\!] &= \bigsqcup\{\bar{F}_i \circ \phi_{\infty,i} \circ \tilde{E}[\![\varepsilon]\!] \mid i \geq 0\} \\ &= \bigsqcup\{\tilde{F}\circ syn_i \circ \phi_{\infty,i} \circ \tilde{E}[\![\varepsilon]\!] \mid i \geq 0\}, \text{ since } \tilde{E}[\![EXP]\!] \subset Low(E_\infty). \\ &= \tilde{F}[\![\varepsilon]\!], \text{ by } (*) \text{ above.}\end{aligned}$$

-Finally,

$$\begin{aligned}\bar{F}\circ E[\![\varepsilon]\!] &= \bar{F}(\bigsqcup\{\tilde{E}[\![\varepsilon']\!] \mid \varepsilon \xrightarrow{\beta} \varepsilon'\}) \\ &= \bigsqcup\{\bar{F}\circ\tilde{E}[\![\varepsilon']\!] \mid \varepsilon \xrightarrow{\beta} \varepsilon'\} \\ &= \bigsqcup\{\tilde{F}[\![\varepsilon']\!] \mid \varepsilon \xrightarrow{\beta} \varepsilon'\} \\ &= F[\![\varepsilon]\!].\end{aligned}$$

†

7.2:COR:-

(i) $\langle F, F\rangle$ is a continuous semantics if and only if it is continuously derivable from $\langle E_\infty, E\rangle$. Thus, $\langle E_\infty, E\rangle$ is the minimal continuous semantics.

(ii) In a continuous semantics, the well-behaved approximate is unique.

(iii) In a continuous semantics, all the Y_i combinators are equivalent.

(iv) In a continuous Scott-model, the Y combinator behaves like the minimal fixed-point operator -

$$D[\![Y(\varepsilon)]\!](\rho) = \mu D[\![\varepsilon]\!](\rho).$$

<u>Proof:-</u>

(i) -By 2.2(iv), 7.0(ii) and 7.1(ii).

(ii) -Suppose that $\tilde{F}$ and $\tilde{G}$ are well-defined approximates of F.
-Construct $\overline{F}$ and $\overline{G}$ as in 7.1. By 7.1(ii), $\overline{F}$ and $\overline{G}$ are the same when restricted to $E[\![\mathrm{EXP}]\!]$.
-But, $\tilde{E}[\![\mathrm{EXP}]\!] \subset E[\![\mathrm{EXP}]\!]$. Hence, $\tilde{F} = \tilde{G}$.

(iii) -By 3.3(ii) and 7.1(ii).

(iv) -A Scott-model has the form $<[\mathrm{ENV} \to \mathrm{D}], \mathcal{D}>$, where the continuous function space, $[\mathrm{D} \to \mathrm{D}]$, is a projection of D. One of its properties is that the value of a combinator (i.e. a λ-expression with no free variables) is independent of the environment used :-

e.g. $\mathcal{D}[\![\Delta\Delta]\!](\rho) = \mathcal{D}[\![\Delta\Delta]\!](\rho')$, for all $\rho, \rho' \in \mathrm{ENV}$.

-Suppose $<[\mathrm{ENV} \to \mathrm{D}], \mathcal{D}>$ is continuous.
-Then, $\mathcal{D}[\![\varepsilon]\!] = \mathcal{D}[\![\Delta\Delta]\!] \sqsubseteq \mathcal{D}[\![x]\!]$, for all $\varepsilon \in \mathrm{INSOL}$, by 2.2(v).
-Choose $\rho \in \mathrm{ENV}$ such that $\rho(x) = \bot$.
-Then, $\mathcal{D}[\![\Delta\Delta]\!](\rho) \sqsubseteq \mathcal{D}[\![x]\!](\rho) = \bot$ and, so, $\mathcal{D}[\![\Delta\Delta]\!] = \bot$, by above remarks.
-Hence all unsolvable expressions are $\bot$ (N.B. in the Park-pathological-model, [4], $\mathcal{D}[\![\Delta\Delta]\!] \neq \bot$, and so it cannot be continuous).
-Now, construct $\overline{\mathcal{D}}$ as in 7.1.
-Thus, $\mathcal{D}[\![Y(\varepsilon)]\!](\rho) = \overline{\mathcal{D}}\circ E[\![Y(\varepsilon)]\!](\rho) = \overline{\mathcal{D}}(\mu E[\![\varepsilon]\!])(\rho)$, by 3.3(iii).

$$= \overline{\mathcal{D}}(\bigsqcup\{E[\![\varepsilon^n(\Delta\Delta)]\!] \mid n \geq 0\})(\rho), \text{ by the above.}$$
$$= \bigsqcup\{\overline{\mathcal{D}}\circ E[\![\varepsilon^n(\Delta\Delta)]\!] \mid n \geq 0\}(\rho)$$
$$= \bigsqcup\{\mathcal{D}[\![\varepsilon^n(\Delta\Delta)]\!](\rho) \mid n \geq 0\}$$
$$= \mu\mathcal{D}[\![\varepsilon]\!](\rho).$$

ǂ

8: References:-

[0] Barendregt,H.P. : "Some Extensional Term Models for Combinatory Logics and λ-Calculi" : Ph.D. Thesis, Utrecht (1971).

[1] Curry,H.B., Feys,R. : "Combinatory Logic - Volume I" : North-Holland, Amsterdam (1958).

[2] Lévy,J-J. : "Another Syntactic Model of the λ-K-β-Calculus" : Symposium on λ-Calculus and Computer Sciences Theory, Roma (1975)

[3] Lévy,J-J. : "Réductions Sures dans le Lambda-Calcul" : Diplome de Docteur de 3e Cycle, Universite Paris VII (1974).

[4] Park,D.M.R. : "The Y-Combinator in Scott's Lambda-Calculus Models" : Unpublished notes, University of Warwick (1970).

[5] Reynolds,J.C. : "Notes on a Lattice-Theoretic Approach to the Theory of Computation" : Systems and Information Science, Syracuse University (1972).

[6] Scott,D. : "Continuous Lattices" : Technical Monograph PRG-7, Oxford University Computing Laboratory, Programming Research Group (1970).

[7] Scott,D. : "Data Types as Lattices" : Unpublished lecture notes, Oxford (1973).

[8] Scott,D. : "The Lattice of Flow Diagrams" : PRG-3 (1970).

[9] Wadsworth,C.P. : "Semantics and Pragmatics of the λ-Calculus" : Ph.D. Thesis, Oxford University (1971).

[10] Wadsworth,C.P. : "Typed λ-Expressions" : Unpublished notes, Oxford (1972).

[11] Welch,P.H. : "The Minimal Continuous Semantics of the Lambda-Calculus" : Ph.D. Thesis (Submitted), Warwick University (1974).

An algebraic interpretation of the λβK-calculus and a labelled λ-calculus

Jean-Jacques LEVY
IRIA-LABORIA
78150-Rocquencourt
France

Introduction : A wide range of λ-calculus models has been proposed by Scott[10,11] In these interpretations, the interconvertibility relation among λ-expressions is extended by mainly equating the unsolvable terms (i.e. expressions M such that, for any arguments $N_1, N_2, \dots N_k$, the expression $MN_1N_2 \dots N_k$ has no normal form). This extension has been shown consistent by Barendregt [1] and Wadsworth [13]. Wadsworth [13] showed the adequacy of most of Scott's models from a computational point of view ; more precisely, each expression is equal to the limit of its approximations in these models. We will try to go in the reverse direction, in the first part of this paper, and to define the value of an expression from its set of approximations. Then we prove that, as usual, our interpretation defines (using Milner's words [7]) a congruence on the language of λ-expressions. For this, we follow Welch [14] who stated a conjecture about the completeness, in the reducibility sense, of "inside-out reductions". This conjecture is proved in the second part of this paper by introducing a "labelled λ-calculus", which the author believes to be a useful tool for some λ-calculus problems. The results in this paper are related to the ones in Hyland [4] and Welch [15]. The definition of our interpretation is very similar to the one of Nivat [9] and Vuillemin [12] used for systems of recursively defined functions. Most results appeared in the author's thesis [5].

Syntax : We consider the set Λ of λ-expressions, built from an infinite alphabet V of variables, which is the minimal set containing :

(1) x (variable)

(2) (λxM) (abstraction)

(3) (MN) (application)

where x is in V and M,N are already in Λ. And we will use the standard abbreviations where :

$$MNN_1N_2 \ldots N_k \text{ stands for } (\ldots(((MN)N_1)N_2)\ldots N_k)$$
$$(\lambda x_1x_2\ldots x_m.M) \text{ " " } (\lambda x_1(\lambda x_2\ldots(\lambda x_mM)\ldots))$$

and M,N,N_i are expressions in Λ, x_i are variables. We shall also omit the outermost parenthesis of an expression. The usual notions of free and bound variables are assumed defined and we note $M[x\backslash N]$ for the substitution of N for the free occurrences of x in M. We consider only two rules of conversion : the α and β rules. If M derives from M by an α-conversion, we write $M \underset{\alpha}{\to} N$. Similarly we have $M \underset{\beta}{\to} N$, and a reduction (possibly of length zero) using only α-conversion from M to N is written $M \overset{*}{\underset{\alpha}{\to}} N$. Hence we note $M \overset{*}{\underset{\beta}{\to}} N$ and $M \xrightarrow[\alpha,\beta]{*} N$ for β-reduction or any sequence of α and β conversions from M to N. We often forget α-conversions and $M \to N$ or $M \overset{*}{\to} N$ are understood as $M \underset{\beta}{\to} N$ or $M \xrightarrow[\alpha,\beta]{*} N$. Equality must also be considered as equality modulo some α-conversions. We will try to use the usual terminology (residuals, standard, reductions ...) defined in [2,3]. We also make use of the context notation (See [8,13]).

Let us first remark that Λ can also be considered as the smallest set containing :

(i) $\lambda x.M$ (abstraction)

(ii) $xM_1M_2\ldots M_n$ (head normal form)

(iii) $(\lambda x.M)NM_1M_2\ldots M_n$

if x is a variable and M,N,M_i are expressions of Λ. More generally, a head normal form is any expression of the form $\lambda x_1x_2\ldots x_m.xM_1M_2\ldots M_n$ where $m,n \geq o$ (See[13]). Others expressions are of the form $\lambda x_1x_2\ldots x_m.(\lambda x.M)NM_1M_2\ldots M_n$ and have a head redex $(\lambda x.M)N$. If $M \overset{*}{\to} N$ and N is an abstraction (respectively a head normal form) we say that M has an abstraction form (respectively a head normal form).

Proposition 1 : If M has an abstraction form, then M has a minimal abstraction form $\lambda x.N_0$, i.e. we have $M \overset{*}{\to} \lambda x.N_0$ and, for any $\lambda x.N$ such that $M \overset{*}{\to} \lambda x.N$, we have $\lambda x.N_0 \overset{*}{\to} \lambda x.N$.

Proof : M can be only of form (i) or (iii). In the first case, we have $M = \lambda x.N_0$. Otherwise for any, $\lambda x.N$ such that $M \overset{*}{\to} \lambda x.N$, by the standardization theorem, there is a standard reduction :

$$M = M_0 \overset{R_1}{\to} M_1 \overset{R_2}{\to} M_2 \overset{R_3}{\to} \ldots \overset{R_n}{\to} M_n = \lambda x.N$$

from M to $\lambda x.N$. Let M_k be the first M_i which is an abstraction. Then, since the reduction is standard, the redexes R_j contracted between M_{j-1} and M_j are the head redexes of M_{j-1} for $1 \leq j \leq k$. So each standard reduction from M to some $\lambda x.N$ has a common initial part :

$$M_0 \overset{R_1}{\to} M_1 \overset{R_2}{\to} M_2 \overset{R_3}{\to} \dots \overset{R_k}{\to} M_k$$ □

Proposition 2 : If M has a head normal form, then M has a minimal one.

The proof is very similar to the preceding one. In both cases, the minimal form is obtained by contracting head redexes until an expression of the desired form is reached.

Approximations : We still follow Wadsworth [13] and define the direct approximation $\phi(M)$ of an expression M by :

$$\begin{cases} \phi(\lambda x.M) = \lambda x.\phi(M) \\ \phi(xM_1M_2\dots M_n) = x(\phi(M_1))(\phi(M_2))\dots(\phi(M_n)) \\ \phi(\lambda x.M)NM_1M_2\dots M_n) = \Omega \end{cases}$$

where Ω is an extra constant. Basically, $\phi(M)$ is obtained from M by replacing all (outermost) redexes of M by Ω and substituting ΩM by Ω until normal form. If Ω is understood as "undefined", $\phi(M)$ is the information we have from M without contracting its redexes. There is a slight modification from the Wadsworth's definition, because we do not want to identify Ω and $\lambda x.\Omega$.

We define N as $N = \phi(\Lambda)$. Obviously, N is the set of expressions in ω-β normal forms. More precisely N is the minimal set containing :

$$\begin{cases} \Omega \\ \lambda x.a \\ xa_1a_2\dots a_n \end{cases}$$

if x is a variable and a, a_i are already in N. By considering Ω as a minimal element in N and extending by monotony, we get the following partial order $<$ in N :

$$\begin{cases} \Omega < a \\ \lambda x.a < \lambda x.b \qquad \text{if } a < b \\ xa_1a_2\dots a_n < xb_1b_2\dots b_n \qquad \text{if } a_i < b_i \text{ for } 1 \leq i \leq n \end{cases}$$

where a, b, a_i are expressions of N, x is a variable and $n \geq 0$. Here, we must take care of α-conversion and the order $<$ is, in fact, an order between equivalence classes defined on N by the α-interconvertibility. So, if $a \overset{*}{\underset{\alpha}{\to}} a'$ and $b \overset{*}{\underset{\alpha}{\to}} b'$, we have $a < b$ iff $a' < b'$. Moreover, we notice that $a<b$ iff there are M,N in Λ such that $\phi(M) = a$, $\phi(N) = b$ and $M \overset{*}{\to} N$.

Proposition 3 : The set N is a semi-lattice where every directed subset (*) is a lattice. More precisely :

1) N has a minimal element Ω.

2) for any pair a,b of elements in N, there exists a greatest lower bound $a \sqcap b$. (meet operation)

3) for any pair a,b of elements in N which are dominated by a common upper bound, there exists a least upper bound $a \sqcup b$. (join operation)

The proof is trivial, and obviously we can give the inductive definitions of $a \sqcup b$ and $a \sqcap b$ (up to some α-conversions as for the definition of < above) :

$$\begin{cases} (\lambda x.a) \sqcap (\lambda x.b) = \lambda x.(a \sqcap b) \\ (xa_1a_2\ldots a_n) \sqcap (xb_1b_2\ldots b_n) = x(a_1 \sqcap b_1)(a_2 \sqcap b_2)\ldots(a_n \sqcap b_n) \\ a \sqcap b = \Omega \quad \text{otherwise} \end{cases}$$

and

$$\begin{cases} \Omega \sqcup a = a \sqcup \Omega = a \\ (\lambda x.a) \sqcup (\lambda x.b) = \lambda x.(a \sqcup b) \\ (xa_1a_2\ldots a_n) \sqcup (xb_1b_2\ldots b_n) = x(a_1 \sqcup b_1)(a_2 \sqcup b_2)\ldots(a_n \sqcup b_n) \\ a \sqcup b \quad \text{is not defined otherwise} \end{cases}$$

where x is a variable and a,b,a_i,b_i are expressions of N and $n \geq 0$. The set N is also complete for the $\sqcap$ operation, i.e. each subset X of N has a greatest lower bound $\sqcap X$ in N. Moreover, the order < is well-founded in N and we have no infinite stricly decreasing chains in N.

To any expression M, Wadsworth [13] associates a set of approximations $A(M)$, which is the set of direct approximations of all expressions reducible from M :

$$A(M) = \{\phi(N) \mid M \overset{*}{\rightarrow} N\}$$

We briefly review some descriptive properties.

Proposition 4 : The set $A(M)$ of approximations of any λ-expression M is a sublattice of N (with the same meet and join operations than in N)

Proof : We only need to show that $\sqcap$ and $\sqcup$ are closed in $A(M)$. Suppose a,b are in $A(M)$.

(*) If (D,<) is a partial order structure, a directed subset X of D is such that for any a,b in X, there is c in X such that a<c, b<c. (Notion similar to ascending chains if D is denumerable). See Scott [10].

1) $a \sqcap b$ is in $A(M)$ by induction on the size $\|a \sqcap b\|$ of $a \sqcap b$. There are three cases :

1.1. If $a \sqcap b = \Omega$, then $a \neq \Omega$ and $b \neq \Omega$ is impossible, because of propositions 1 and 2. Therefore, $a = \Omega$ or $b = \Omega$ which implies $\phi(M) = \Omega$, and then $a \sqcap b$ is in $A(M)$.

1.2. If $a \sqcap b = \lambda x.c_1$, then $a = \lambda x.a_1$ and $b = \lambda x.b_1$. Hence, M has an abstraction form and, by proposition 1, a minimal one $\lambda x.M_0$. As a_1 and b_1 are in $A(M_0)$, we know by induction that $a_1 \sqcap b_1$ is in $A(M_0)$. Thus, $\lambda x.c_1 = \lambda x.(a_1 \sqcap b_1)$ is in $A(\lambda x.M_0)$, and $a \sqcap b$ is in $A(M)$.

1.3. If $a \sqcap b = xc_1c_2 \ldots c_n$, we have the same proof and we use proposition 2.

2) The Church-Rosser theorem shows the existence of c such that $a < c$ and $b < c$. Hence, $a \sqcup b$ is defined and a similar proof, based on an induction on the the size $\|a \sqcup b\|$, shows that $a \sqcup b$ is in $A(M)$. □

Proposition 5 : For any a in $A(M)$, there is a minimal λ-expression N_a such that :

i) $M \overset{*}{\rightarrow} N_a$

ii) $\phi(N_a) = a$

iii) for any N such that $M \overset{*}{\rightarrow} N$ and $a < \phi(N)$, then $N_a \overset{*}{\rightarrow} N$

Proof : by induction on the size of a

1) $a = \Omega$. Then $\phi(M) = \Omega = a$ and $N_a = M$

2) $a = \lambda x.a_1$. Then if $M \overset{*}{\rightarrow} N$ and $a < \phi(N)$, we have $N = \lambda x.N_1$, $\phi(N) = \lambda x.\phi(N_1)$ and $a_1 < \phi(N_1)$. By proposition 1, there is a minimum abstraction form $\lambda x.M_0$ of M. Hence $M_0 \overset{*}{\rightarrow} N_1$ and by induction there is N_{a_1} minimum for a_1 and reducible from M_0. Hence, if $N_a = \lambda x.N_{a_1}$, we have :

$$M \overset{*}{\rightarrow} \lambda x.M_0 \overset{*}{\rightarrow} N_a = \lambda x.N_{a_1} \overset{*}{\rightarrow} \lambda x.N_1 = N.$$

and $\phi(N_a) = \lambda x.\phi(N_{a_1}) = \lambda x.a_1 = a$.

3) $a = xa_1a_2 \ldots a_n$. Same proof but we need now proposition 2. □

Hence, generalizing propositions 1 and 2, we have a minimal expression N_a for any approximation a. We reach it by head reductions (leftmost outermost reductions) until a head normal form (if necessary) and repeating this process on arguments of the head normal form (when necessary). We notice also that if a,b are in some $A(M)$, then $a < b$ iff for any N', such that $M \overset{*}{\rightarrow} N'$ and

$\phi(N') = b$, there is N such that $M \overset{*}{\to} N$ and $\phi(N) = a$ and $N \overset{*}{\to} N'$.

<u>The interpretation domain</u> : In the set Λ, some λ-expressions have a finite set of approximations ; we call them expressions of finite information. The other expressions have an infinite information and, in order to be able to speak of their value, we will complete the set N of ω-β normal forms, by adding infinite points. Let $\bar{N}$ be the set of all directed subsets of N :

$$\bar{N} = \{ S \mid S \text{ directed}, S \subset N \}$$

We can extend the relation $<$ to $\bar{N}$ by defining for S and S' in $\bar{N}$:

$$S \subseteq S' \text{ iff } \forall a \in S. \exists b \in S'. a<b$$

In order to keep an ordering, we define a quotient set :

$$\hat{N} = \bar{N}/\equiv$$

where :

$$S \equiv S' \text{ iff } S \subseteq S' \subseteq S$$

Hence, if we note by $[S]$ the equivalence class of S in $\hat{N}$, we have in $\hat{N}$:

$$[S] \subseteq [S'] \text{ iff } S \subseteq S'$$

<u>Proposition 6</u> : The set $\hat{N}$ is a semi-lattice where every directed subset is a lattice. More precisely :

1) $\hat{N}$ has a minimal element $[\{\Omega\}]$
2) any pair of elements $[S],[S']$ in $\hat{N}$ has a greatest lower bound $[S] \sqcap [S']$
3) any pair of elements $[S],[S']$ in $\hat{N}$, which is dominated by a common upper bound, has a least upper bound $[S] \sqcap [S']$.

The proof is obvious and the definitions of $\sqcap$ and $\sqcup$ in $\hat{N}$ are given by :

$$[S] \sqcap [S'] = [\{ a \sqcap b \mid a \in S, b \in S' \}]$$
$$[S] \sqcup [S'] = [\{ a \sqcup b \mid a \in S, b \in S' \}]$$

But $\hat{N}$ has a richer structure. Using Scott's terminology (see for instance [10]), we have :

<u>Proposition 7</u> : The domain $\hat{N}$ is :

1) complete for directed subsets of $\hat{N}$,
2) continuous,
3) algebraic since $\hat{N}$ admits a denumerable basis of isolated elements $[\{a\}]$ where $a \in N$.

This means that every directed subset X of $\hat{N}$ has a least upper bound $\sqcup X$ and that each element of $\hat{N}$ is the least upper bound of the finite information points $[\{a\}]$ (where $a \in N$) which are below it. The proof follows from the construction of $\hat{N}$ and we skip it. The method we use for the completion of $\hat{N}$ is equivalent to the one of Vuillemin [12].

Interpretation of λ-expressions : We associate to any expression M an element in $\hat{N}$ by the following equation :

$$J(M) = [A(M)]$$

and we can thus induce a partial preorder on Λ defined by : $M \subseteq M'$ iff $J(M) \subseteq J(M')$. By the definition of $\hat{N}$, we have :

$$M \subseteq M' \text{ iff } \forall N \text{ s.t. } M \overset{*}{\to} N.\ \exists N' \text{ s.t. } M' \overset{*}{\to} N' \text{ and } \phi(N) < \phi(N').$$

We write $M \equiv M'$ for $M \subseteq M' \subseteq M$ and we expect the usual properties for our interpretation J. Moreover, if X is a directed subset of λ-expressions, $\sqcup X$ means $\sqcup J(X)$.

Theorem 1 : The β-rule of conversion is valid in J, i.e. if $M \overset{*}{\to} M'$ then $M \equiv M'$.

Proof : Since $M \overset{*}{\to} M'$, we have $A(M') \subset A(M)$ and then $M' \subseteq M$. Now, suppose $M \overset{*}{\to} N$; then as $M \overset{*}{\to} M'$, we know, by the Church-Rosser theorem, the existence of an N' such that $M \overset{*}{\to} N'$ and $M' \to N'$, and hence $\phi(N) < \phi(N')$. So we have $M \subseteq M'$. □

Let the set Λ be now extended by allowing the constant Ω whenever a free variable is possible. So Λ is now the set of λ-Ω expressions and we consider not only β-reductions but also an ω-rule of conversion defined by replacing any subexpression of the form ΩM by Ω. We note this kind of reduction $N \underset{\omega}{\overset{*}{\to}} N'$. Let $C[\ ]$ denote any context (see [8]), i.e a λ-Ω expression with one subexpression missing, and $C[M]$ be the corresponding expression where M stands at the place of the previously missing subexpression.

Proposition 8 : $C[\Omega] \subseteq C[M]$ for any context M and any expression M.

Proof : The set of expressions reducible from $C[\Omega]$ is isomorphic to a subset of the one reducible from $C[M]$. Moreover, $\phi(N[x\backslash\Omega] < \phi(N[x\backslash M])$ for any λ-Ω expressions M,N. Hence, $C[\Omega] \subseteq C[M]$ by definition of $\subseteq$. □

Theorem 2 : The ω-rule of conversion is valid in J, i.e. if $M \underset{\omega}{\overset{*}{\to}} M'$ then $M \equiv M'$.

Proof : By the above proposition, we already know that $M' \subseteq M$. Now, suppose $M \overset{*}{\rightarrow} N$, we can show easily (by an induction on the pair $< \ell, \ell' >$ if ℓ and ℓ' are the length of the reductions $M \overset{*}{\rightarrow} N$ and $M \overset{*}{\underset{\omega}{\rightarrow}} M'$) the existence of N' such that $N \overset{*}{\underset{\omega}{\rightarrow}} N'$ and $M' \overset{*}{\rightarrow} N'$. Hence $\phi(N) = \phi(N')$ and then $M \subseteq M'$. □

We turn now to the main point of this paper, i.e. we show that, for any context C[ ], we have $C[M] \subseteq C[M']$ if $M \subseteq M'$. So, using the definition of $\subseteq$, we need to show that if $C[M] \overset{*}{\rightarrow} N$, then there is an N' such that $C[M'] \overset{*}{\rightarrow} N'$ and $\phi(N) < \phi(N')$. But all we know is that, for any approximation of M, we can have a better one for M'. Therefore, in order to compare C[M] and C[M'], we try to point out the approximation of M needed by any reduction from C[M] to some N. That is the Welch' conjecture about inside-out reductions ([14]), which we prove later. Using the Welch's notations, let $C[M] \underset{M}{\overset{*}{\not\rightarrow}} N$ designate any reduction :

$$C[M] = M_0 \overset{R_1}{\rightarrow} M_1 \overset{R_2}{\rightarrow} M_2 \overset{R_3}{\rightarrow} \cdots \overset{R_n}{\rightarrow} M_n = N$$

where, for all i ($1 \leq i \leq n$), the redex R_i contracted between M_{i-1} and M_i is not a residual of a redex internal to the subexpression M in M_0. Similarly, if F is a set of redexes of M_0, we write $M_0 \underset{F}{\overset{*}{\not\rightarrow}} M_n$ if none of the R_i is a residual of a redex of F. Hence, if F is the set of all redexes of M, we have $C[M] \underset{M}{\overset{*}{\not\rightarrow}} N$ iff $C[M] \underset{F}{\overset{*}{\not\rightarrow}} N$. Moreover, let $M[F \backslash \Omega]$ denote the substitution in M of all redexes of F by the constant Ω. Now, we can state what we want. (The proof is postponed until the next section.)

Proposition 9 : For any context C[ ] and any expression M, if $C[M] \overset{*}{\rightarrow} N$, then there are expressions M' and N', such that $M \overset{*}{\rightarrow} M'$, $N \overset{*}{\rightarrow} N'$ and $C[M'] \underset{M}{\overset{*}{\not\rightarrow}} N'$.

Lemma 1 : Given a set of redexes F in an expression M, if $M \underset{F}{\overset{*}{\rightarrow}} M'$ and iff F' is the set of the residuals of the redexes of F in M', then $M[F \backslash \Omega] \overset{*}{\rightarrow} M'[F' \backslash \Omega]$.

Proof : by induction on the length of the reduction from M to M', which is of the form $M \overset{R}{\rightarrow} M_1 \underset{F_1}{\overset{*}{\not\rightarrow}} M'$, where the redex R, first contracted, is not one of the redexes of F and F_1 is the set of residuals of the redexes of F in M_1. Then, depending on the relative position of R and F, we obviously have $M[F \backslash \Omega] \overset{*}{\rightarrow} M_1 [F_1 \backslash \Omega]$. □

Lemma 2 : If a,b are in the set N of ω-β normal forms and such that a<b, then $C[a] \sqsubseteq C[b]$ for any context C[ ].

Proof : Corollary of proposition 8, since, if a<b, then a matches b except for some Ω's. □

Lemma 3 : For any context C[ ] and expression M :

$$C[M] \equiv \sqcup \{ C[a] \mid a \in A(M) \}$$

Proof : Let $X = \{C[a] \mid a \in A(M)\}$. By lemma 2, as $A(M)$ is a directed set, X also is directed and $\sqcup X$ exists, since $\hat{N}$ is complete for directed subsets.

First, if a is in $A(M)$, there is an expression N such that $M \overset{*}{\rightarrow} N$ and $a = \phi(N)$. So, we have, by proposition 8, $C[a] \sqsubseteq C[N]$ since a matches N except in some Ω's. But, as $M \overset{*}{\rightarrow} N$, we have too $C[M] \overset{*}{\rightarrow} C[N]$ and $C[M] \equiv C[N]$ by theorem 1. Hence, we get $C[a] \sqsubseteq C[M]$ for any a in $A(M)$. Therefore C[M] is an upper bound of X and $\sqcup X \sqsubseteq C[M]$

Conversely, if $C[M] \overset{*}{\rightarrow} N$, there are M' and N' such that $M \overset{*}{\rightarrow} M'$, $N \overset{*}{\rightarrow} N'$ and $C[M'] \overset{*}{\underset{M'}{\not\rightarrow}} N'$ (by proposition 9). Let F be the set of all redexes of M' ; we have $C[M'] \overset{*}{\underset{F}{\not\rightarrow}} N'$. By lemma 1, if F' is the set of the residuals of the redexes of F in N', we have $C[M'][F\backslash\Omega] \overset{*}{\rightarrow} N'[F'\backslash\Omega]$. Moreover, $C[M'][F\backslash\Omega] \underset{\omega}{\overset{*}{\rightarrow}} C[\phi(M')]$ and since ω and β-conversions are valid, we get :

$$\phi(N)<\phi(N') = \phi(N'[F'\backslash\Omega]) \sqsubseteq N'[F'\backslash\Omega] \equiv C[M'][F\backslash\Omega] \equiv C[\phi(M')]$$

And then, for any N such that $C[M] \overset{*}{\rightarrow} N$, there is an M' such that $M \overset{*}{\rightarrow} M'$ and $\phi(N) \sqsubseteq C[\phi(M')]$. Since $C[M] \equiv \sqcup \{ a \mid a \in A(C[M]) \}$, we have $C[M] \sqsubseteq \sqcup X$. □

Theorem 3 : If $M \sqsubseteq M'$, then $C[M] \sqsubseteq C[M']$ for any context C[ ].

Proof : Since $M \sqsubseteq M'$, for any a in $A(M)$, there is a b in $A(M')$ such that a<b. Hence, by lemma 2, $C[a] \sqsubseteq C[b]$. Since $\hat{N}$ is complete, we get :

$$\sqcup \{ C[a] \mid a \in A(M) \} \sqsubseteq \sqcup \{ C[b] \mid b \in A(M') \}$$

and, by lemma 3, $C[M] \sqsubseteq C[M']$. □

Inside-out reductions : In order to show proposition 9, we follow Welch [14] and define an inside-out reduction as any reduction :

$$M = M_0 \overset{R_1}{\to} M_1 \overset{R_2}{\to} M_2 \overset{R_3}{\to} \cdots \overset{R_n}{\to} M_n = M'$$

where, for all i,j such that i<j and $1 \le i \le n-1$, $2 \le j \le n$, the redex R_j (contracted between M_{j-1} and M_j) is not a residual of a redex internal to R_i in M_{i-1}. Let $M \xrightarrow[i.o.]{*} M'$ be a notation for any such inside-out reduction from M to M'. Welch conjectured the completeness of inside-out reductions, i.e. if $M \overset{*}{\to} N$, then there exists N' such that $N \overset{*}{\to} N'$ and $M \xrightarrow[i.o.]{*} N'$. So, as pointed out by Welch :

Proposition 10 : If inside-out reductions are complete, then proposition 9 is true.

Proof : Let $C[M] \overset{*}{\to} N_1$, there is N such that $N_1 \overset{*}{\to} N$ and $C[M] \xrightarrow[i.o.]{*} N$, as inside-out reductions are complete. Let us prove, by induction on the size of the context C[ ], that if $C[M] \xrightarrow[i.o.]{*} N$, there is an M' such that $M \to M'$ and $C[M'] \overset{*}{\underset{M'}{\not\nearrow}} N$.

1) If C[ ] = [ ], then C[M] = M and M' = C[M'] = N.

2) If C[ ] = $\lambda x.C_1$[ ], the induction works easily.

3) If C[ ] = $M_1 C_1$[ ], as the reduction $C[M] \xrightarrow[i.o.]{*} N$ is inside-out, we have $M_1 \xrightarrow[i.o.]{*} M'_1$, $C_1[M] \xrightarrow[i.o.]{*} N'$ and $M'_1 N' \overset{*}{\underset{N'}{\not\nearrow}} N$. Hence, by induction there is M' such that $M \overset{*}{\to} M'$ and $C_1[M'] \overset{*}{\underset{M'}{\not\nearrow}} N'$. Thus $M \overset{*}{\to} M'$ and $C[M'] = M_1 C_1[M'] \overset{*}{\underset{M'}{\not\nearrow}} N$.

4) If C[ ] = C_1[ ] M_1, we have the same proof. □

A labelled λ-calculus : The problem is now to keep track of the redexes contracted in some reduction $M \overset{*}{\to} N$ in order to be able to reorder them in an inside-out way, and to show the inside-out completeness. We do this by introducing a new set of λ-expressions (Λ') defined on a set of labels L as follows. Let L_0 = { a,b,c ... } be an infinite set of letters. We consider the set L of all strings of characters formed on L_0, with any level of overlining and underlining. So expressions of L are :

$$\begin{cases} a & \text{if } a \in L_0 \\ \alpha\beta & \text{if } \alpha,\beta \in L \\ \overline{\alpha} & \text{"} \quad \text{"} \\ \underline{\alpha} & \text{"} \quad \text{"} \end{cases}$$

and expressions of Λ' are :

$$\begin{cases} x^{\alpha} & \text{if} \quad \alpha \in L \quad \text{and} \quad x \in V \\ (\lambda x.M)^{\alpha} & \text{if} \quad " \quad \text{and} \quad M,N \in \Lambda' \\ (MN)^{\alpha} & \text{if} \quad " \quad " \quad " \end{cases}$$

Thus, the labelled λ-expressions are like usual λ-expressions except that every subexpression has an arbitrary label. This λ-calculus is a generalization of the one of Wadsworth [13] since, instead of considering integers as exponents, we have strings of characters. For any labël α and expression M of Λ', we define $\alpha.M$ as :

$$\begin{cases} \alpha.x^{\beta} = x^{\alpha\beta} \\ \alpha.(\lambda x.M)^{\beta} = (\lambda x.M)^{\alpha\beta} \\ \alpha.(MN)^{\beta} = (MN)^{\alpha\beta} \end{cases}$$

and the substitution operation is defined by :

$$\begin{cases} x^{\alpha}[x\backslash N] & = \alpha_{\circ}N \\ y^{\alpha}[x\backslash N] & = y^{\alpha} \\ (\lambda y.M)^{\alpha}[x\backslash N] & = (\lambda y.M[x\backslash N])^{\alpha} \\ (MM')^{\alpha}[x\backslash N] & = (M[x\backslash N]\ M'[x\ N])^{\alpha} \end{cases}$$

where we forget the difficulties due to α-conversion. Then the β-rule is defined (by monotony) from :

$$((\lambda x.M)^{\alpha} N)^{\beta} \rightarrow \beta\bar{\alpha}.M[x\backslash\underline{\alpha}_{\circ}N]$$

(We do not care for the precedences between the . and substitution operators because they commute). Furthermore we will allow this reduction iff some predicate $P(\alpha,\beta)$ is verified. So, for instance, using a graph notation for λ-expressions (see Morris [8] where nodes ⓛ and ⓨ corresponds to abstraction and application), we have figure 1 if we suppose $P(\alpha,\beta)$ always true. In fact, we can restrict our attention to λ-expressions labelled by a set L' of labels defined as containing :

$$\begin{cases} a & \text{if} \quad a \in L_0 \\ \alpha\bar{\beta}\gamma & \text{if} \quad \alpha,\beta,\gamma \in L' \\ \alpha\underline{\beta}\gamma & \text{if} \quad " \quad " \end{cases}$$

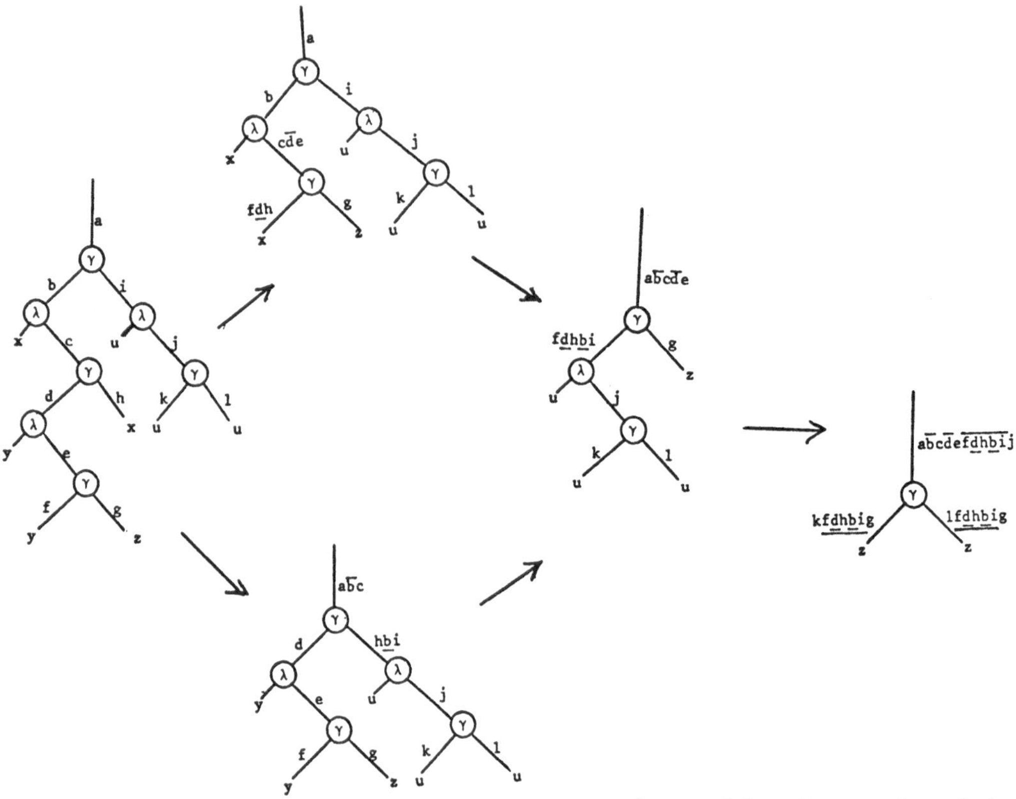

figure 1: Reductions from $((\lambda x.(\lambda y.(y^f z^g)^e)^d x^h)^c)^b(u.(u^k u^e)^j)^i)^a$ and it is clear that expressions labelled by L' keep their labels in L' after some β-reductions. We remark too that other λ-calculus languages are obtainable from this one by some homomorphism : for instance Wadsworth's typed λ-calculus and Morris' definition of descendants. Let the heigth h(α) of a label α of L be defined by :

$$\begin{cases} h(a) = o & \text{if } a \in L_0 \\ h(\alpha\beta) = \lceil h(\alpha), h(\beta)\rceil & \text{if } \alpha,\beta \in L \\ h(\underline{\alpha}) = h(\bar{\alpha}) = 1 + h(\alpha) & \text{if } \alpha \in L. \end{cases}$$

and let the degree of a redex be the label of its abstraction part. Hence, $\text{degree}(((\lambda x.M)^\alpha N)^\beta) = \alpha$.

Proposition 1' : The residuals of a redex R have the same degree as R.

Proof : Suppose $M \overset{R}{\rightarrow} N$ and R is a redex, in M, of the form $R = ((\lambda x.P)^\alpha Q)^\beta$. If $S = ((\lambda y.T)^\gamma U)^\delta$ is another redex of M, we show by cases that residual(s) of S in N have the same degree γ than S in M.

1) If R and S are 2 disjoint expressions, it is obvious.
2) If S is in $\mathcal{R}$, then S is in P or Q and the contraction of R may have only an effect on the external label δ of S.
3) If R is in S, then R is in T or U and the contraction of R has no effect on the degree α of S. □

<u>Proposition 2'</u> : If $P(\alpha,\beta)$ implies $P(\alpha,\gamma\beta)$ for any labels α,β,γ of L, then the β-rule of the labelled calculus is Church-Rosser.

<u>Proposition 3'</u> : If

1) $P(\alpha,\beta)$ implies $P(\alpha,\gamma\beta)$ for any labels α,β,γ of L
2) the set $\{ h(\alpha) \mid P(\alpha,\beta) \text{ is true} \}$ is bounded

then any labelled λ-expression strongly normalizes (i.e. any reduction in this labelled calculus has a finite length).

The proofs of these propositions are given in the appendix, by the usual techniques. We go back to the inside-out completeness and we will use letters as M,N to designate expressions of Λ and U,V for labelled λ-expressions of Λ'.

<u>Theorem 4</u> : If $M \overset{*}{\rightarrow} N$, then there is an expression N' such that $M \xrightarrow[\text{i.o.}]{*} N'$ and $N \overset{*}{\rightarrow} N'$.

<u>Proof</u> : Let U be the labelled λ-expression obtained from M by labelling all the subexpressions of M with a different letter of L_0. We can associate to the reduction $M \overset{*}{\rightarrow} N$, an isomorphic labelled reduction $U \overset{*}{\rightarrow} V$. More precisely, this reduction can be written :

$$U = U_0 \xrightarrow{R_1} U_1 \xrightarrow{R_2} U_2 \xrightarrow{R_3} \cdots \xrightarrow{R_n} U_n = V$$

Let us now consider the predicate $P(\alpha,\beta)$ defined on labels by :

$$P(\alpha,\beta) \text{ is true iff } \alpha = \text{degree } (R_i) \text{ for some } i \ (1 \le i \le n)$$

The two assumptions of proposition 3' are verified and, hence, U strongly normalizes. Let V' be the normal form of V, then $V \overset{*}{\rightarrow} V'$, because the Church-Rosser assumption is true and for instance, any innermost reduction reaches the normal form V'. Let

$$U = V_0 \xrightarrow{S_1} V_1 \xrightarrow{S_2} V_2 \xrightarrow{S_3} \cdots \xrightarrow{S_m} V_m = V'$$

be such an innermost reduction. (We then have, for all i, degree (S_i) = degree (R_j) for some j between 1 and n). We claim that this reduction is inside-out. Suppose i<j for some i,j between 1 and m and suppose S_j is a residual of a redex S'_j internal to S_i in V_{i-1}. By proposition 1', we have degree (S'_j) = degree (S_j) and then, as the predicate P is true for S_j, P is also true for S'_j. The reduction from U to V' is thus not an innermost reduction and we have a contradiction. Let N' be the λ-expression obtained by erasing the labels of V'. As an isomorphic reduction of Λ corresponds to any labelled reduction, we have $N \overset{*}{\rightarrow} N'$ and $M \xrightarrow[i.o.]{*} N'$. □

In fact, with the same method, we have, if $M \overset{*}{\rightarrow} M'$ and $M \overset{*}{\rightarrow} M''$, the existence of an N such that $M' \xrightarrow[i.o.]{*} N$ and $M'' \xrightarrow[i.o.]{*} N$.

Conclusion: The interpretation $\langle \mathcal{I},\hat{\mathcal{N}}\rangle$,although strongly inspired by Scott's theory of computation,is purely algebraic.Here,we do not have a definition of application as in Scott [10,11] or Welch [15] .But with the help of the labelled calculus, any expression can be considered as the limit of expressions having a normal form.If we think of λ-expressions as programs, the interpretation $\langle \mathcal{I},\hat{\mathcal{N}}\rangle$ seems to be the minimal one to consider.Thus we expect that $\langle \mathcal{I},\hat{\mathcal{N}}\rangle$ is some kind of free interpretation.This is proved by Welch for "continuous semantics",i.e. roughly speaking for interpretations where the Wadsworth theorem is true.Welch did it for his model but his interpretation seems to be equivalent to the one we used here.Hyland [4],who independently considered also the same interpretation, proved that there is an extensional equivalence relation corresponding to equality in $\mathcal{I}$.Furthermore, he showed for any λ-expressions M,M' that $M \underline{\subseteq} M'$ iff $M \subseteq_{P\omega} M'$ where Pω is Scott's model [11] .Another question is to take into account extensionality and build an algebraic interpretation where the η-rule is valid.This is done by Hyland [4] Finally,the labelled λ-calculus seems interesting in itself [6],since we can capture the history of any reduction in the labels.

Aknowledgements: to D.Park and P.Welch without whom I would not have thought of the inside-out reductions; to G.Plotkin,D. van Daalen and R. de Vrijer without whom I could not have done the appendix; to G.Kahn and G.Huet for their constant help.

Appendix 1: The Church-Rosser property in the labelled calculus (by the Taït-Martin Löf method)

Let $C(M)$ be defined by:

$$\begin{cases} C(x^{\alpha}) = x^{\alpha} \\ C((\lambda x.M)^{\alpha}) = (\lambda x.C(M))^{\alpha} \\ C(((\lambda x.M)^{\alpha}N)^{\beta}) = \beta\overline{\alpha}.C(M)[x\backslash\underline{\alpha}.C(N)] & \text{if } P(\alpha,\beta) \\ C((MN)^{\alpha}) = (C(M)C(N))^{\alpha} & \text{otherwise} \end{cases}$$

Let $M \to M'$ denote a parallel step of reduction and be defined by the following inference rules and axiom:

-I- $\quad x^{\alpha} \to x^{\alpha}$

-II- $\quad \dfrac{M \to M'}{(\lambda x.M)^{\alpha} \to (\lambda x.M')^{\alpha}}$

-III- $\quad \dfrac{M \to M', N \to N'}{(MN)^{\alpha} \to (M'N')^{\alpha}}$

-IV- $\quad \dfrac{M \to M', N \to N'}{((\lambda x.M)^{\alpha}N)^{\beta} \to \beta\overline{\alpha}.M'[x\backslash\underline{\alpha}.N']}$ $\quad$ if $P(\alpha,\beta)$

Moreover, we suppose $P(\alpha,\beta)$ implies $P(\alpha,\gamma\beta)$ for any labels α,β,γ.

First, we notice that the associativity of concatenation implies:

$$\alpha.(\beta.M) = \alpha\beta.M$$

Lemma 1: $\quad \alpha.(M[x\backslash N]) = (\alpha.M)[x\backslash N]$

Proof: by cases on M. The only problem is when $M = x^{\beta}$. Then the associativity of concatenation gives the answer. □

Lemma 2: If $x \neq y$ and x is not free in N', then:

$$M[x\backslash N][y\backslash N'] = M[y\backslash N'][x\backslash N[y\backslash N']]$$

Proof: by induction on the size of M. The only problem is when $M = x^{\alpha}$. Then we apply lemma 1. □

Lemma 3: If $M \to M'$, then $\alpha.M \to \alpha.M'$

Proof: by cases on the rule or axiom used for $M \to M'$. The only interesting case is when:

$$M = ((\lambda x.M_1)^{\gamma}M_2)^{\beta} \to M' = \beta\overline{\gamma}.M'_1[x\backslash\underline{\gamma}.M'_2]$$

with $M_1 \to M'_1$, $M_2 \to M'_2$ and $P(\gamma,\beta)$. Then:

$$\alpha.M = ((\lambda x.M_1)^{\gamma}M_2)^{\alpha\beta} \quad \text{and} \quad \alpha.M' = \alpha.(\beta\overline{\gamma}.M'_1[x\backslash\underline{\gamma}.M'_2])$$

As $P(\gamma,\beta)$ implies $P(\gamma,\alpha\beta)$, we have by rule IV:

$$\alpha.M = ((\lambda x.M_1)^{\gamma}M_2)^{\alpha\beta} \rightarrow \alpha\beta\bar{\gamma}.M_1'[x\backslash\underline{\gamma}.M_2']$$

and by the associativity of concatenation $\alpha.M \rightarrow \alpha.M'$.□

<u>Lemma 4</u>: If $M \rightarrow M'$ and $N \rightarrow N'$,then $M[x\backslash N] \rightarrow M'[x\backslash N']$

<u>Proof</u>:by induction on the size of M.

1) M is a variable:

a) $M = x^{\alpha} = M'$ .Then we use lemma 3.

b) $M = y^{\alpha} = M'$ .Then obvious by axiom I.

2) M is not a variable and we have several cases according to the rule used for $M \rightarrow M'$ .The only interesting one is when:

$$M = ((\lambda y.M_1)^{\alpha}M_2)^{\beta} \rightarrow M' = \beta\bar{\alpha}.M_1'[y\backslash\underline{\alpha}.M_2']$$

with $M_1 \rightarrow M_1'$, $M_2 \rightarrow M_2'$ and $P(\alpha,\beta)$.Then,ignoring α-conversions,we have:

$$\begin{aligned}
M[x\backslash N] &= ((\lambda y.M_1[x\backslash N])^{\alpha}M_2[x\backslash N])^{\beta} \\
M'[x\backslash N'] &= (\beta\bar{\alpha}.M_1'[y\backslash\underline{\alpha}.M_2'])[x\backslash N'] \\
&= \beta\bar{\alpha}.(M_1'[y\backslash\underline{\alpha}.M_2'][x\backslash N']) && \text{(by lemma 1)} \\
&= \beta\bar{\alpha}.(M_1'[x\backslash N'][y\backslash(\underline{\alpha}.M_2')[x\backslash N']]) && \text{(by lemma 2)} \\
&= \beta\bar{\alpha}.(M_1'[x\backslash N'][y\backslash\underline{\alpha}.(M_2'[x\backslash N'])]) && \text{(by lemma 1)}
\end{aligned}$$

By induction,we know that $M_1[x\backslash N] \rightarrow M_1'[x\backslash N']$ and $M_2[x\backslash N] \rightarrow M_2'[x\backslash N']$,and using rule IV,we have $M[x\backslash N] \rightarrow M'[x\backslash N']$.□

<u>Lemma 5</u>: If $M \rightarrow M'$,then $M' \rightarrow C(M)$.

<u>Proof</u>:by induction on the size of M .There are two interesting cases:

1)When:

$$M = ((\lambda x.M_1)^{\alpha}M_2)^{\beta} \rightarrow M' = (M_3'M_2')^{\beta}$$

with $(\lambda x.M_1)^{\alpha} \rightarrow M_3'$, $M_2 \rightarrow M_2'$ and $P(\alpha,\beta)$ .Then it is clear that $M_3' = (\lambda x.M_1')^{\alpha}$ and $M_1 \rightarrow M_1'$ .Hence we have by induction $M_1' \rightarrow C(M_1)$ and $M_2' \rightarrow C(M_2)$ .Then,using rule IV:

$$M' = ((\lambda x.M_1')^{\alpha}M_2')^{\beta} \rightarrow \beta\bar{\alpha}.C(M_1)[x\backslash\underline{\alpha}.C(M_2)] = C(M)$$

2)When:

$$M = ((\lambda x.M_1)^{\alpha}M_2)^{\beta} \rightarrow M' = \beta\bar{\alpha}.M_1'[x\backslash\underline{\alpha}.M_2']$$

with $M_1 \rightarrow M_1'$, $M_2 \rightarrow M_2'$ and $P(\alpha,\beta)$ .Then we have by induction $M_1' \rightarrow C(M_1)$ and $M_2' \rightarrow C(M_2)$ .Hence,by lemma 3: $\underline{\alpha}.M_2' \rightarrow \underline{\alpha}.C(M_2)$,and by lemma 4 and 3:

$$M' \rightarrow \beta\bar{\alpha}.C(M_1)[x\backslash\underline{\alpha}.C(M_2)] = C(M) \quad .\square$$

<u>Lemma 6</u>: If $M \rightarrow M'$ and $M \rightarrow M''$, then there is an N such that $M' \rightarrow N$ and $M'' \rightarrow N$.

Proof: We take $N = C(M)$ and use lemma 5. □

Proposition: If $M \to M'$ and $M \to M''$, then there is an N such that $M' \to N$ and $M'' \to N$. □

Proof: by induction on the sum of length of the reductions $M \to M'$ and $M \to M''$. □

Appendix 2: Strong normalization in the labelled λ-calculus by a method due to D. van Daalen. We suppose:

(1) $P(\alpha,\beta)$ implies $P(\alpha,\gamma\beta)$

(2) $\{h(\alpha) \mid P(\alpha,\beta) \text{ is true}\}$ is bounded

Hence, we have the Church-Rosser property. Let write $\tau(N)$ for the external label of N. So $\tau(x^\alpha) = \tau((\lambda x.M)^\alpha) = \tau((MN)^\alpha) = \alpha$ and we call SN the set of strongly normalizable labelled λ-expressions.

Lemma 1: If $(\dots((MN_1)^{\beta_1}N_2)^{\beta_2}\dots N_n)^{\beta_n} \overset{*}{\to} (\lambda x.N)^\alpha$, then we have $h(\tau(M)) \le h(\alpha)$.

Proof: by induction on n. If $n = 0$, this is clearly true. Otherwise, we must have:

$$(\dots((MN_1)^{\beta_1}N_2)^{\beta_2}\dots N_{n-1})^{\beta_{n-1}} \overset{*}{\to} (\lambda y.P)^\gamma$$

and:

$$((\lambda y.P)^\gamma N_n)^{\beta_n} \to \beta_n\overline{\gamma}.P[y\backslash \underline{\gamma}.N_n] \overset{*}{\to} (\lambda x.N)^\alpha$$

Hence, we get $h(\tau(M)) \le h(\gamma)$ by induction. We also have:

$$h(\gamma) < h(\overline{\gamma}) \le h(\tau(\beta_n\overline{\gamma}.P[y\backslash \underline{\gamma}.N_n])) \le h(\alpha) \quad .□$$

Lemma 2: If $M[x\backslash N] \overset{*}{\to} (\lambda y.P)^\alpha$, we only have two cases:

1) $M \overset{*}{\to} (\lambda y.M')^\alpha$ and $M'[x\backslash N] \overset{*}{\to} P$

or

2) $M \overset{*}{\to} M' = (\dots((x^\beta M_1)^{\beta_1}M_2)^{\beta_2}\dots M_n)^{\beta_n}$ and $M'[x\backslash N] \overset{*}{\to} (\lambda y.P)^\alpha$

Proof: application of propositions 1 and 2. □

Lemma 3: If M,N are in SN, then $M[x\backslash N]$ is in SN.

Proof: Let m be the upper bound of the set $\{h(\alpha) \mid P(\alpha,\beta) \text{ is true}\}$, which exists by assumption (2), and $prof(M)$ be the maximal length of reductions starting from M in SN. We do an induction on the triple:

$$< h(\tau(N)) - m, prof(M), \|M\| >$$

where $\|M\|$ is the size of M.

The only interesting case is when :

$$M = (M_1M_2)^{\alpha} \quad \text{and} \quad M_1[x\backslash N] \overset{*}{\to} (\lambda y.P_1)^{\alpha_1}$$

We know by induction that $M_1[x\backslash N]$ and $M_2[x\backslash N]$ are in *SN* ,but we wonder if $((\lambda y.P_1)^{\alpha_1}M_2[x\backslash N])^{\alpha}$ is in *SN* ,i.e. if $M' = \alpha\overline{\alpha_1}.P_1[y\backslash \underline{\alpha_1}.M_2[x\backslash N]]$ is in *SN* .Then lemma 2 tells us there are two subcases:

1) $M_1 \overset{*}{\to} (\lambda y.M_1')^{\alpha_1}$ and $M_1'[x\backslash N] \overset{*}{\to} P_1$ .Then by lemmas 2,3,4 of the Church-Rosser proof,we get:

$$\alpha\overline{\alpha_1}.M_1'[y\backslash \underline{\alpha_1}.M_2][x\backslash N] = \alpha\overline{\alpha_1}.M_1'[x\backslash N][y\backslash \underline{\alpha_1}.M_2[x\backslash N]]$$
$$\overset{*}{\to} \alpha\overline{\alpha_1}.P_1[y\backslash \underline{\alpha_1}.M_2[x\backslash N]] = M'$$

But $M = (M_1M_2)^{\alpha} \overset{*}{\to} ((\lambda y.M_1')^{\alpha_1}M_2)^{\alpha} \to \alpha\overline{\alpha_1}.M_1'[x\backslash \underline{\alpha_1}.M_2]$ .Hence:

$$\text{prof}(\alpha\overline{\alpha_1}.M_1'[x\backslash \underline{\alpha_1}.M_2]) < \text{prof}(M)$$

and by induction M' is in *SN*.

2) $M_1 \overset{*}{\to} Q_1 = (\ldots((x^{\beta}N_1)^{\beta_1}N_2)^{\beta_2}\ldots N_n)^{\beta_n}$ and $Q_1[x\backslash N] \overset{*}{\to} (\lambda y.P_1)^{\alpha_1}$ .As $M_1[x\backslash N]$ is in *SN* and $M_1[x\backslash N] \overset{*}{\to} Q_1[x\backslash N]$,we have P_1 in *SN* .Moreover:

$$Q_1[x\backslash N] = (\ldots(((\beta.N)N_1')^{\beta_1}N_2')^{\beta_2}\ldots N_n')^{\beta_n}$$

where $N_i' = N_i[x\backslash N]$ for all i.Hence,using lemma 1:

$$h(\tau(N)) \le h(\tau(\beta.N)) \le h(\alpha_1) < h(\underline{\alpha_1}) \le h(\tau(\underline{\alpha_1}.M_2[x\backslash N]))$$

and by induction M' is in *SN* .□

<u>Proposition</u>: If *P* verifies assumptions (1) and (2), every expression M strongly normalizes.

<u>Proof</u>:by induction on the size of M and application of lemma 3.□

<u>References</u>:

[1] Barendregt H.P.,"Some extensional term models for combinatory logics and lambda-calculi",PhD thesis,Utrecht(1971)

[2] Church A.,"The calculi of lambda conversion",Annals of Math.Studies,n°6, Princeton(1941)

[3] Curry H.B.,Feys R.,"Combinatory logic",Vol 1,North Holland(1958)

[4] Hyland M.,to appear in the Rome conference(1975)

[5] Lévy J-J.,"Réductions sures dans le lambda calcul",3° cycle,Univ of Paris (1974)

[6] Lévy J-J.,"Réductions sures et optimales dans le lambda calcul",to appear

[7] Milner R.,"Processes;A model of computing agents",Univ. of Edinburgh,Internal report(1973)

[8] Morris J.H.,"Lambda calculus models of programming languages",PhD thesis, MIT,Cambridge(1968)

[9] Nivat M.,"Sur l'interprétation des shémas de programmes monadiques",Rapport IRIA-LABORIA (1972)

[10] Scott D.,"Continuous lattices",Technical monograph,PRG-7,Oxford(1971)

[11] Scott D.,"Data types as lattices",to appear in Springer Lectures Notes

[12] Vuillemin J.,"Syntaxe,sémantique et axiomatique d'un langage de programmation simple",Thèse d'état,Univ. Paris 7(1974)

[13] Wadsworth C.P.,"Semantics and pragmatics of the lambda calculus",PhD thesis, Oxford (1971)

[14] Welch P.,"Another problem of lambda calculus",Univ. of Kent,Private communication (1973)

[15] Welch P.,to appear in the Rome conference (1975)

LES MODELES INFORMATIQUES DES λ-CALCULS [1]

Louis NOLIN
Université Paris VII
U.E.R. de Mathématiques
2, place Jussieu
75230 - PARIS - France -

Nous nous proposons de dire ce que peut être l'interprétation d'un λ-calcul dans une collection d'algorithmes, notion que nous avons déjà définie [2], et quelles en sont les propriétés essentielles.

Les résultats sont présentés dans la première partie de ce mémoire et démontrés dans la troisième ; nous rappelons dans la seconde partie la définition des collections d'algorithmes et certaines de leurs propriétés qui sont utilisées par la suite ; nous terminons par quelques commentaires et un hommage aux auteurs dont l'oeuvre nous a inspiré ou qui nous paraissent avoir obtenu des résultats voisins des nôtres.

1. RESULTATS.

Une collection d'algorithmes, $\mathcal{A}$, est de l'un des trois types suivants :

1er type. Pas d'ensembles de base : les seuls algorithmes sont T (l'ensemble total), $\emptyset$, les algorithmes propres tels

$$\underline{I} = \cap\{FXX/X \in \mathcal{A}\},$$

$$P^1_2 = \cap\{FX(FYX)/X,\ Y \in \mathcal{A}\},$$

$$\underline{S} = \cap\{FX(FY(FZ((X[Z])\ [Y[Z]])))/X,\ Y,\ Z \in \mathcal{A}\}.$$

Tous ces algorithmes sont atomiques.

2ème type. Tous les ensembles de base sont atomiques ; ce sont, par exemple, les singletons {VRAI}, {FAUX}, {i} pour $i \in \mathbb{N}$; en plus des algorithmes précédents, $\mathcal{A}$ contient des algorithmes propres (toujours atomiques) tels que :

---

(1) Ce travail a été grandement facilité par l'aide que nous a apportée le Centre National de la Recherche Scientifique (A T P 7110).

$\cap \{F\{i\}\{i+1\}/i \in \mathbb{N}\}$,

$\cap \{F\{i\}(F\{j\}\{i+j\})/i,\ j \in \mathbb{N}\}$,

$(\cap \{F\{i\}(F\{j\}\{VRAI\})/i \leqslant j \in \mathbb{N}\}) \cap (\cap \{F\{i\}(F\{j\}\{FAUX\})/i > j \in \mathbb{N}\})$.

qui donnent la signification usuelle des procédures ≪ successeur ≫, ≪ somme ≫ et ≪ plus petit que ≫.

3ème type. Quelques ensembles de base ne sont pas atomiques ; tels sont, par exemple, les sous-ensembles récursivement énumérables de $\mathbb{N}$ qui ne sont ni vides ni singletons, ou encore l'ensemble {VRAI, FAUX} ; ainsi l'algorithme

$$\{VRAI, FAUX\} \cap Z,$$

qui donne une partie de la signification de la déclaration ≪ BOOLEAN Z ≫ d'ALGOL 60.

Nous notons X[Y] l'image que donne de Y l'algorithme X ; ainsi, en désignant par $\underline{suc}$ et $\underline{+}$ les algorithmes ≪ successeur ≫ et ≪ somme ≫, on a :

$\underline{suc}\,[\{0\}] = \{1\}$, $\underline{+}\,[\{1\}] = \cap \{F\{j\}\{1+j\}/j \in \mathbb{N}\} = \underline{suc}$,

$(\underline{+}\,[\{1\}])\,[\{3\}] = \{4\}$, $\underline{suc}\,[\{0,\ 1\}] = \underline{suc}\,[\{0\}]\ \bar{\cup}\ \underline{suc}\,[\{1\}] = \{1,\ 2\}$

($X\ \bar{\cup}\ Y$ est le plus petit élément de $\mathcal{A}$ dans lequel sont inclus X et Y),

$\underline{+}\,[\{0,\ 1\}] = (\cap\{F\{j\}\{j\}/j \in \mathbb{N}\})\ \bar{\cup}\ \underline{suc} = \cap \{F\{j\}\{j,\ j+1\}/j \in \mathbb{N}\}$,

$(\underline{+}\,[\{0,\ 1\}])\,[\{3,\ 4\}] = \{3,\ 4\}\ \bar{\cup}\ \{4,\ 5\} = \{3,\ 4,\ 5\}$.

DEFINITION 1. Λ est le langage défini par la grammaire suivante :

$$\Lambda \to V_i \ ;\ \Lambda \to (\Lambda\Lambda)\ ;\ \Lambda \to (\lambda V_i.\Lambda) \quad \text{pour } i \in \mathbb{N}.$$

Les éléments terminaux V_i $(i \in \mathbb{N})$ en sont les variables ; les notions d'occurrence libre ou liée d'une variable et de variable libre ou liée sont définies comme d'ordinaire ; l'opération de substitution dont le résultat est noté $\mathrm{Sub}_B^{V_i}/A$ est définie comme dans [1].

DEFINITION 2. Soit $\mathcal{A}$ une collection d'algorithmes ; $\varphi : \Lambda \to \mathcal{A}$ est une interprétation de Λ dans $\mathcal{A}$ ssi :

$\varphi V_i \in \mathcal{A}$,

$\varphi(AB) = (\varphi A)\ [\varphi B]$,

$\varphi(\lambda V_i.A) = \cap \{F(\psi V_i)\ (\psi A)/$ toute interprétation ψ telle que $\psi V_j = \varphi V_j$

si $j \neq i\}$.

DEFINITION 3. Soient A et B deux mots de Λ ; alors $A \equiv_\nu B$ ssi $\varphi A = \varphi B$ pour toute interprétation φ de Λ dans toute collection d'algorithmes de type ν ($\nu = 1, 2, 3$).

Chacune de ces relations est évidemment une équivalence et même une congruence : $A \equiv_\nu A'$ et $B \equiv_\nu B'$ entraînent $(AB) \equiv_\nu (A'B')$ et $(\lambda V_i.A) \equiv_\nu (\lambda V_i.A')$; de plus, si $\nu \geqslant \nu'$, alors $A \equiv_\nu A'$ entraîne $A \equiv_{\nu'} A'$.

Les résultats annoncés s'énoncent ainsi :

PROPOSITION.

(α) $(\lambda V_i.A) \equiv_3 (\lambda V_j.\ \mathrm{Sub}_{V_j}^{V_i} / A)$ si V_j n'a pas d'occurrence dans A ;

(β) $((\lambda V_i.A)B) \equiv_2 \mathrm{Sub}_B^{V_i}/A$;

(η') $(\lambda V_j.((\lambda V_i.A)V_j)) \equiv_3 (\lambda V_i.A)$ si V_j n'est pas libre dans A ;

(η) $(\lambda V_j.(AV_j)) \equiv_1 A$ si V_j n'est pas libre dans A.

Un λ-calcul est pour nous le quotient du langage Λ par une relation d'équivalence. Ainsi :

Un λ-calcul où valent (α), (β) et (η) ne peut avoir pour modèles que des collections d'algorithmes de type 1 ; on n'a donc affaire, ici, qu'aux seuls <u>algorithmes universels</u> ;

un λ-calcul avec (α), (β) et (η') admet en outre des modèles de type 2 ; il s'agit, en tout cas, d'<u>algorithmes déterministes</u>, comme ceux qu'on décrit dans des programmes qui ne contiennent pas de déclarations ;

la règle (β) n'est plus valable en général si on accepte des modèles de type 3, modèles dans lesquels on peut avoir affaire à des <u>algorithmes non-déterministes</u>, par exemple un sous-ensemble non vide de $\mathbb{N}$ qui n'est pas un singleton.

2. RAPPELS SUR LES COLLECTIONS D'ALGORITHMES.

Soit T une collection non vide, $\mathcal{A}$ une collection de parties de T contenant $\emptyset$ et T comme éléments, close pour l'intersection infinie.

Dans tout ce qui suit, les lettres V, W, X, Y, Z, affectées au besoin d'indices, désignent des éléments de $\mathcal{A}$, sauf indication contraire explicite.

DEFINITION 4.

1. $\bar{\bigcup}_i X_i$ est le plus petit majorant des X_i $(i \in I \neq \emptyset)$;

2. X est atomique ssi $X \subseteq \bigcup_i Y_i$ $(i \in I \neq \emptyset)$ entraîne $X \subseteq Y_j$ pour un j au moins dans I ;

3. $\mathcal{F}$ est la collection des applications $f : \mathcal{A} \to \mathcal{A}$ qui sont normales ; i.e. telles que, pour tout X, $f(X) = \bar{\bigcup}\{f(Y)/Y$ atomique $\subseteq X\}$ et, pour un X au moins, $f(X) \neq \emptyset$ [2] ;

4. $\mathcal{F}XT = T$ pour tout X ; si $Y \neq T$ alors $\mathcal{F}XY$ est la sous-collection de $\mathcal{F}$ telle que $f \in \mathcal{F}XY$ ssi $f(X) \subseteq Y$;

5. $\mathcal{A}_F$ est la plus petite collection qui contient tous les $\mathcal{F}XY$ où $Y \neq T$ et qui est close pour l'intersection infinie ;

6. Si $X' \in \mathcal{A} \cup \mathcal{A}_F$, l'image de Y par X', en abrégé $X'[Y]$ est $\bigcap\{Z/X' \subseteq \mathcal{F}YZ\}$.

DEFINITION 5. Soit T une collection non vide ; une collection $\mathcal{A}$ de parties de T est une collection d'algorithmes ssi elle remplit les conditions suivantes :

1. il existe une collection $\mathcal{B}$ de parties de T qui contient $\emptyset$ et T comme éléments, telle que $\mathcal{A}$ est la plus petite sous-collection de $\mathcal{P}(T)$ qui contient $\mathcal{B}$ et qui est close pour l'intersection infinie et pour l'opération : $X, Y \mapsto \mathcal{F}XY$;

2. soit $\mathcal{A}_B$ la fermeture de $\mathcal{B}$ par l'intersection infinie, amputée de $\emptyset$ et de T ; si $X \in \mathcal{A}_B$ et si $Y \in \mathcal{A}_F$, alors $X \cap Y = \emptyset$;

3. $\bigcup\{Z/Z$ atomique $\subseteq X\} = X$;

4. T est atomique.

Notons que $\{\mathcal{A}_B$ (les ensembles de base), $\mathcal{A}_F$ (les algorithmes propres), $\{T\}\}$ est une partition de $\mathcal{A}$ (si $\mathcal{A}_B \neq \emptyset$), et que les éléments non atomiques, s'il y en a, sont dans $\mathcal{A}_B$.

---

(2) C'est là la seule différence avec la version exposée dans [2] ; il en résulte que $\mathcal{F}T\emptyset = \emptyset$.

Nous utilisons dans la troisième partie les propriétés suivantes :

A_1. $FXY \subseteq FX_1Y_1$ ssi $Y_1 = T$ ou $X_1 \subseteq X$ et $Y \subseteq Y_1$.

A_2. $X \subseteq X_1$ et $Y \subseteq Y_1$ entraînent $X[Y] \subseteq X_1[Y_1]$

(monotonie de l'opération : $X, Y \rightsquigarrow X[Y]$).

A_3. Tout algorithme propre, X, a une base unique, $\hat{X} = \{FY_i\ Z_i / i \in I \neq \emptyset\}$ telle que $X = \cap\ \hat{X}$ et, pour tous $i, j \in I$:

. Y_i est atomique,

. si Y est atomique et si $Y \subseteq Y_i$ $(i \in I)$, il existe $j \in I$ tel que $Y = Y_j$,

. $Y_i \neq Y_j$ si $i \neq j$,

. $Y_i \subseteq Y_j$ entraîne $Z_i \subseteq Z_j$,

. $Z_i \neq T$.

A_4. Soit X un algorithme propre, de base $\hat{X} = \{FY_i\ Z_i / i \in I \neq \emptyset\}$; alors, pour tout V, $X[V] = (\,\overline{\cup}\,\{Z_i\,/\,Y_i \subseteq V\})\ \overline{\cup}$ (si l'un des éléments atomiques inclus dans V n'est pas l'un des X_i, alors T sinon $\emptyset$).

A_5. Si $Z \in \mathcal{A}_F \cup \{T\}$ alors $\cap\,\{FX(Z[X])/X \in \mathcal{A}\} = Z$.

A_6. $FX(\cap\{Y_i/i \in I \neq \emptyset\}) = \cap\{FXY_i/i \in I\}$.

A_7. Soient $\underline{I} = \cap\,\{FXX/X \in \mathcal{A}\}$, $P_2^1 = \cap\,\{FX(FYX)/X, Y \in \mathcal{A}\}$,

$\underline{S} = \cap\,\{FX(FY(FZ((X[Z])\,[Y[Z]])))/X, Y, Z \in \mathcal{A}\}$; alors, si Z est atomique,

$\underline{I}\,[X] = X$, $FTX = P_2^1[X]$, $(P_2^1[X])\,[Y] = X$, $((\underline{S}\,[X])\,[Y])\,[Z] = (X[Z])[Y[Z]]$.

3. DEMONSTRATION DES RESULTATS.

Dans toute cette partie, les lettres A, B, affectées au besoin d'indices, désignent des mots du langage Λ.

DEFINITION 6. La <u>variable</u> V_i a une <u>occurrence libre</u> dans V_i ; elle n'a pas d'<u>occurrence liée</u> dans V_i et aucune occurrence, libre ou liée dans V_j si $j \neq i$; les occurrences libres (resp^t. liées) de V_i dans (AB) sont les occurrences libres (resp^t. liées) de V_i dans A ou dans B ; les occurrences libres (resp^t. liées) de V_j dans $(\lambda V_i.A)$ sont les mêmes que dans A, si $j \neq i$; toutes les occurrences de V_i dans $(\lambda V_i.A)$ sont liées. La variable V_i est libre (resp^t. liée) dans A ssi elle a une occurrence libre (resp^t. liée) dans ce mot.

<u>Remarque.</u> Une interprétation est entièrement déterminée par $\mathcal{A}$ et par les images qu'elle donne des variables V_i.

LEMME 1. Pour toute interprétation φ, $\varphi(\lambda V_i . V_i) = \underline{I}$.

Car $\varphi(\lambda V_i . V_i) = \cap \{F(\psi V_i)\ (\psi V_i) / \psi V_j = \varphi V_j \text{ si } j \neq i\} = \cap \{FXX / X \in \mathcal{A}\} = \underline{I}\ (A_7)$.

LEMME 2. Soient φ et ψ deux interprétations telles que $\varphi V_k = \psi V_k$ si $k \neq i$; alors, pour tout $A \in \Lambda$, $\varphi A = \psi A$ si V_i n'est pas libre dans A.

C'est vrai si $A = V_j$ $(j \neq i)$; supposons qu'il en soit de même pour A_1 et A_2 ; alors :

. si $A = (A_1 A_2)$, $\varphi(A_1 A_2) = (\varphi A_1)\ [\varphi A_2] = (\psi A_1)\ [\psi A_2] = \psi(A_1 A_2)$;

. si $A = (\lambda V_i . A_1)$, $\varphi(\lambda V_i . A_1) = \cap \{F(\theta V_i)\ (\theta A_1) / \theta V_k = \varphi V_k \text{ si } k \neq i\}$
$= \cap \{F(\theta V_i)\ (\theta A_1) / \theta V_k = \psi V_k \text{ si } k \neq i\} = \psi(\lambda V_i . A_1)$;

. si $A = (\lambda V_j . A_1)$ avec $j \neq i$, $\varphi(\lambda V_j . A_1) = \cap \{F(\theta V_j)\ (\theta A_1) / \theta V_k$
$= \varphi V_k \text{ si } k \neq j\} = \cap \{F(\theta' V_j)\ (\theta' A_1) / \theta' V_k = \theta V_k \text{ si } k \neq i\}$
$= \cap \{F(\theta' V_j)\ (\theta' A_1) / \theta' V_k = \psi V_k \text{ si } k \neq j\} = \psi(\lambda V_j . A_1)$.

COROLLAIRE. Si V_i n'est pas libre dans A, alors $\varphi(\lambda V_i . A) = P^1_2\ [\varphi A]$.
Car alors $\varphi(\lambda V_i . A) = \cap \{F(\theta V_i)\ (\theta A) / \theta V_k = \varphi V_k \text{ si } k \neq i\} = \cap \{F(\theta V_i)\ (\varphi A) / \theta V_k = \varphi V_k \text{ si } k \neq i\}$ (lemme 2) $= \cap \{FX(\varphi A) / X \in \mathcal{A}\} = FT(\varphi A)\ (A_1) = P^1_2\ [\varphi A]\ (A_7)$.

LEMME 3. Soient φ et ψ comme dans le lemme 2 ; si $\varphi V_i \subseteq \psi V_i$ alors, pour tout $A \in \Lambda$, $\varphi A \subseteq \psi A$.

C'est vrai si V_i n'est pas libre dans A (lemme 2) et si $A = V_i$; pour les autres cas, on procède comme dans la démonstration du lemme 2 en utilisant A_1 et A_2.

COROLLAIRE. $\{F(\theta V_k)\ (\theta A) / \theta V_k = \varphi V_k \text{ si } k \neq i,\ \theta V_k \text{ atomique},\ \theta A \neq T\}$ est la base de $\varphi\ (\lambda V_i . A)$.

Car tout θV_k atomique tel que $\theta A \neq T$ figure dans la collection des $F(\theta V_k)$ (θA) ; et $\theta V_k \subseteq \theta' V_k$ entraîne $\theta A \subseteq \theta' A$ (lemme 3) ; ainsi les conditions énoncées en A_3 sont satisfaites.

COROLLAIRE. Si φV_i est atomique, alors $\varphi((\lambda V_i.A)\ V_i) = \varphi A$.

C'est évident, eu égard au Lemme 2 et au corollaire précédent, car la seule interprétation convenable est $\theta = \varphi$.

LEMME 4. Soient A sans occurrence de V_1 ; A' identique à A à ceci près, que toute occurrence libre de V_i est remplacée par une occurrence (libre) de V_1 ; soient φ et ψ des interprétations telles que $\psi V_k = \varphi V_k$ si $k \neq 1$ et $\psi V_1 = \varphi V_i$; alors $\varphi A = \psi A'$.

C'est vrai si V_i n'est pas libre dans A car alors $A = A'$ et $\varphi A = \psi A'$ (lemme 2) ; vrai aussi si $A = V_i$ car $\varphi A = \psi V_1 = \psi A'$; supposons qu'il en est de même pour A_1 et A_2 ; alors :

- si $A = (A_1 A_2)$, $\varphi(A_1 A_2) = (\varphi A_1)\ [\varphi A_2] = (\psi A_1')\ [\psi A_2'] = \psi(A_1' A_2') = \psi A'$;
- si $A = (\lambda V_j.A_1)$ avec $j \neq i$, $\varphi A = \cap\ \{F(\theta V_j)\ (\theta A_1)\ /\ \theta V_k = \varphi V_k$ si $k \neq j\}$ $= \cap\ \{F(\theta' V_j)\ (\theta' A_1')\ /\ \theta' V_k = \psi V_k$ si $k \neq j\} = \psi(\lambda V_j.A_1') = \psi A'$.

COROLLAIRE. Soient A et A' comme dans le Lemme 4 ; alors,

$$\varphi(\lambda V_i.A) = \varphi(\lambda V_1.A') .$$

Car $\varphi(\lambda V_i.A) = \cap\ \{F(\theta V_i)\ (\theta A)\ /\ \theta V_k = \varphi V_k$ si $k \neq i\} = \cap\ \{F(\theta' V_1)\ (\theta' A')\ /$ $\theta' V_k = \psi V_k$ si $k \neq 1\}$ (en vertu du Lemme 2, avec φ et ψ définis comme dans le Lemme 4) $= \cap\ \{F(\theta' V_1)\ (\theta' A')\ /\ \theta' V_k = \varphi V_k$ si $k \neq 1\} = \varphi(\lambda V_1.A')$.

LEMME 5. Soient φ et ψ deux interprétations telles que $\psi V_k = \varphi V_k$ si $k \neq i$ et $\psi V_i = \varphi B$; alors $\varphi((\lambda V_i.A)B) = \psi A$ si φB est atomique.

C'est vrai si V_i n'est pas libre dans A car alors $\varphi((\lambda V_i.A)B) =$ $(\varphi(\lambda V_i.A))\ [\varphi B] = \left(P_2^1\ [\varphi A]\right)\ [\varphi B]$ (Corollaire du Lemme 2) $= \varphi A$ (A_7) ; c'est vrai aussi si $A = V_i$, car $\varphi((\lambda V_i.V_i)B) = (\varphi(\lambda V_i.V_i))\ [\varphi B] = \underline{I}\ [\varphi B]$ (Lemme 1) $= \varphi B$ $(A_7) = \psi A$; supposons qu'il en est de même pour A_1 et A_2 ; alors :

- si $A = (A_1 A_2)$, $(\varphi(\lambda V_i.(A_1 A_2)))\ [\varphi B] = (\cap\{F(\theta V_i)\ (\theta A_1[\theta A_2])\ /\ \theta V_k = \varphi V_k$ si $k \neq i\})\ [\varphi B] = \cap\ \{(\theta A_1)\ [\theta A_2]\ /\ \theta V_k = \varphi V_k$ si $k \neq i$ et $\theta V_i = \varphi B\}$ (Corollaire du Lemme 3, car φB est atomique et A_4) $= (\psi A_1)\ [\psi A_2] = \psi(A_1 A_2) = \psi A$;

. si $A = (\lambda V_j . A_1)$ avec $j \neq i$, $(\varphi(\lambda V_i A)) = \cap \{F(\theta V_i)\ (\theta A) / \theta V_k = \varphi V_k$ si $k \neq i\} = \cap \{F(\theta V_i)\ (\cap\{F(\tau V_j)\ (\tau A_1) / \tau V_1 = \theta V_1$ si $1 \neq j\}) / \theta V_k = \varphi V_k$ si $k \neq i\} = \cap \{F(\theta V_i)\ (F(\tau V_j)\ (\tau A_1)) / \tau V_1 = \theta V_1$ si $1 \neq j$, $\theta V_k = \varphi V_k$ si $k \neq i\}$ (A_6) ; ainsi $\varphi((\lambda V_i . A)B) = (\varphi(\lambda V_i . A))\ [\varphi B] = \cap \{F(\tau V_j)\ (\tau A_1) /$ $\tau V_1 = \psi V_1$ si $1 \neq j\} = \psi(\lambda V_j . A_1) = \psi A$.

COROLLAIRE. Si φB est atomique, alors $\varphi((\lambda V_i . (A_1 A_2))B) =$

$$\underline{S}\ [\varphi(\lambda V_i . A_1),\ \varphi(\lambda V_i . A_2),\ \varphi B].$$

Car le second membre est identique à $((\varphi(\lambda V_i . A_1))\ [\varphi B])\ \ [(\varphi(\lambda V_i . A_2))\ [\varphi B]]$ $(A_7) = (\psi A_1)\ [\psi A_2]$ (en vertu du Lemme 4, avec φ et ψ définis comme dans le Lemme 5) $= \varphi((\lambda V_i . (A_1 A_2))B)$.

DEFINITION 7. $\mathrm{Sub}_B^{V_i} / V_i = B$; $\mathrm{Sub}_B^{V_i} / V_j = V_j$ si $j \neq i$;

$\mathrm{Sub}_B^{V_i} / (A_1 A_2) = ((\mathrm{Sub}_B^{V_i} / A_1)\ (\mathrm{Sub}_B^{V_i} / A_2))$;

$\mathrm{Sub}_B^{V_i} / (\lambda V_i . A_1) = (\lambda V_i . A_1)$;

$\mathrm{Sub}_B^{V_i} / (\lambda V_j . A_1) = (\lambda V_j .\ \mathrm{Sub}_B^{V_i} / A_1)$ si $i \neq j$ et si (V_j non libre dans B ou V_i non libre dans A_1) ;

$\mathrm{Sub}_B^{V_i} / (\lambda V_j . A_1) = (\lambda V_k .\ \mathrm{Sub}_B^{V_i} / (\mathrm{Sub}_{V_k}^{V_j} / A_1))$ si $i \neq j$, V_j libre dans B, V_i libre dans A_1 et si k est le plus petit entier tel que V_k n'a pas d'occurrence dans $(A_1 B)$.

LEMME 6. $\mathrm{Sub}_B^{V_i} / A = A$ si V_i n'est pas libre dans A .

Car $\mathrm{Sub}_B^{V_i} / A = A$ si $A = V_j$ $(j \neq i)$; supposons qu'il en est de même pour A_1 et A_2 ; alors

. $\mathrm{Sub}_B^{V_i} / (A_1 A_2) = ((\mathrm{Sub}_B^{V_i} / A_1)\ (\mathrm{Sub}_B^{V_i} / A_2)) = (A_1 A_2)$;

. $\mathrm{Sub}_B^{V_i} / (\lambda V_i . A_1) = (\lambda V_i . A_1)$;

. $\mathrm{Sub}_B^{V_i} / (\lambda V_j . A_1)$ (avec $j \neq i$) $= \lambda V_j . (\mathrm{Sub}_B^{V_i} / A_1) = \lambda V_j . A_1$.

LEMME 7. Soient A et A' comme dans le Lemme 4 ; si V_1 n'a pas d'occurrence dans A alors $\mathrm{Sub}_{V_1}^{V_i} / A = A'$.

C'est vrai si V_i n'est pas libre dans A (Lemme 6) et si $A = V_i$:

$\mathrm{Sub}_{V_1}^{V_i} / V_i = V_1 = A'$; supposons qu'il en est de même pour A_1 et A_2 ; alors :

. $\mathrm{Sub}_{V_1}^{V_i} / (A_1A_2) = ((\mathrm{Sub}_{V_1}^{V_i} / A_1)\ (\mathrm{Sub}_{V_1}^{V_i} / A_2)) = (A_1'A_2') = (A_1A_2)'$;

. $\mathrm{Sub}_{V_1}^{V_i} / (\lambda V_j.A_1)$ (avec $j \neq i$) $= \lambda V_j.(\mathrm{Sub}_{V_1}^{V_i} / A_1)$

(car $V_j \neq V_1$) $= (\lambda V_j.A_1') = (\lambda V_j.A_1)'$.

COROLLAIRE. Si V_1 n'a pas d'occurrence dans A alors $\varphi(\lambda V_i.A) = \varphi(\lambda V_1.\mathrm{Sub}_{V_1}^{V_i} / A)$.

C'est évident en vertu des Lemmes 3 et 7;

LEMME 8. Soient φ et ψ des interprétations telles que $\psi V_k = \varphi V_k$ si $k \neq i$ et $\psi V_i = \varphi B$; alors $\varphi(\mathrm{Sub}_B^{V_i} / A) = \psi A$.

C'est vrai si V_i n'est pas libre dans A (Lemmes 6 et 2) et si $A = V_i$ car $\psi V_i = \varphi B = \varphi(\mathrm{Sub}_B^{V_i} / A)$; supposons qu'il en est de même pour A_1 et A_2 ; alors :

. si $A = (A_1A_2)$, $\psi(A_1A_2) = (\psi A_1)\ [\psi A_2] =$

$(\varphi(\mathrm{Sub}_B^{V_i} / A_1))\ \left[\varphi(\mathrm{Sub}_B^{V_i} / A_2)\right] = \varphi((\mathrm{Sub}_B^{V_i} / A_1)\ (\mathrm{Sub}_B^{V_i} / A_2)) = \varphi(\mathrm{Sub}_B^{V_i} / (A_1A_2))$;

. si $A = (\lambda V_j.A_1)$ avec $j \neq i$ et V_j non libre dans B ,

$\varphi(\mathrm{Sub}_B^{V_i} / (\lambda V_j.A_1)) = \varphi(\lambda V_j.\ (\mathrm{Sub}_B^{V_i} / A_1))$

$= \cap \{F(\theta V_j)\ (\theta(\mathrm{Sub}_B^{V_i} / A_1)) / \theta V_k = \varphi V_k \text{ si } k \neq j\}$

$= \cap \{F(\theta' V_j)\ (\theta' A_1) / \theta' V_k = \theta V_k \text{ si } k \neq i \text{ et } \theta' V_i = \theta B\}$

$= \cap \{F(\theta' V_j)\ (\theta' A_1) / \theta' V_k = \psi V_k \text{ si } k \neq i\}$ (Lemme 2) $= \psi(\lambda V_j.A_1)$;

. si $A = (\lambda V_j.A_1)$ avec $j \neq i$ et V_j libre dans B ,

$\varphi(\mathrm{Sub}_B^{V_i} / (\lambda V_j.A_1)) = \varphi(\lambda V_1.\ \mathrm{Sub}_B^{V_i} / (\mathrm{Sub}_{V_1}^{V_j} / A_1))$

(avec V_1 sans occurrence dans (A_1B), donc $V_1 \neq V_i$ et $V_1 \neq V_j$)

$= \cap \{F(\theta V_1)\ (\theta(\mathrm{Sub}_B^{V_i} / (\mathrm{Sub}_{V_1}^{V_j} / A_1))) / \theta V_k = \varphi V_k \quad \text{si} \quad k \neq 1\}$

$= \cap \{F(\theta' V_1)\ (\theta'(\mathrm{Sub}_{V_1}^{V_j} / A_1)) / \theta' V_k = \theta V_k \quad \text{si} \quad k \neq i \quad \text{et} \quad \theta' V_i = \theta B\}$

$= \cap \{F(\theta' V_1)\ (\theta'(\mathrm{Sub}_{V_1}^{V_j} / A_1)) / \theta' V_k = \psi V_k \quad \text{si} \quad k \neq i\} = \psi(\lambda V_1 .\ \mathrm{Sub}_{V_1}^{V_j} / A_1) =$

$\psi(\lambda V_j . A_1)$ (Corollaire du Lemme 7).

COROLLAIRE. Si φB est atomique, alors $\varphi((\lambda V_i . A)B) = \varphi(\mathrm{Sub}_B^{V_i} / A)$.

Cela résulte des Lemmes 5 et 8.

LEMME 9. Si V_j n'est pas libre dans A alors $\varphi(\lambda V_j . ((\lambda V_i . A)V_j)) = \varphi(\lambda V_i . A)$.

Car $\varphi(\lambda V_j . ((\lambda V_i . A)V_j)) = \cap \{F(\psi V_j)\ ((\psi(\lambda V_i . A))\ [\psi V_j]) / \psi V_k = \varphi V_k \quad \text{si} \quad k \neq j\}$

$= \cap \{F(\psi V_j)\ ((\varphi(\lambda V_i . A))\ [\psi V_j]) / \psi V_k = \varphi V_k \quad \text{si} \quad k \neq j\}$ (Lemme 2) =

$= \cap \{FX\ ((\varphi(\lambda V_i . A))\ [X]) / X \in \mathcal{A}\} = \varphi(\lambda V_i . A)$ (A_5, car $\varphi(\lambda V_i . A)$ est T ou un algorithme propre).

La proposition s'ensuit aisément ; en effet, α, β et η' résultent immédiatement des corollaires des lemmes 7 et 8 ou du lemme 9 ; et η est un corollaire de η' puisque tous les éléments d'une collection d'algorithmes de type 1 sont T ou un algorithme propre.

4. COMMENTAIRES.

Dans tous les cas, on peut étendre le langage Λ (qui devient alors un λ-calcul appliqué) en adjoignant les règles de grammaire :

$$\Lambda \to C_j \qquad (j \in J \neq \emptyset)$$

(les constantes C_j étant différentes de λ , du point, des parenthèses et de V_i pour tout $i \in \mathbb{N}$), et en précisant qu'aucune variable n'est liée ni libre dans C_j . On considère alors, plus particulièrement, les interprétations φ pour lesquelles φC_j est un élément fixe d'une collection d'algorithmes $\mathcal{A}$.

On peut avoir ainsi les constantes I , K et S telles que $\varphi I = \underline{I}$, $\varphi K = P_2^1$, $\varphi S = \underline{S}$ pour tout φ ; ou encore les constantes VRAI, FAUX, i (pour $i \in \mathbb{N}$) avec $\varphi(\mathrm{VRAI}) = \{\mathrm{VRAI}\}$, $\varphi(\mathrm{FAUX}) = \{\mathrm{FAUX}\}$, $\varphi i = \{i\}$ pour tout φ ; et peut-être aussi les constantes INTEGER , BOOLEAN , [m : n],... (ALGOL 60).

Un λ-calcul appliqué est donc, en somme, un langage de programmation du genre LISP ; et la façon la plus simple de définir la sémantique d'un langage de programmation quelconque nous semble être de le réduire à un λ-calcul par des voies purement syntaxiques.

Nos sources sont, pour l'essentiel, les travaux des spécialistes de la Logique Combinatoire, magistralement exposés dans les ouvrages, qui défient le temps, de H.B. CURRY, R. FEYS et H.B. CURRY, J.R. HINDLEY, J.P. SELDIN, et agréablement résumés dans le petit livre de J.R. HINDLEY, B. LERCHER, J.P. SELDIN [1].

Nous avons exposé notre point de vue sur la sémantique des langages de programmation dans divers articles (Colloque de l'IRIA, 1972 ; Colloque de Rome, 1973 ; International summer school de Capri, 1973) ; et après quelques variations nous nous en tenons pour l'instant à la version suivante :

L. NOLIN - Algorithmes universels. R.A.I.R.O., 1974, 5 - 18 [2].

Deux thèses de doctorat ont montré tout le parti qu'on pouvait en tirer pour clarifier certains problèmes d'Informatique :

B. ROBINET - Contribution à l'étude de réalités informatiques, Paris, 1974.

G. RUGGIU - De l'organigramme à la formule, Paris, 1974.

Nos modèles sont assez proches de ceux, bien connus, de D. SCOTT, voire des URS (E.G. WAGNER, H.R. STRONG, M. VENTURINI ZILLI) qu'on aurait tort, selon nous d'oublier.

Parmi les langages de programmation les plus proches des λ-calculs, nous mentionnerons tout particulièrement LISP (J. Mc CARTHY) et CUCH (C. BÖHM).

Nous ne saurions citer tous ceux qui ont eu l'idée de ramener les langages de programmation à des λ-calculs, depuis P. LANDIN, J.H. MORRIS, R.J. ORGASS et D. PARK ; tout le monde les connaît, d'ailleurs.

Disons pour terminer que nos travaux semblent converger vers le même point que des recherches d'inspiration toute différente, celles de M. NIVAT en particulier (voir notre communication conjointe au Colloque de Sarrebrück, 1974).

ON THE DESCRIPTION OF TIME VARYING SYSTEMS IN λ-CALCULUS

Giorgio Ausiello

Centro di Studio dei Sistemi di Controllo e Calcolo Automatici

Roma, Italy

Abstract

The properties of λ-calculus and related formal systems as description languages of programs and machines have been studied by several authors. In this paper we examine how λ-calculus can be used to describe the behaviour of time varying systems; for this purpose a particular kind of typed calculus is introduced. The problem of synchronization is then considered and an application to rewriting and developmental systems is shown.

Summary

1. On the intuitive description of time variations

The use of λ-calculus and related algorithmic languages such as CUCH [1] as languages for describing machines and programs has been widely studied in the past (a non exaustive list of applications is referred: [2][3][4][5][6][7][12]).More recently the interest for stating and proving properties of programs such as correctness, parallelism, existence of deadlocks, dynamic storage optimization, has drawn considerable attention on formal systems where the notion of time can be explicitly expressed throughout the use of a special type of variables and relations, or by introducing modality in the language [8][9] [10].

Developing ideas of Fitch [12], a first step toward the explicit introduction of time in λ-calculus was made in [4]. The aim was to provide a programming and simulation language for sequential-analog machines and the presence of time was needed to take care both of time varying components (such as integrators) and of time varying connections in the network. In both cases the effect of time was conveyed through logical variables ϕ_i taking as values the constants $\underline{0}$ and $\underline{K}$ (*), depending on whether the value of the time variable t to which they were applied was in a certain domain or not. For example the time varying network

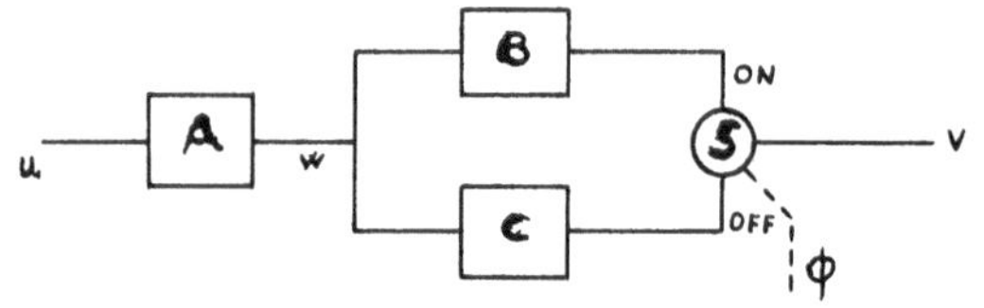

where u,v,w are variables, A,B,C are boxes, ϕ is a logical variable, S is a switch ("on" in odd instants) was represented by the following system of equations:

$$v = \lambda x \lambda y[Sxy](\lambda x[Bx]w)(\lambda x[Cx]w)$$

$$w = \lambda x[Ax]u$$

$$\lambda t S = \lambda s \lambda t[\underline{I}st]\phi$$

which β-reduced (η-reductions were not allowed) to the system:

$$v = S(Bw)(Cw)$$

$$w = Au$$

$$S = \underline{I}(\phi\bar{t})$$

if $\bar{t}$ was a specific instant at which we wanted to know the state of the network. If $\bar{t}$ was in the domain $D_\phi = \{2n \mid n \geq 0\}$ then $\phi\bar{t} = \underline{0}$ and $v = C(Au)$, otherwise $\phi\bar{t} = \underline{K}$ and $v = B(Au)$.

The main obstacle in developping this naïf way of dealing with time varying objects is in the fact that "time" β-reductions (that we need to fix the value of the time variables) and "space" β-reductions (that we need, for example, to connect systems) cannot be ordered "a priori". So that, for example, if $\lambda u \lambda x[A(x,u)]$ and $\lambda v \lambda y[B(y,v)]$ are formulae de-

(*) We underline the names of combinators.

scribing two time varying systems with input functions x and y and time variables u and v respectively, we want to be able of performing both the following series of reductions (denoted by $\geq$)

i) $\lambda u \lambda x[A(x,u)]n \geq \lambda x[A(x,n)]$

$\lambda v \lambda y[B(y,v)]m \geq \lambda y[B(y,m)]$

$\underline{B}(\lambda x[A(x,n)])(\lambda y[B(y,m)]) \geq \lambda z[A(B(z,m),n)]$

where we first fix the configuration of the two systems and then we make the connection,

and

ii) $\underline{B}_{\tau}(\lambda u \lambda x[A(x,u)])(\lambda v \lambda y[B(y,v)]) \geq \lambda u \lambda v \lambda z[A(B(z,v)u)]$

(where $\underline{B}_{\tau} \equiv \lambda s_1 \lambda s_2 \lambda n \lambda m[\underline{B}(s_1 n)(s_2 m)]$)

$\lambda u \lambda v \lambda z[A(B(z,v),u)]n\ m \geq \lambda z[A(B(z,m),n)]$

where we first make the connection and then we fix the value of the time variables as two independent parameters.

Clearly, the operator that we need to connect two static systems ($\underline{B}$) must be different from the operator that we need to connect two time varying systems ($\underline{B}_{\tau}$). In fact if we would use $\underline{B}$ also in the second case we would come out with a type error because we would assign the output of the system B in correspondence with the abstraction u that represents the time variancy of the system A.

In another example we may consider an integrator I which is in the state "operate" at time t_o and gives the output 0 if $t \leq t_o$ and $\int_{t_o}^{t} y(u)$ if $t \geq t_o$. Also in this case we want to be able either of first assigning the input y and then specifying the time parameter t_o or of first specifying the parameter and then assigning the input; this means that the result of applying $\lambda t_o \lambda y[I(y,t_o)]$ to z (input) and n (time) must be equal to the result of applying it first to n and then to z, that is equal to $I(z,n)$.

What we need, hence, is the ability of saturating the abstractions with respect to time and to signals independently from their order, but only according to their type as if the abstractions were made on two different axes of a two dimensional space.

In the more general case of a time varying system realized by connecting (in a time varying way) several (time varying) subsystems, we are faced with the problem of expressing that the output signal (as a function of time) depends on the input signals and on time itself. To describe this type of systems, hence, we need a calculus where "time" β-reductions and "space" β-reductions are performed in whatever order is required, without changing the formulae that describe the single subsystems and the connectors, and where we may choose either of synchronizing the time variancies of the subsystems or, more generally, of keeping several independent time parameters (as if any subsystem had its own time reference).

In this paper we introduce a calculus that satisfy the said requirements. After sketching in §2 and §3 the basic definitions of the language and of the reduction rules, in §4 we introduce the slight variations which allow to describe the synchronization of all components of a system and in §5 we show, mainly through examples, how the rules of the calculus (both in the asynchronous and in the synchronous form) are suitable for describing the derivation of sentential forms in various classes of rewriting systems.

2. *A "two dimensional" typed language*

In order to clarify the basic ideas we will show how we formalize the notion of time varying system and how we define a calculus for describing this kind of systems.

The objects of different type we want to deal with are, for example:
Time, $t : T$,
signal, $x : X$,
that we may consider the *elementary types*, even though what we call signals are mappings from T to the set of the possible input/output values (integers, reals, ecc.)
Some of the *types of higher order are:* (*)
system, $B : S \equiv X^n \to X$

(*) Here α^n means $\underbrace{\alpha \times \ldots \times \alpha}_{n \text{ times}}$ or $\underbrace{\alpha \to (\alpha \ldots (\alpha \to}_{n \text{ times}}$ indifferently.

m-time varying system: $\mathcal{B}_\tau : S_\tau \equiv T^m \to (X^n \to X)$

static connector, $\Phi : S^n \to S$; $\Phi_\tau : S_\tau^n \to S_\tau$

m-time varying connector, $\Phi^m : T^m \to (S^n \to S)$; $\Phi_\tau^m : T^m \to (S_\tau^n \to S_\tau)$

etc.

The idea of a two dimensional calculus is based on the formalization of the notion of "time" as a type with peculiar properties. In terms of types, infact, what we need to describe the time variancy of systems and connectors are the following equivalences: if T is time and α and γ are types different from time then

$$(T\to\alpha)\to\gamma \equiv \alpha\to(T\to\gamma) \equiv T\to(\alpha\to\gamma)$$

As a consequence, for example, if S is the type of systems

$$(T\to S)\to(T\to S) \equiv T\to(T\to(S\to S)) \equiv T^2\to(S\to S)$$

which means that mapping 1-time varying systems into 1-time varying systems is the same of using 2-time varying connectors. Similarly, the type of a 1-time varying connector of two 1-time varying systems,

$$T\to((T\to(X\to X))\to((T\to(X\to X))\to(T\to(X\to X))))$$

is equivalent to the type of a 4-time varying connector of two systems

$$T\to(T\to(T\to(T\to((X\to X)\to((X\to X)\to(X\to X))))))$$

The *basic objects* of the language are variables and constants over different types: $t, t_0, t_1, \ldots$ are time variables, $x, y, z, \ldots$ are input/output variables, $\mathcal{A}, \mathcal{B}, \ldots$ system variables (boxes), $\Phi, \Psi, \ldots$ connector variables, ecc. An example of connector constants are $\underline{B}$ and $\underline{B}_\tau$ that are used in the preceding paragraph.

The *terms* are defined in the following way:

i) all basic objects are terms,

ii) if u is a term of type $\alpha\to\beta$ and v is a term of type α $(\alpha\neq T)$ then $(u\ v)_\lambda$ is a term of type β,

iii) if u is a term of type $T\to\beta$ and v is a term of type T then $(u\ v)_\tau$ is a term of type β,

iv) if u is a term of type β and x is a variable over type $\alpha\neq T$ then

$\lambda x[u]$ is a term of type $\alpha\to\beta$,

v) if u is a term of type β and t is a time variable then $\tau t[u]$ is a term of type $T\to\beta$.(*)

As far as the interpretation is concerned we have the following (obvious) examples:

- abstraction: if $x_1,\ldots,x_n$ are signal variables and if $t_1,\ldots,t_m$ are time variables

$$\tau t_1 \ldots \tau t_m \lambda x_1 \ldots \lambda x_n [B_\tau(x_1,\ldots,x_n,t_1,\ldots,t_m)]$$

  is interpreted as a time varying system which operates on n input signals depending on m time parameters;
- application: if N is an object of type time and if M has a time variability, $(M\ N)_\tau$ is interpreted as the configuration of M at time N; if N is an object of type signal and M has an input, $(M\ N)_\lambda$ is interpreted as the result of feeding M with the signal N on the input.

3. The reduction rules in the asynchronous case

i) Globalization of time variables: all time variables are made as global as possible and renamed to avoid conflicts;
   - if t does not occur neither free nor bound in B

   $$(\tau t[A(\ldots t\ldots)]B)_\lambda \geq \tau t[(A(\ldots t\ldots)B)_\lambda]$$

   - if t occurs free in B and t_1 does not occur in B

   $$(\tau t[A(\ldots t\ldots)]B)_\lambda \geq \tau t_1[(A(\ldots t_1\ldots)B)_\lambda]$$

   - if t occurs bound in B and t_1 does not occur in B

   $$(\tau t[A(\ldots t\ldots)]B)_\lambda \geq \tau t[(A(\ldots t\ldots)B[t/t_1])_\lambda]$$

   - if t does not occur neither free nor bound in A

   $$(A\tau t[B(\ldots t\ldots)])_\lambda \geq \tau t[(AB(\ldots t\ldots))_\lambda]$$

   - if t does occur free in A and t_1 does not occur in A

   $$(A\tau t[B(\ldots t\ldots)])_\lambda \geq \tau t_1[(AB(\ldots t_1\ldots))_\lambda]$$

(*) As usual we will assume the left associativity of λ-applications.

finally:

- $\lambda x[\tau t[\ldots t\ldots]] \geq \tau t[\lambda x[\ldots t\ldots]]$

ii) β-reduction and application of the rules of the constants: after all time variables are globalized we may perform β-reductions and apply the rules of the constants both in λ-applications: $(\lambda x[A(\ldots x\ldots)]B)_\lambda$ where B is, for example, of type signal, and in τ-applications: $(\tau t[A(\ldots t\ldots)]B)_\tau$ where B is of type time; in both cases β-reductions are performed according to the usual definition and in the usual left-most outermost way (*).

iii) η-reductions are not allowedin order to preserve explicit type declarations.

Clearly the globalization of time variables is allowed by the stated equivalences among types, for which we may choose a particular standard form and put all terms in that form. The bidimensionality of the calculus is due to the fact that time abstractions freely sweep outward the formulae looking for the innermost object of type "time" able to saturate them.

Besides it is also clear that in the particular case that explicit references to time variancy are completely absent we have an instance of the usual typed λ-calculus.

If we consider again our example we see that we can compose the two time varying sistems $\tau u\lambda x[A(x,u)]$ and $\tau u\lambda y[B(y,u)]$ by the connector B so that both the following sequences of reductions can be realized:

$$(\tau u\lambda x[A(x,u)]n)_\tau \geq \lambda x[A(x,n)]$$

$$(\tau u\lambda y[B(y,u)]m)_\tau \geq \lambda y[B(y,m)]$$

$$((\underline{B}\lambda x[A(x,n)])_\lambda \lambda y[B(y,m)])_\lambda \geq \lambda z[A(B(z,m),n)]$$

and

$$((((\underline{B}\tau u\lambda x[A(x,u)])_\lambda \tau u\lambda y[B(y,u)])_\lambda n)_\tau m)_\tau$$

$$\geq(((\tau u[(\underline{B}\lambda x[A(x,u)])_\lambda]\tau u\lambda y[B(y,u)])_\lambda n)_\tau m)_\tau$$

(*) The order of execution of the rules have to be fixed because as a consequence of the introduction of the globalization rules the calculus is not Church-Rosser.

$$\geq((\tau u[((\underline{B}\lambda x[A(x,u)])_{\lambda}\tau t\lambda y[B(y,t)])_{\lambda}]n)_{\tau}m)_{\tau}$$

$$\geq((\tau u[\tau t[((\underline{B}\lambda x[A(x,u)])_{\lambda}\lambda y[B(y,t)])_{\lambda}]]n)_{\tau}m)_{\tau}$$

$$\geq((\tau u\tau t[((\underline{B}\lambda x[A(x,u)])_{\lambda}\lambda y[B(y,t)])_{\lambda}]n)_{\tau}m)_{\tau}$$

by globalizations and

$$\geq\lambda z[A(B(z,m),n)]$$

by β-reductions on λ and τ-applications.

4. The problem of synchronization

In most applications the connection between two time varying systems implies a well defined relation between the temporal variables of the two systems. In a fully synchronous system the "time" is always the same in all parts of the system and is often measured on an external clock on which also the input signal is synchronized. In other systems we have to take care of delays among subsystems and these delays may also be constant or time varying. In connecting, for example, the two systems $\tau t\lambda x[A(x,t)]$ and $\tau u\lambda y[B(y,u)]$ we may want the resulting system to be

$$\tau t\lambda z[A(B(D_{-1}z,t-1),t)]$$

(where D denotes a delay and, in general, $D_f zt = z(f(t))$)) if the output of B at time t is a function of the state of B and of the input of B at time t-1, or to be

$$\tau t\lambda z[A(B(D_{/2}\ z,\ t/2),t)]$$

if the delay of B increases with t.

In terms of types the synchronization process requires that if α is any type

$$T\to(T\to\alpha)\ \equiv\ T\to\alpha$$

In correspondence of this equivalence we have to modify the rules of globalization of time variables by adding the following rule:

$$-\tau t[\tau t_1[\ldots t\ldots t_1\ldots]]\ \geq\ \tau t[\ldots t\ldots t_1/t]$$

which makes all time abstractions to coincide so that the full synchronization of the system is achieved.

5. An application to grammars

The calculi that have been introduced in the preceding paragraphs were mainly motivated by the purpose of describing time varying systems such as analog sequential computers or, more generally, systems with an input-output relation depending on several time parameters.

In this paragraph, very informally, we want to show an application of the same concepts, (in particular of the same reduction rules) to rather different systems, such as the developmental systems that have been introduced [13] with the purpose of modelling particular biological behaviours. Relations among classes of developmental systems and rewriting systems have been extensively studied by several authors (see, for example, papers in [15] and [16]) and in [11] Salomaa shows how classes of developmental systems and classes of rewriting systems can be characterized by introducing various degrees of synchronization of the application of rewriting rules in grammars and by using the concept of level grammar which has been introduced in [14].

Here we are interested in showing how we may use a formalism similar to the two dimensional calculus for describing the generation of sentences by context free productions and how, by using asynchronous or synchronous reduction rules, we get CF languages or EOL(*) languages respectively.

Given a finite set of context free productions (over a non terminal alphabet $N \subseteq V$ and a terminal alphabet V-N) of the type

$$\alpha ::= \alpha_{i_1} \dots \alpha_{i_n} \qquad \text{where } \alpha, \alpha_{i_1}, \dots, \alpha_{i_n} \in N$$

$$\alpha ::= x \qquad \text{where } x \in (V-N)^*$$

a CF language is defined as the set of words in $(V-N)^*$ that can be derived from the axiom $S \in N$ by repeatedly applying the rule $u\alpha_i v \to u\alpha_{i_1} \dots \alpha_{i_n} v$ (where $u, v \in V^*$) while an EOL language is defined as the set of words in

(*) "extended Lindenmayer languages of type 0"

$(V-N)^*$ that can be derived from the axiom $W \varepsilon N^*$ by repeatedly applying the rule $w_1 \to w_2$ where $w_1, w_2 \varepsilon N^*$ and w_2 is obtained from w_1 by simultaneously replacing each non terminal α in w_1 whith the corresponding string of non terminals $\alpha_{i_1} \ldots \alpha_{i_n}$ or with the corresponding terminal word x.

Clearly the essential difference between CF languages and EOL languages is based on the fact that in EOL derivations (such as it is in all Lindernmayer systems) the productions (which simulate cells development) are applied synchronously through out a sentential form (often called a "filament"). The classical example of an EOL language which is not CF is $\{a^{2^n} | n \geq 0\}$ which is obtained with the productions S::=SS,S::=a.

To give a representation of the derivation of words by CF rewriting systems and EOL developmental systems we assume that a nonterminal behaves like a time varying system whose configuration at a "certain time" is given by the sentential form that is achieved after a "certain number" of production rules have been applied, starting from the given nonterminal.

We act in such a way that, for example, given the above EOL system, asking for the state of the nonterminal S at time 3 we pass through the following steps:

$$(S3)_\tau \ldots ((S2)_\tau (S2)_\tau)_\lambda \ldots (((S1)_\tau (S1)_\tau)_\lambda ((S1)_\tau (S1)_\tau)_\lambda)_\lambda \ldots$$

$$\ldots ((((S0)_\tau (S0)_\tau)_\lambda ((S0)_\tau (S0)_\tau)_\lambda)_\lambda (((S0)_\tau (S0)_\tau)_\lambda ((S0)_\tau (S0)_\tau)_\lambda)_\lambda)_\lambda \ldots$$

$$\ldots (((aa)_\lambda (aa)_\lambda)_\lambda ((aa)_\lambda (aa)_\lambda)_\lambda)_\lambda$$

Remark 1 - Since we use a leftmost outermost reduction rule the indicated expressions are not actually achieved during the reduction process.

Remark 2 - The parenthesis of λ-applications will always visualize the (binary) derivation tree of the achieved sentential form.

Remark 3 - In the case of CF systems, since the productions are allowed to act asynchronously we will need many time parameters to characterize the state corresponding to a sentential form. For example, given the grammar S::=ASA S::=b A::=a the state of the axiom S at times 1,2,1,1,1 is achieved through the steps

$(((((S1)_\tau 2)_\tau 1)_\tau 1)_\tau 1)_\tau \ldots ((((A0)_\tau (S1)_\tau (A0)_\tau)_\lambda 1)_\tau 1)_\tau \ldots$

$\ldots (((a(S1)_\tau a)_\lambda 1)_\tau 1)_\tau \ldots$

$\ldots (((a((A0)_\tau (S0)_\tau (A0)_\tau)_\lambda a)_\lambda \ldots$

$\ldots (a(aba)_\lambda a)_\lambda$

Remark 4 - As it is possible to realize by looking at the example the time parameters correspond to the lenghts of the branches in the derivation tree (reduced by 1). Note that the new branches are added in order of appearence of the correspondent nonterminal.

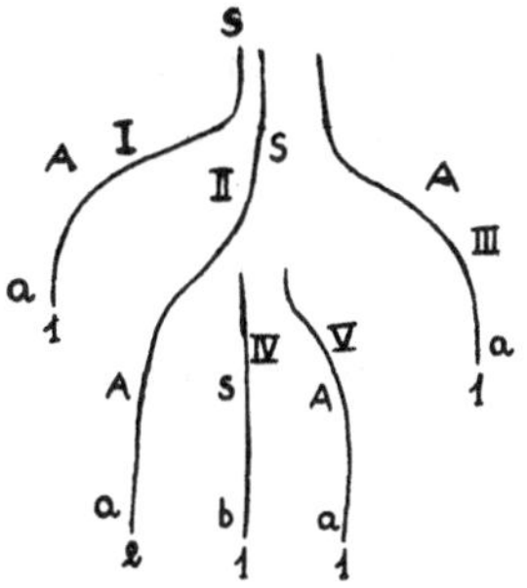

Remark 5 - According to the preceding point the time will be represented by integers (over any chosen numbering system).

Let $\alpha ::= \alpha_{i_1} \ldots \alpha_{i_n} | \alpha_{j_1} \ldots \alpha_{j_n} | \ldots | x_i | x_j | \ldots$ be the general form of the productions whose left members are the non terminal α. To the non terminal α we associate the formulae:

$$\tau t_1 [(\Delta \alpha_{i_1} t_1)_\tau \tau t_2 [(\Delta \alpha_{i_2} t_2)_\tau \tau t_3 [\ldots \tau t_n [(\Delta \alpha_{i_n} t_n)_\tau] \ldots]]]$$

where $(\Delta xy)_\tau$ is " if y=0 then x else $(x(y-1))_\tau$

$$\tau t_1 [(\Delta \alpha_{j_1} t_1)_\tau \tau t_2 [(\Delta \alpha_{j_2} t_2)_\tau \tau t_3 [\ldots \tau t_n [(\Delta \alpha_{j_n} t_n)] \ldots]]]$$

...

$$\tau t_1 [(\Delta' x_i t_1)_\tau] \quad ; \quad \tau t_1 [(\Delta' x_j t_1)_\tau]$$

where $(\Delta' xy)_\tau$ is " if y=0 then x else $(xy)_\tau$"

The reduction rule correspondent to α consists of replacing an occurrence of α by all associated formulae, non deterministically.

Then we characterize the languages according to the following definitions:

- the state of a sentential form $\beta_1 \ldots \beta_m \varepsilon N^+$ at time $n_1, \ldots, n_k$ is the parenthesized sentential form which may be obtained by asynchronously reducing $(\ldots(\beta_1 \ldots \beta_m n_1)_\tau \ldots n_k)_\tau$ (in some cases the normal form which is achieved may not be a sentential form)
- a CF language is the set of parenthesized sentential forms in $(V-N)^+$ which are the states of the axiom $S \varepsilon N$ in correspondence of all possible finite time parameter sets;
- an EOL language is the set of parenthesized sentential forms in $(V-N)^+$ which are the states of the axiom $W \in N^+$ achieved by synchronous reductions in correspondence of all possible time instants.

Example 1. Let S::=SS|a be the productions of an EOL system. To S the following formulae correspond:

$$S \equiv \tau t_1 [(\Delta S t_1)_\tau \tau t_2 [(\Delta S t_2)_\tau]]$$

$$S \equiv \tau t_1 [(\Delta' a t_1)_\tau]$$

Among all possible non deterministic reductions of $(S3)_\tau$ we give the following:

$(S3)_\tau \geq (\tau t_1 [(\Delta S t_1)_\tau \tau t_2 [(\Delta S t_2)_\tau]] 3)_\tau \geq \ldots$ (by globalization and synchronization of variables) $\ldots \geq (\tau t_1 [(\Delta S t_1)_\tau (\Delta S t_1)_\tau] 3)_\tau \geq$

$\geq ((\Delta S3)_\tau (\Delta S3)_\tau)_\lambda \geq$

$((S2)_\tau (\Delta S3)_\tau)_\lambda \geq (((\tau t_1 [(\Delta S t_1)_\tau \tau t_2 [(\Delta S t_2)_\tau]] 2)_\tau (\Delta S3)_\tau)_\lambda \geq \ldots$

$(((S1)_\tau (\Delta S2)_\tau)_\lambda (\Delta S3)_\tau)_\lambda \geq \ldots$

$((((S0)_\tau (\Delta S1)_\tau)_\lambda (\Delta S2)_\tau)_\lambda (\Delta S3)_\tau)_\lambda \geq \ldots$

$((((\tau t_1 [(\Delta' a t_1)_\tau] 0)_\tau (\Delta S1)_\tau)_\lambda (\Delta S2)_\tau)_\lambda (\Delta S3)_\tau)_\lambda \geq \ldots$

$(((a (S0)_\tau)_\lambda (\Delta S2)_\tau)_\lambda (\Delta S3)_\tau)_\lambda \geq \ldots$

$(((a\ a)_\lambda (a\ a)_\lambda)_\lambda ((a\ a)_\lambda (a\ a)_\lambda)_\lambda)_\lambda$

which corresponds to the derivation

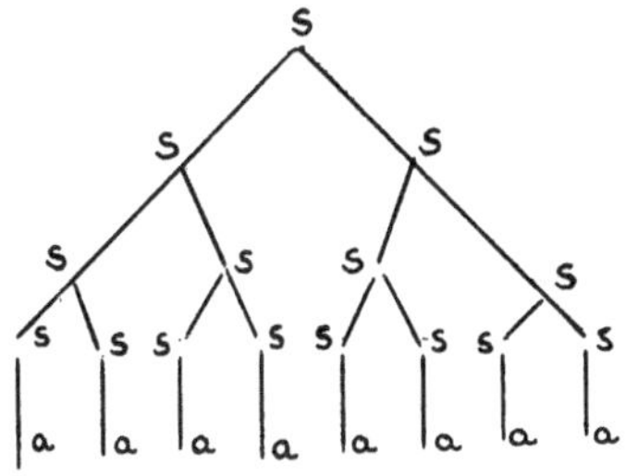

Example 2. Let S::=ASA|b A::=a be the productions of a CF language. To S and A the following formulae correspond:

$S \equiv \tau t_1[(\Delta A t_1)_\tau \tau t_2[(\Delta S t_2)_\tau \tau t_3[(\Delta A t_3)_\tau]]]$

$S \equiv \tau t_1[(\Delta' b t_1)_\tau]$

$A \equiv \tau t_1[(\Delta' a t_1)_\tau]$

Among all possible non deterministic reductions of $(((((S1)_\tau 2)_\tau 1)_\tau 1)_\tau 1)_\tau$ we give the following

$(((((S1)_\tau 2)_\tau 1)_\tau 1)_\tau 1)_\tau \geq$

$(((((\tau t_1[(\Delta A t_1)_\tau \tau t_2[(\Delta S t_2)_\tau \tau t_3[(\Delta A t_3)_\tau]]]1)_\tau 2)_\tau 1)_\tau 1)_\tau 1)_\tau \geq \ldots$

...(by asynchronous globalization)...$\geq$

$(((((\tau t_1 \tau t_2 \tau t_3[(\Delta A t_1)_\tau (\Delta S t_2)_\tau (\Delta A t_2)_\tau]1)_\tau 2)_\tau 1)_\tau 1)_\tau 1)_\tau \geq \ldots$

$((((\Delta A1)_\tau (\Delta S2)_\tau (\Delta A1)_\tau)_\lambda 1)_\tau 1)_\tau \geq \ldots$

$((((A0)_\tau (\Delta S2)_\tau (\Delta A1)_\tau)_\lambda 1)_\tau 1)_\tau \geq \ldots$

$((((\tau t_1[(\Delta' a t_1)_\tau]0)_\tau (\Delta S2)_\tau (\Delta A1)_\tau)_\lambda 1)_\tau 1)_\tau \geq \ldots$

$(((a(S1)_\tau (\Delta A1)_\tau)_\lambda 1)_\tau 1)_\tau \geq \ldots$

$(((a(\tau t_1[(\Delta A t_1)_\tau \tau t_2[(\Delta S t_2)_\tau \tau t_3[(\Delta A t_3)_\tau]]]1)_\tau (\Delta A1)_\tau)_\lambda 1)_\tau 1)_\tau \geq \ldots$

$(((a(\tau t_1 \tau t_2 \tau t_3[(\Delta A t_1)_\tau (\Delta S t_2)_\tau (\Delta A t_3)_\tau]1)_\tau (\Delta A1)_\tau)_\lambda 1)_\tau 1)_\tau \geq \ldots$

$(((a\ \tau t_2 \tau t_3[(\Delta A1)_\tau (\Delta S t_2)_\tau (\Delta A t_3)_\tau](\Delta A1)_\tau)_\lambda 1)_\tau 1)_\tau \geq \ldots$

$((\tau t_2 \tau t_3[a((\Delta A1)_\tau (\Delta S t_2)_\tau (\Delta A t_3)_\tau)_\lambda (\Delta A1)_\tau]1)_\tau 1)_\tau \geq \ldots$

$(a((\Delta A1)_\tau (\Delta S1)_\tau (\Delta A1)_\tau)_\lambda (\Delta A1)_\tau)_\lambda \geq \ldots$

$(a(a\,b\,a)_\lambda a)_\lambda$

which corresponds to the derivation

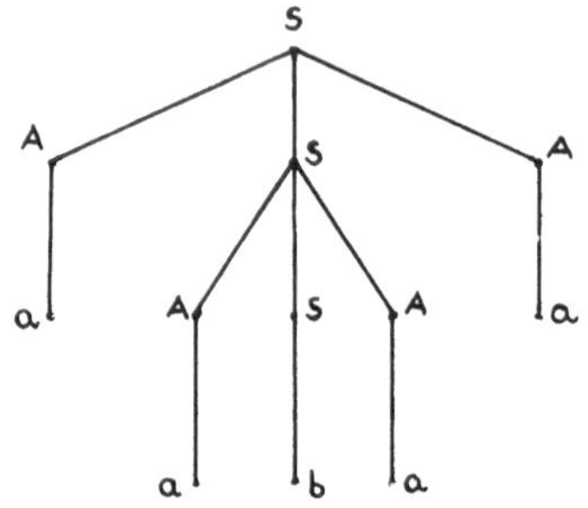

6. *Bibliography*

[1] Böhm, C.; W. Gross, Introduction to the CUCH, in Automata theory, Caianiello Ed., Academic Press, 1966.

[2] Böhm, C., The CUCH as a formal and description language, in Formal languages description languages for computer programming, North Holland, 1966.

[3] Landin, P.J., A correspondence between ALGOL 60 and Church's lambda notation, Communications ACM, 8, 1965

[4] Ausiello, G.; C. Böhm, Applicazione del linguaggio CUCH alla programmazione delle macchine analogiche sequenziali, Atti del IX Convegno della Automazione e Strumentazione, FAST, 1966.

[5] Morris, J.H., Lambda calculus models of programming languages, Ph. D. Thesis, MIT, 1968.

[6] Orgass, R.J.; F.B. Fitch, A theory of computing machines; A theory of programming languages, Studium Generale, 22, 1969.

[7] Wegner, P., Programming languages, information structures and machine organization, Mc Graw-Hill, 1968.

[8] Rescher, N.; A. Urquart, Temporal logic, Springer Verlag, 1971.

[9] Burstall, R.M., Program proving as hand simulation with a little induction, Proceedings of IFIP 74, 1974.

[10] Batini, C., Time structures with a root and a sink: a completeness proof, to be published in Pubblicazioni dell'Istituto di Automatica, Univ. di Roma, 1974.

[11] Salomaa, A., Parallelism in rewriting systems, in Automata, Languages and Programming, Lecture Notes in Computer Science 14, Springer Verlag, 1974

[12] Fitch, F., Representation of sequential circuits in combinatory logic, Phil. of Science, 25, 1958

[13] Lindenmayer, A., Developmental systems without cellular interactions, their languages and grammars, J. Theoretical Biology, 30, 1971

[14] Skyum, S., On extensions of ALGOL-like languages, DAIMI Publications, University of Aarhus, 1974

[15] Open House in Unusual Automata Theory, DAIMI Publications, University of Aarhus, 1973

[16] Salomaa, A., L-systems, Lecture Notes in Computer Science, Springer Verlag, 1974.

UNIFICATION IN TYPED LAMBDA CALCULUS

Gérard HUET

IRIA-LABORIA

Abstract :

This paper discusses the problem of finding common instances to terms in typed λ-calculus. It is shown that here the notion of most general unifier must be extended. Complete sets of unifiers are defined, and their structure with respect to substitution composition is studied.

Introduction :

We are concerned here with the problem of studying the set of common instances of two formulas, by substitution for their free variables. More precisely, we want to know the structure of the set of substitutions that unify these two formulas into a common instance. This problem is relevant to symbolic formula manipulation and automatic theorem proving.

In first-order logic, for any two terms e_1 and e_2 having a common instance, there exists a substitution σ called a most general unifier (MGU) of e_1 and e_2, such that for any unifier ρ (i.e. a substitution verifying $\rho e_1 = \rho e_2$) there exists some substitution η, with

$$\rho = \eta\sigma$$

The unification algorithm described by J.A. Robinson [17] computes for every pair $<e_1,e_2>$ a MGU, if there exists one, or returns a negative answer. This MGU is unique, up to composition with some substitution effecting a permutation on the set of variables.

For instance, taking

$$e_1 = P(F(x),x,u)$$

and $$e_2 = P(y,A,z),$$

then $\sigma = \{x \leftarrow A,\ y \leftarrow F(A),\ u \leftarrow z\}$ is a MGU,

$\sigma' = \{x \leftarrow A,\ y \leftarrow F(A),\ z \leftarrow u\}$ is another MGU,

but $\rho = \{x \leftarrow A,\ y \leftarrow F(A),\ u \leftarrow B,\ z \leftarrow B\}$ is a unifier less general than σ or σ'.

The unification algorithm is the basic tool needed for the inference rules used in automatic theorem proving, such as resolution [17], factoring and paramodulation. The generalization of these rules to higher-order logic is very complex, mainly because of the hard unification problems discussed here.

First we give an overview of the typed lambda calculus, used in formulations of higher-order logic such as the simple theory of types of Church [2].

I. A typed lambda-calculus.

1) Types

Every expression in the language possesses a unique type, defining its position in a functional hierarchy. We suppose given a finite set T_0 of elementary types, and we define the set T of types as the smallest superset of T_0 closed by the operation :

$$\alpha,\beta \in T \Rightarrow (\alpha\to\beta) \in T.$$

The composite type $(\alpha\to\beta)$ is the type of functions of domain elements of type α and of range elements of type β. We shall denote the types by the Greek symbols $\alpha,\beta,\gamma,\ldots$

2) Terms

The terms are the well-formed formulas of our language. We have three categories of terms :

a-atoms, either variables or constants.

We suppose given a denumerable set V_α of variables, for every type α, and an at most denumerable set C of constants of any type.

We impose $\forall\alpha,\beta \in T \ \alpha \neq \beta \Rightarrow V_\alpha \cap V_\beta = \emptyset$

and denoting $V = \bigcup_{\alpha \in T} V_\alpha$: $V \cap C = \emptyset$.

Variables will be denoted by lower case letters x,y,...,f,g,... and constants by upper case letters A,B,...,F,G,... Atoms in general will be denoted by @,@',...

b-applications

For any term e_1 of type $(\alpha\to\beta)$ and for any term e_2 of type α, we define $(e_1 e_2)$ as a term of type β.

c-abstractions

For any term e of type β and any variable u of type α (i.e. in V_α), we define $\lambda u.e$ as a term of type $(\alpha\to\beta)$.

More precisely, we can now define the set of terms as the smallest set containing $V \cup C$ and closed by the operations of application and abstraction. We shall denote by $\tau(e)$ the type of term e.

We shall use the context notation : $E[e]$ denotes a term in which we distinguish an occurrence of subterm e. $E[e']$ denotes the same term, in which the distinguished occurrence is replaced by e', provided of course that $\tau(e) = \tau(e')$.

Let $E = E[\lambda u.e]$. All the occurrences of u in $\lambda u.e$ are said to be *bound* in E. Any occurrence of a variable which is not bound is said to be *free*. We denote by $F(E)$ the set of all variables having some free occurrence in term E.

We define $\langle e_1/u \rangle e_2$ as the term obtained from e_2 by substituting every free occurrence of u by e_1, provided $\tau(e_1) = \tau(u)$.

3) λ-conversion

We have the usual rules of α-conversion and β-reduction.

a- α-conversion

Let $e_1 = E[\lambda u.e_2]$, and let v be a variable which does not occur in e_2, with $\tau(v) = \tau(u)$. We say that $e_3 = E[\lambda v.<v/u>e_2]$ derives from e_1 by α-conversion, and we write $e_1 \underset{\alpha}{\rightarrow} e_3$.

b- β-reduction

Let $e_1 = E[(\lambda u.e_2\ e_3)]$. If no variable in $F(e_3)$ occurs bound in e_2, we say that $e_4 = E[<e_3/u>e_2]$ derives from e_1 by β-reduction, and we write $e_1 \underset{\beta}{\rightarrow} e_4$.

c- λ-conversion

λ-conversion is the reflexive and transitive closure of α-conversion and β-reduction.

In general we shall omit α-conversions, which are necessary only to effect the necessary renamings needed by the conditions of application of the β-reduction rule.

4) Normal form

A term is said to be in *normal form* if it does not contain any subterm of the form (λu.e e').

It is well-known that, in typed λ-calculus, every term e can be transformed into a term in normal form, using λ-conversion. By the Church-Rosser property, this term is unique, up to α-conversion. It is called *the normal form* of e.

Actually, a stronger result holds : using α-conversion only when needed to apply β-reduction, every sequence of λ-conversions leads to the normal form of any term. This strong normalization theorem may be found in Sanchis [18] .

We shall use the following abbreviations :

- $(\ldots((e_1 e_2)e_3)\ldots e_n)$ will be written

  $e_1(e_2, e_3, \ldots, e_n)$

  when no ambiguity arises.

- $\lambda u_1.\lambda u_2.\ \ldots\lambda u_n.e$ will be written $\lambda u_1 u_2 \ldots u_n.e$ when $u_1, \ldots, u_n$ are distinct variables.

Any term in normal form may thus be standardized into the general form :

$\lambda u_1 u_2 \ldots u_n . @(e_1, e_2, \ldots, e_p)$ where :

. $@ \in V \cup C$

. $n \geq 0$; if $n = 0$, $\lambda.$ is omitted.

. $p \geq 0$; if $p = 0$, ( ) is omitted.

. $u_1 \ldots u_n$ are distinct variables.

. e_i is a term standardized in the same way, $1 \leq i \leq p$.

We may impose in supplement that no u_i appears bound in any e_j, $1 \leq i \leq n$, $1 \leq j \leq p$.

Remark that no ambiguity is possible. For instance, $\lambda u.F(x)$ is an abbreviation of $\lambda u.(Fx)$ and not of $(\lambda u.Fx)$, which would be β-reduced in its normal form F.

In the following, we shall write $e_1 = e_2$ if and only if e_1 and e_2 have the same normal form (up to α-conversion). We shall denote by T_α the set of terms of type α in normal form, and by $T = \bigcup_{\alpha \in T} T_\alpha$ the set of terms in normal form.

5) Substitutions

A substitution is a finite set of pairs

$$\sigma = \{<x_i, e_i> \mid 1 \leq i \leq n\}$$

such that $\forall_i \leq n \ \tau(x_i) = \tau(e_i)$

and $i \neq j \Rightarrow x_i \neq x_j$.

We shall ignore in σ the pairs $<x,x>$; that is :

$$\sigma = \sigma' \iff [(\sigma - \sigma') \cup (\sigma' - \sigma)] \subset \{<x,x> \mid x \in V\} .$$

In other words, a substitution σ is a type-preserving mapping from V to T, equal to the identity almost everywhere. this mapping is extended to T as follows.

We define the application of substitution $\sigma = \{<x_i, e_i> \mid 1 \leq i \leq n\}$ to term e, written σe, as the normal form of the term :

$$(\lambda x_1 \ldots x_n . e)(e_1, \ldots, e_n).$$

Since the conflicts of variables are automatically resolved when reducing to normal form, σe does not depend on the order in which we take the x_i's in σ, as can be proved easily.

Let us denote by S the set of substitutions.

We shall now state without proofs a few easy lemmas on substitutions.

Lemma 1 $\forall \sigma, \rho \in S$

$$\sigma = \rho \iff \forall x \in V \quad \sigma x = \rho x \iff \forall e \in T \quad \sigma e = \rho e$$

Definition

We define the *composition* of substitutions σ and ρ, written $\rho\sigma$, as the substitution :

$$\rho\sigma = \{\langle x, \rho(\sigma x)\rangle \mid x \in V\}$$

This is exactly the composition of the corresponding mappings, as shown by the next lemma.

Lemma 2

$$\forall \sigma, \rho \in S \quad \forall e \in T \quad (\rho\sigma)e = \rho(\sigma e).$$

From this we get immediately the associativity of substitution composition :

Lemma 3

$$\forall \sigma, \rho, \eta \in S : (\sigma\rho)\eta = \sigma(\rho\eta).$$

The last two lemmas allow us to suppress parentheses, and to write $\rho\sigma e$ and $\sigma\rho\eta$.

We call *domain* of substitution σ the finite set :

$$\mathcal{D}(\sigma) = \{x \in V \mid \sigma\, x \neq x\}.$$

We end this section with one last easy lemma :

Lemma 4

$$\forall e \in T, \forall \sigma \in S : \mathcal{D}(\sigma) \cap F(e) = \emptyset \Rightarrow \sigma e = e.$$

II. The unification problem.

1 Unifiers and complete sets of unifiers.

Definitions.

We call *unifier* of two terms e_1 and e_2 of the same type any substitution σ such that $\sigma e_1 = \sigma e_2$. We write :

$$\mathcal{U}(e_1,e_2) = \{\sigma \in \mathcal{S} \mid \sigma e_1 = \sigma e_2\}.$$

e_1 and e_2 are said to be *unifiable* iff $\mathcal{U}(e_1,e_2) \neq \emptyset$.

Of course, if $\sigma \in \mathcal{U}(e_1,e_2)$, then for every $\eta \in \mathcal{S}$ we also have $\eta\sigma \in \mathcal{U}(e_1,e_2)$. We say that $\eta\sigma$ is a unifier *less general* than σ.

We call *most general unifier*, or MGU, of e_1 and e_2 any substitution σ such that :

$$\begin{cases} \sigma \in \mathcal{U}(e_1,e_2) \\ \forall\rho \in \mathcal{U}(e_1,e_2)\ \exists\, \eta \in \mathcal{S} : \rho = \eta\sigma. \end{cases}$$

Our first result is to shown that most general unifiers do not always exist.

Proposition 1.

Certain pairs of unifiable terms do not possess a most general unifier.

Proof

Let $\begin{cases} e_1 = f(A) \\ e_2 = A \end{cases}$ with $\begin{cases} \tau(A) = \alpha \\ \tau(f) = (\alpha\to\alpha). \end{cases}$

Let us consider :

$$\sigma_1 = \{\langle f,\lambda u.A\rangle\} \in \mathcal{U}(e_1,e_2)$$

$$\sigma_2 = \{\langle f,\lambda u.u\rangle\} \in \mathcal{U}(e_1,e_2)$$

$\mathcal{F}(\lambda u.A) = \emptyset$, and by lemma 4

$\forall\eta \in \mathcal{S}\ \ \eta(\lambda u.A) = \lambda u.A \neq \lambda u.u$. Therefore, there does not exist η such that $\sigma_1 = \eta\sigma_2$. In the same way, there is no η such that $\sigma_2 = \eta\sigma_1$.

Finally, it is easy to show that every unifier of e_1 and e_2 must contain either σ_1 or σ_2, which concludes the proof. □

The situation is therefore very different from first-order unification. Moreover, we have here the need to introduce "new" variables, in order to describe the unifiers of two terms. For instance, let us consider $e_1 = f(A)$ and $e_2 = f(B)$, with $\tau(f) = (\alpha \to \alpha)$. Any substitution that substitutes to f a constant function, for instance $\lambda u.C$, is in $U(e_1,e_2)$. Conversely, this is the only way to unify e_1 and e_2. We would therefore like to say that $\sigma = \{<f,\lambda u.x>\}$ is a MGU of e_1 and e_2. However, this is not quite true since, taking $\rho = \{<f,\lambda u.C>\}$ and $\eta = \{<x,C>\}$, we have

$$\eta\sigma = \rho \cup \eta \neq \rho.$$

However, we have $\rho e = \eta\sigma e$ for every term e such that $x \notin F(e)$, and this is all that really matters. This leads us to the definitions below.

Definitions

Let V be a finite set of variables, σ a substitution. We call *restriction of* σ *to* V the substitution :

$$\sigma{\restriction}V = \{<x_i,e_i> \mid \sigma x_i = e_i \ \& \ x_i \in V\}.$$

For every V, we define an equivalence $\underset{V}{=}$ between substitutions by ;

$$\sigma \underset{V}{=} \sigma' \iff \sigma{\restriction}V = \sigma'{\restriction}V.$$

We can now define a relation $\underset{V}{\leq}$ by :

$$\sigma \underset{V}{\leq} \sigma' \iff \exists \eta \in S : \sigma \underset{V}{=} \eta\sigma'.$$

We say that σ is *less general than* σ' *on* V.

Using lemma 4, we have $\sigma e = (\sigma{\restriction}F(e))e$ for every σ and e, and we can write the analogue of lemma 1 as :

Lemma 5

$\forall\sigma, \rho \in S, \forall e \in T$:

$$\sigma \underset{F}{=} \rho \iff \forall x \in F \ \sigma x = \rho x \implies \sigma e = \rho e, \text{ where } F = F(e).$$

However, the last converse does not hold here, taking for instance $e = f(x)$, $\sigma = \{<f,\lambda u.u>\}$, $\rho = \{<f,\lambda u.x>\}$.

Lemma 6

If $\rho \underset{V}{=} \rho'$, then for every σ in S $\sigma\rho \underset{V}{=} \sigma\rho'$.

Proof

$\rho \underset{V}{=} \rho'$ implies $\forall x \in V \quad \rho x = \rho' x$.

$\sigma\rho \restriction V = \{<x,\sigma e> \mid \rho x = e \ \& \ x \in V\} = \sigma\rho' \restriction V$. □

Corollary $\underset{V}{\leq}$ is a transitive relation. (Immediate).

Remark however that it is not true that $\rho \underset{V}{=} \rho'$ implies $\rho\sigma \underset{V}{=} \rho'\sigma$. Consider for instance $V = \{x\}, \sigma = \{<x,z>\}$, $\rho = \emptyset$, $\rho' = \{<z,A>\}$.

$\underset{V}{\leq}$ is a reflexive and transitive relation, defining a preorder structure on S. It is not antisymmetric, even on the quotient by $\underset{V}{=}$. For instance, taking

$\sigma = \{<x,f(A)>\}$, $\sigma' = \{<x,y>\}$ and $V = \{x\}$,

we have $\sigma \underset{V}{=} \rho\sigma'$ with $\rho = \{<y,f(A)>\}$

and $\sigma' \underset{V}{=} \rho'\sigma$ with $\rho' = \{<f,\lambda u.y>\}$,

but $\sigma \underset{V}{\neq} \sigma'$. We therefore need to define a new equivalence :

$$\sigma \underset{V}{\equiv} \sigma' \iff \sigma \underset{V}{\leq} \sigma' \ \& \ \sigma' \underset{V}{\leq} \sigma.$$

We are now able to state our main definition.

Definition.

Let e_1 and e_2 be two terms of the same type, and V be a finite set of variables containing $F(e_1)$ and $F(e_2)$. We call *complete set of unifiers* (CSU) *of* e_1 *and* e_2 *on* V any set of substitutions Σ such that :

1) $\Sigma \subset U(e_1,e_2)$ (consistence)

2) $\forall\rho \in U(e_1,e_2) \ \exists\sigma \in \Sigma : \rho \underset{V}{\leq} \sigma$ (completeness)

3) $\forall\sigma_1,\sigma_2 \in \Sigma \ \sigma_1 \underset{V}{\not\equiv} \sigma_2$ (non-duplication)

$\sigma_1 \neq \sigma_2$

We shall write $\Sigma \in CSU(e_1,e_2,V)$. Remark that, for every e_1,e_2 and V, the set $CSU(e_1,e_2,V)$ is never empty, since it always contains at least $\mathcal{U}(e_1,e_2) \,/\, \overset{\equiv}{V}$.

The first question we ask concerning CSU's regards finiteness : if e_1 and e_2 are unifiable, do they always possess a finite CSU ? The next section will answer by the negative.

2. Non-finiteness of CSU's

Proposition 2. Certain pairs of unifiable terms do not possess a finite CSU. I.e., there exist $e_1,e_2 \in \mathcal{T}_\alpha$ with $\mathcal{U}(e_1,e_2) \neq \emptyset$ and $V \subset \mathcal{V}$, with $\mathcal{F}(e_1) \cup \mathcal{F}(e_2) \subset V$, such that every Σ in $CSU(e_1,e_2,V)$ is infinite.

Proof :

$$\text{Let} \begin{cases} e_1 = f(F(A)) \\ e_2 = F(f(A)) \\ V = \{f\} \end{cases} \quad \text{with} \begin{cases} \tau(A) = \alpha \\ \tau(f) = \tau(F) = (\alpha\to\alpha) \end{cases}$$

and let us consider

$$\Sigma = \{\{<f,F>\}\} \cup \{\{<f,\lambda u.F^n(u)>\} \mid n \geq 0\}, \text{ where}$$

$$F^0(u) = u, \text{ and } F^n(u) = F(F^{n-1}(u)).$$

First, we want to show that Σ is a CSU of e_1 and e_2 on V.

The first condition is easy to check, we verify that every σ in Σ unifies e_1 and e_2.

For the completeness condition, let $\rho \in \mathcal{U}(e_1,e_2)$. There must exist in ρ some pair <f,e>, by lemma 4; let us write $\rho = \{<f,e>\} \cup \bar{\rho}$.

Case 1. If e is not an abstraction, then

$$\rho e_1 = \{<f,e>\}e_1 \quad \text{by lemma 5}$$
$$= (e(FA))^*.$$

* We do not use the standardized form here, because it depends on term e. For instance, if e = B, then it is B(F(A)), whereas, if e = (CD) then it is C(D,F(A)).

Same : $\rho e_2 = (F(eA))$.

Since $\rho e_1 = \rho e_2$, we must have $e = F$, and therefore

$$\rho = \bar{\rho}\{<f,F>\}.$$

Case 2. If e is an abstraction : $e = \lambda u.e'$, with $\tau(u) = \tau(e') = \alpha$. Let p be the largest integer such that $e' = F^p(e'')$, with $e'' \in T_\alpha$.

Then $\rho e_1 = F^p(<F(A)/u>e'')$

and $\rho e_2 = F^{p+1}(<A/u>e'')$ and since e" does not start with an F, in order to have $\rho e_1 = \rho e_2$ it must start with a u. Since $\tau(e'') = \tau(u)$, this imposes $e'' = u$, and we have

$$\rho = \bar{\rho}\{<f,\lambda u.F^p\ (u)>\}.$$

It is easy to check that all the substitutions in Σ are independent, i.e., that :

(1) $\forall \sigma_1,\sigma_2 \in \Sigma \quad \sigma_1 \not\leq_V \sigma_2$, since for every σ in Σ :
$\sigma_1 \neq \sigma_2$

$<f,e> \in \sigma \Rightarrow F(e) = \emptyset$. This concludes the proof that Σ is a CSU of e_1 and e_2 on V.

Now let Σ' be any finite CSU of e_1 and e_2 on V. This implies that there exist σ_1 and σ_2 in Σ, $\sigma_1 \neq \sigma_2$, such that $\exists \rho \in \Sigma : \sigma_1 \leq_V \rho$ and $\sigma_2 \leq_V \rho$. But, since Σ is a CSU, $\exists \sigma \in \Sigma : \rho \leq_V \sigma$ and, by transitivity of $\leq_V$:

(2) $\sigma_1 \leq_V \sigma$

and (3) $\sigma_2 \leq_V \sigma$.

If $\sigma \neq \sigma_1$ (2) contradicts (1), otherwise (3) contradicts (1). This shows that there cannot exist any finite set in $CSU(e_1,e_2,V)$. □

3. Redundancy of CSU's

Although we sometimes must consider infinite CSU's, we may wonder whether or not one may impose on CSU's a non-redundancy condition stronger than the condition (3) of non-duplication. More precisely, we would like to replace (3), in the definition of a CSU, by :

$$(3')\quad \forall \sigma_1,\sigma_2 \in \Sigma \quad \sigma_1 \not\leq_V \sigma_2 \qquad \text{(non redundancy).}$$
$$\sigma_1 \neq \sigma_2$$

A CSU that verifies such a condition will be called a *complete set of maximal unifiers* (CSMU).

We may wonder whether such a CSMU always exists. The next proposition answers this question negatively, answering a conjecture of Plotkin [16].

Proposition 3.

Certain pairs of unifiable terms do not possess a CSMU.

Proof.

Let
$$\begin{cases} e_1 = f(x,A) \\ e_2 = f(x,B) \\ V = \{x,f\} \end{cases} \quad \text{with} \quad \begin{cases} \tau(A) = \tau(B) = \alpha \\ \tau(x) = (\alpha\to\alpha) \\ \tau(f) = ((\alpha\to\alpha) \to (\alpha\to\alpha)) \end{cases}$$

We consider :

$$\rho = \{<f,\lambda uv.h(u)>\},$$
$$\sigma_0 = \{<f,\lambda u.u>,<x,\lambda v.z>\}$$

and
$$\sigma_n = \{<f,\lambda uv.g_n(u,u(h_1^n\ (u,v)),\dots,u(h_n^n(u,v)))>,<x,\lambda v.z>\}\ (n>0)$$

with
$$\begin{cases} \tau(v) = \tau(z) = \alpha \\ \tau(u) = (\alpha\to\alpha) \\ \tau(h) = ((\alpha\to\alpha) \to \alpha) \\ \tau(h_i^j) = ((\alpha\to\alpha) \to (\alpha\to\alpha)), \\ \tau(g_i) = ((\alpha\to\alpha) \to \underbrace{(\alpha\to \dots(\alpha\to\alpha)\dots)}_{i}) \end{cases}$$

and let $\Sigma = \{\sigma_i \mid i\geq 0\} \cup \{\rho\}$.

The proof follows from a number of lemmas, like for proposition 2.

(a) Σ is a CSU of e_1 and e_2 on V.

First, we check that for every $\sigma\epsilon\Sigma$, $\sigma e_1 = \sigma e_2$.

Let us now consider an arbitrary unifier σ of e_1 and e_2.

Let $e_f = \sigma f$

and $e_x = \sigma x$.

We first show by cases on the structure of e_f that σ is less general than some member of Σ.

<u>a-1.</u> e_f is not an abstraction.

Then $\sigma e_1 = e_f(e_x,A) \neq e_f(e_x,B) = \sigma e_2$, and therefore this case cannot arise.

<u>a-2.</u> $e_f = \lambda u.e$, where e is not an abstraction.

Then, denoting by E the normal form of $<e_x/u>e$, we have :

$$\sigma e_1 = E(A)$$

$$\text{and } \sigma e_2 = E(B).$$

Therefore $\sigma e_1 = \sigma e_2$ iff $E = \lambda v.E'$, where $v \notin \mathcal{F}(E')$. Since e is not an abstraction, there are three cases :

<u>a-2-1.</u> $e = u$, then $e_x = \lambda v.E'$, and therefore $\sigma = \{<z,E'>\}\sigma_o$.

<u>a-2-2.</u> $e = z$; this is impossible, because we would have $E = z$.

<u>a-2-3.</u> $e = (e'e'')$, where e' is not an abstraction. Then

$$E = (<e_x/u>e' \quad <e_x/u>e'') = \lambda v.E'$$

imposes that $<e_x/u>e'$ be an abstraction, and therefore that $e' = u$ or $e' = (u\ e''')$. But this is impossible because of the types, and thus this case cannot arise.

a-3. $e_f = \lambda uv.e$.

The proof proceeds by cases on the structure of e.

a-3-1. $v \notin F(e)$.

Then $\sigma \underset{V}{=} \{\langle h,\lambda u.e\rangle, \langle x,e_x\rangle\}\rho$.

a-3-2. If a contains a free occurrence of v which is not inside the argument to an u, the we prove that, for every term e_x, v occurs free in the normal form E of $\langle e_x/u\rangle e$. The proof goes by induction on the structure of e.

Let us write $e = \lambda w_1 \ldots w_n . @(E_1, \ldots, E_p)$. If $@ = v$ (with p = o) this is trivial. Otherwise, by hypothesis we have $@ \neq u$ and, neglecting possible renamings we have :

$$\langle e_x/u\rangle e = \lambda w_1 \ldots w_n . @(\ldots, \langle e_x/u\rangle E_i, \ldots).$$

Since at least one of the E_i's must satisfy the induction hypothesis, we get easily the result.

Now

$\sigma e_1 = \langle A/v\rangle E \neq \langle B/v\rangle E = \sigma e_2$, contrary to the hypothesis, which proves that this case cannot arise.

a-3-3. Otherwise, it is possible to write e under the form :

$$e = E[u,u,\ldots,u,u(E_1),u(E_2),\ldots,u(E_n)]$$

where all occurrences of u and v in e appear between the brackets. Since $v \in F(e)$, this imposes $\exists k \leq n \;\; v \in F(E_k)$. In this notation, we are supposing that the occurrences of u appearing isolated do not appear in the left part of an application (in which case we would have written instead the corresponding term $u(E_i)$).

We have now two cases, according to the structure of e_x.

a-3-3-1. If $e_x = \lambda v.E'$, with $v \notin F(E')$,

then $\sigma \underset{V}{=} \eta\sigma_n$, with

$$\eta = \{\langle g_n, \lambda uw_1 \ldots w_n . E[u,u,\ldots,u,w_1,w_2,\ldots,w_n]\rangle, \langle z,E'\rangle\} \cup \{\langle h_i^n, \lambda uv.E_i\rangle \mid 1 \leq i \leq n\}.$$

<u>a-3-3-2</u>. In all other cases, it is easy to show that v appears free in the normal form of $<e_x/u>u(E_k)$, and therefore also in the normal form of $<e_x/u>e$. We can conclude as in a-3-2 that this case cannot arise.

This concludes the proof of the completeness of Σ. It remains to show its non-duplication. In fact we are going to study in more detail the structure of Σ with respect to $\underset{V}{\leq}$.

(b) $\forall_i > o \qquad \sigma_i \underset{V}{\leq} \sigma_{i+1}$.

Let us consider :

$$\eta_i = \{<g_{i+1}, \lambda uv_1 v_2 \ldots v_{i+1}.g_i(u,v_1,\ldots,v_i)>\} \cup \{<h_j^{i+1}, h_j^i> \mid 1 \leq j \leq i\}.$$

We check that :

$$\sigma_i \underset{V}{=} \eta_i \ \sigma_{i+1}.$$

(c) $\forall_i \geq o \qquad \rho \underset{V}{\not\leq} \sigma_i$.

Let us suppose that there exists η such that $\rho \underset{V}{=} \eta\sigma_i$. In particular, we should have

$$\rho x = x = \eta(\lambda v.z) = \lambda v'.\eta z, \text{ which is impossible.}$$

(d) $\sigma_1 \underset{V}{\not\leq} \rho$

Let us suppose that there exists η such that $\sigma_1 \underset{V}{=} \eta\rho$. In particular, we should have :

$\sigma_1 f = \eta(\lambda uv.h(u)) = \lambda uv.g_1(u(h_1^1(u,v)))$, which is impossible, since

$\eta(\lambda uv.h(u)) = \lambda u'v'.e$, with $v' \notin F(e)$.

(e) $\forall_i > o \quad \sigma_o \underset{V}{\not\leq} \sigma_i$.

In the same way, if $\sigma_o \underset{V}{=} \eta\sigma_i$, we should have :

$$\sigma_o f = \lambda u.u = \eta(\lambda uv.g_i(\ldots)) = \lambda u'v'.e, \quad \text{impossible.}$$

(f) $\sigma_1 \underset{V}{\not\leq} \sigma_o$.

In the same way, if $\sigma_1 \underset{V}{=} \eta\sigma_o$, we should have :

$$\sigma_1 f = \eta(\lambda u.u) = \lambda u.u, \text{ impossible.}$$

(g) $\forall_{i>0}\ \sigma_i \not\leq_V \sigma_0\ \&\ \sigma_i \not\leq_V \rho$.

Immediate from (b), (d), (f) and the transitivity of $\leq_V$.

(h) $\sigma_0 \not\leq_V \rho$.

Proof same as for (e).

(i) $\forall_{i>0}\ \sigma_{i+1} \not\leq_V \sigma_i$.

This proof is rather tedious, essentially for notational reasons. We are going to show that $\sigma_2 \not\leq_V \sigma_1$; the general proof is similar.

Let us suppose that there exists η such that $\sigma_2 =_V \eta\sigma_1$. According to lemma 6 we must have, for every $\xi \in S$: $\xi\sigma_2 = \xi\eta\sigma_1$. Let us choose

$$\xi = \{<g_2,G>,<h_1^2,\lambda uv.C(v)>,<h_2^2,\lambda uv.D(v)>\},$$

with $\tau(G) = \tau(g_2)$, $\tau(C) = \tau(D) = (\alpha\to\alpha)$, and let $\zeta = \xi\eta$. We must have, in particular :

$$\xi\sigma_2 f = \lambda uv.G(u,u(C(v)),u(D(v))) = \zeta(\lambda uv.g_1(u,u(h_1^1(u,v)))).$$

Let us write $e = \zeta g_1$, $e' = \zeta(h_1^1(u,v))$; we may suppose that neither u nor v appears free in ζg_1 and ζh_1^1. We must therefore get $G(u,u(C(v)),u(D(v)))$ as normal form of $e(u,u(e'))$, which u and v not free in e. A study by cases on the structure of e shows that we must have $e = \lambda w_1 w_2.G(E_1,E_2,E_3)$. Now, by cases on E_2, we have two possibilities :

1) $E_2 = w_1(C(\bar{E}_2))$. In this case we must have

$$v = <u/w_1><u(e')/w_2>\bar{E}_2 \quad \text{with} \quad v \notin F(\bar{E}_2),$$

which is impossible.

2) $E_2 = w_2$. In this case we must have $e' = C(v)$, and by the same reasoning applied to E_3, $e' = D(v)$, which is impossible.

This complete the proof of $\sigma_2 \not\leq_V \sigma_1$.

(j) We have shown so far that Σ is a CSU of e_1 and e_2 on V having the following structure :

$$\rho \quad \sigma_0 \quad \sigma_1<\sigma_2<\ldots<\sigma_n<\ldots$$

where $\sigma<\sigma'$ denotes $\sigma \leq_V \sigma'$ and $\sigma' \not\leq_V \sigma$.

Let Σ' be any CSMU of e_1 and e_2 on V. Since Σ' is a CSU, $\exists\mu\in\Sigma' : \sigma_1 \underset{V}{\leq} \mu$. But, Σ being a CSU too, $\exists\sigma\in\Sigma : \mu \underset{V}{\leq} \sigma$. By transitivity $\sigma_1 \underset{V}{\leq} \sigma$, and therefore $\sigma \neq \rho$ by (d), and $\sigma \neq \sigma_o$ by (f). There must therefore exist $k\geq 1$ such that $\sigma = \sigma_k$. Now using again the fact that Σ' is a CSU, $\exists\mu'\in\Sigma' : \sigma_{k+1} \underset{V}{\leq} \mu'$. Using (b) and the transitivity of $\underset{V}{\leq}$, we get $\mu \underset{V}{\leq} \mu'$. But Σ' being a CSMU, this implies $\mu = \mu'$, and now we get by transitivity of $\underset{V}{\leq} : \sigma_{k+1} \underset{V}{\leq} \sigma_k$. This is impossible, by (i), which contradicts the existence of Σ'.

This completes the proof of proposition 3. □

Jensen and Pietrzykowski give in [19] an algorithmic procedure to generate a CSU for two terms e_1 and e_2. This procedure may not stop, by proposition 2, and it is impossible to avoid generating redundant solutions, by proposition 3. In contrast, it is shown in [12] that is possible to test whether two terms are unifiable or not without doing redundant computation. However, it is only possible to give a semi-decision algorithm to test unifiability, as shown by the next proposition.

4. Undecidability of unification.

Proposition 4.

Unifiability is undecidable. In order words, the set

$$\{\langle e_1,e_2\rangle \in \bigsqcup_{\alpha\in T} \mathcal{T}_\alpha \times \mathcal{T}_\alpha \mid \mathcal{U}(e_1,e_2) = \emptyset\}$$

is not recursive.

Proof.

Let us consider the alphabet of two variables $\Delta = \{u,v\}$, with $\tau(u) = \tau(v) = (\alpha\to\alpha)$. With every word $\xi\in\Delta^*$ we associate the term $\tilde{\xi}\in\mathcal{T}_{(\alpha\to\alpha)}$ defined recursively by :

$$\begin{cases} \tilde{\Lambda} = \lambda z.z & (\Lambda \text{ is the empty word in } \Delta^*) \\ \widetilde{u\xi} = \lambda z.u(\tilde{\xi}(z)) & \\ \widetilde{v\xi} = \lambda z.v(\tilde{\xi}(z)) & \text{where } \tau(z) = \alpha. \end{cases}$$

With every Post correspondence problem $\Pi = \begin{pmatrix} X_1 & X_2 & \dots & X_n \\ Y_1 & Y_2 & \dots & Y_n \end{pmatrix}$

over Δ, we associate the unifiability problem for the terms :

$$\begin{cases} e_1 = \lambda uvw.w(f(\widetilde{X}_1,\widetilde{X}_2,\ldots,\widetilde{X}_n),f(u,u,\ldots,u)) \\ e_2 = \lambda uvw.w(f(\widetilde{Y}_1,\widetilde{Y}_2,\ldots,\widetilde{Y}_n),u(g(u))) \end{cases}$$

where

$$\begin{cases} \tau(f) = \underbrace{((\alpha\to\alpha) \to ((\alpha\to\alpha) \to (\ldots\to((\alpha\to\alpha)}_{n} \to \alpha)\ldots))) \\ \tau(g) = ((\alpha\to\alpha) \to \alpha) \\ \tau(w) = (\alpha \to (\alpha\to\alpha)). \end{cases}$$

It is shown in [10] that $\mathcal{U}(e_1,e_2) \neq \emptyset$ if and only if Π has a solution. This proves that unifiability is undecidable in our language. □

5. Discussion and open problems

Let us define a function $\pi : T \to \mathbb{N}$ by :

$$\pi(\alpha) = \begin{cases} 1 \text{ if } \alpha\in T_0 \\ \max(\pi(\beta)+1,\ \pi(\gamma)) \text{ if } \alpha=(\beta\to\gamma). \end{cases}$$

We call *order* of a term e the number $\pi(\tau(e))$. This notion corresponds to the usual notion of order : at order 1 terms represent individuals, at order 2 functions, at order 3 fonctionals, etc...

Let us call *language of order* n a language where we restrict ourselves to variables of order at most n. The full language is said to be of order ω. Note that for us languages of order 1 correspond with the usual terminology of sorted first-order languages (types play the role of arities and sorts).

Looking back at the examples used in the proofs above, we see that propositions 1 and 2 are true over any language of order ≥ 2, and that propositions 3 and 4 are true over any language of order ≥ 3. The four propositions are false in first-order languages, since we can define in this case a most general unifier, obtainable with the unification algorithm [17].

Propositions 3 and 4 are still open for second-order languages. Proposition 3 is conjectured to be false. It may be possible for instance to modify the unification algorithm given in [14] in such a way that it generates a CSMU. Proposition 4 is certainly much harder to settle in 2nd order logic.

The unifiability problem generally corresponds to the existence of solutions to sets of equations in some algebraic system. For instance, if we restrict our functions to be of arity 1 (i.e. $T = T_o \cup (T_o \to T_o)$) and if we limit abstraction so that we stay in λ-I-calculus (i.e. $\lambda u.e$ is a term only if $u \in F(e)$), then terms are just strings. The unifiability problem is equivalent in this case to finding a solution to an equation in some free monoid, a well-known open problem. Allowing λ-K-calculus terms amounts to having in supplement a right-zero element. Related problems are treated in [16].

If we add to our λ-calculus an η-rule :

$$\lambda u.e(u) = e \quad \text{provided} \quad u \notin F(e),$$

all the results stay the same.

BIBLIOGRAPHY

[1] Andrews P.B., (1971) : " Resolution in type theory". Journal of Symbolic Logic 36,3 pp. 414-432.

[2] Church A. (1940) : "A formulation of the simple theory of types" Journal of Symbolic Logic 5,1 pp. 56-68.

[3] Church A. (1941) : "The calculi of lambda-conversion." Annals of Mathematical Studies n°6. Princeton University Press.

[4] Curry H.B., Feys R., Craig W. (1958) : Combinatory Logic, vol 1, North Holland.

[5] Darlington J.L., (1971) : "A partial mechanization of second-order logic". Machine Intelligence 6, pp. 91-100. American Elsevier, New York.

[6] Darlington J.L. (1973) : "Automatic program synthesis in second-order logic". Proceedings 3rd IJCAI, Stanford August 73.

[7] Ernst G.W. (1971) : "A matching procedure for type theory". Personal communication.

[8] Gould W.E. (1966) : "A matching procedure for ω-order logic". Scientific report n°4, AFCRL 666-781. Contrat AF 19(628)-3250.AD 646-560.

[9] Guard J.R. (1964) : "Automated logic for semi-automated mathematics". Scientific report n°1, AFCRL-64-411. Contrat AF 19(628)-3250. AD 602 710.

[10] Huet G.P. (1973) : "The undecidability of unification in third order logic". Information and Control 22,3 pp. 257-267.

[11] Huet G.P. (1973) : "A mechanization of type theory". Proceedings of 3rd IJCAI, Stanford August 73.

[12] Huet G.P. (1975) : "A unification algorithm for typed λ-calculus". Theoretical Computer Science 1,1.

[13] Lucchesi C.L. (1972) : "The undecidability of the unification problem for third order languages". Report CSRR 2059, Dept. of Applied Analysis and Computer Science. University of Waterloo.

[14] Pietrzykowski T. (1971) : "A complete mechanization of second order logic". Journal of Assoc. for Comp. Mach. 20,2 pp. 333-364.

[15] Pietrzykowski T. and Jensen D. (1972) : " A complete mechanization of ω-order type theory". Association for computing Machinery National Conference 1972, vol.1, pp. 82-92.

[16] Plotkin G.D. (1972) : "Building-in equational theories". Machine Intelligence 7, pp. 73-90. American Elsevier. New York.

[17] Robinson J.A. (1965) : "A machine-oriented logic based on the resolution principle". Journal of Assoc. for Comp. Mach. 12,1, pp.23-41.

[18] Sanchis L.E. (1967) : "Functionals defined by recursion". Notre Dame Journal of Formal Logic VIII, 3 pp. 161-174.

[19] Jensen D. & Pietrzykowski T. (1973) : "Mechanizing ω-order type theory through unification". Report CS-73-16, Dept. of Applied Analysis and Computer Science, University of Waterloo.

A CONDITION FOR IDENTIFYING TWO ELEMENTS OF WHATEVER MODEL OF COMBINATORY LOGIC

G. Jacopini

Istituto per le Applicazioni del Calcolo, IAC - CNR, Roma.

Summary - The purpose of this work is to show a necessary and sufficient condition by which two combinators (elements of a particular model of combinatory logic) can be identified without introducing contradictions with the axioms of combinatory logic itself.

A model of combinatory logic Γ is constituted by a quaternary:

$$\Gamma = \{E_\Gamma, S_\Gamma, K_\Gamma, \tau_\Gamma\}$$

where E_Γ is a set, S_Γ and K_Γ are two of its elements and τ_Γ is a binary function defined on the elements of E_Γ. ($\tau_\Gamma(X,Y)$ is written (XY) or simply XY. $(XY)Z$ is abbreviated XYZ) since the following properties have to be verified:

1) $S_\Gamma XYZ = XZ(YZ)$

2) $K_\Gamma XY = X$

3) $S_\Gamma B_\Gamma (K_\Gamma I_\Gamma) = I_\Gamma$

4) $B_\Gamma S_\Gamma (B_\Gamma K_\Gamma) = K_\Gamma$

5) $S_\Gamma (D_\Gamma B_\Gamma)_\Gamma (K_\Gamma K_\Gamma) = B_\Gamma (B_\Gamma K_\Gamma) I_\Gamma$

6) $S_\Gamma (D_\Gamma (B_\Gamma S_\Gamma (B_\Gamma S_\Gamma))(K_\Gamma S_\Gamma)) = B_\Gamma (B_\Gamma S_\Gamma)(W_\Gamma I_\Gamma)(B_\Gamma S_\Gamma))$

which are abbreviated in the following way:

$$B_\Gamma = S_\Gamma (K_\Gamma S_\Gamma) K_\Gamma \,, \quad D_\Gamma = B_\Gamma B_\Gamma \,, \quad I_\Gamma = S_\Gamma K_\Gamma K_\Gamma$$

$$C_\Gamma = S_\Gamma (D_\Gamma S_\Gamma)(K_\Gamma K_\Gamma) \,, \quad W_\Gamma = S_\Gamma S_\Gamma (S_\Gamma K_\Gamma)$$

If E_Γ has at least two distinct elements, Γ is said to be a non-trivial model otherwise Γ is trivial ($\Gamma = \Phi$).

For Γ to be banal it is sufficient that it be: $S = K$, in fact this would imply for each $X, Y \in E_\Gamma$

$$S_\Gamma K_\Gamma K_\Gamma K_\Gamma XY = K_\Gamma S_\Gamma K_\Gamma K_\Gamma XY \Longrightarrow X = Y$$

It is said that Γ' is a submodel of Γ if $E_{\Gamma'} \subset E_\Gamma$

$$S_{\Gamma'} = S_\Gamma \,, \quad K_{\Gamma'} = K_\Gamma \,, \quad X, Y \in E_{\Gamma'} \Longrightarrow (XY)_{\Gamma'} = (XY)_\Gamma$$

It is said that Γ is a prime model if it does not have submodels except itself.

Let $\mathcal{L}$ be a set of words in the alphabet: $\{s, k, [,]\}$ thus inductively defined:

$$s \in \mathcal{L} \qquad k \in \mathcal{L}$$
$$P_1 \in \mathcal{L}\,,\ P_2 \in \mathcal{L} \Longrightarrow [P_1 P_2] \in \mathcal{L}$$

Let P be a generic word of $\mathcal{L}$ and Γ be a prime model; by P^{Γ} we shall mean a function from $\mathcal{L}$ in E_{Γ} defined in the following way:

$$s^{\Gamma} = S_{\Gamma}\,,\ k^{\Gamma} = K_{\Gamma}\,,\ [P_1 P_2]^{\Gamma} = (P_1^{\Gamma}, P_2^{\Gamma})_{\Gamma}$$

P^{Γ} is called an interpretation of the word P in the model Γ. By $P_1 \overset{\Gamma}{=} P_2$ it is meant that the words P_1, P_2 have the same interpretation in Γ that is: $P_1^{\Gamma} = P_2^{\Gamma}$

It is said that $\Gamma' < \Gamma$ (Γ' homomorphous to Γ) if for each pair of words P_1 and P_2 belonging to $\mathcal{L}$ it results that:

$$P_1 \overset{\Gamma}{=} P_2 \Longrightarrow P_1 \overset{\Gamma'}{=} P_2$$

We shall call Λ the model in which E_{Λ} is constituted by the set of classes of λ-formulae lacking free variables convertible among themselves; S_{Λ} is the class of formulae which are convertible to $\lambda x \lambda y \lambda z (x z (y z))$; k_{Λ} is the class of formulae which are convertible to $\lambda x \lambda y (x)$ and τ_{Λ} is the simple concatenation of two formulae.

For each Γ we have $\Gamma < \Lambda$. The affirmation is based on the proved equivalence between the axioms of combinatory logic and the rules of λ-conversion for which if two formulae are convertible, they result as being the same on the basis of the axioms of combinatory logic and vice versa. Thus their interpretations are the same in any model.

Let Γ now be any model and $P_1, P_2 \in \mathcal{L}\, P_1^{\Gamma} = U,\ P_2^{\Gamma} = V$ we propose to establish whether or not a non-trivial model homomorphous to Γ exists, in which the elements U and V will be identified in the homomorphism; that is we propose to establish if such a Γ' exists that $\Gamma' < \Gamma,\ \Gamma' \neq \phi$ and $P_1 \overset{\Gamma'}{=} P_2$.

In order to do this let us begin by establishing between the elements X, Y of E_{Γ} a relationship in the following way:

$X \overset{U,V}{\longleftrightarrow} Y$ if such a $Q \in E_{\Gamma}$ exists that:

$$QUV=X \quad e \quad QVU=Y$$

It can be verified that:

1) $X \overset{U,V}{\longleftrightarrow} X$

2) $X \overset{U,V}{\longleftrightarrow} Y \Rightarrow Y \overset{U,V}{\longleftrightarrow} X$

3) $X \overset{U,V}{\longleftrightarrow} Y \Rightarrow X' \overset{U,V}{\longleftrightarrow} Y' \Rightarrow XX' \overset{U,V}{\longleftrightarrow} YY'$

4) $U \overset{U,V}{\longleftrightarrow} V$

In fact:

1) $K_r(K_rX)UV=X$, $K_r(K_rX)VU=X$

2) $C_rQUV=QVU=Y$, $C_rQVU=QUV=X$

3) $Q'UV=X'$, $Q'VU=Y' \Longrightarrow S_r(B_rS_rQ)UV=XX'$

$S_r(B_rS_rQ)VU=YY'$

4) $K_rUV=U$, $K_rVU=V$

Let us now call $\overset{U,V}{\sim}$ the transitive closure of the relationship $\overset{U,V}{\longleftrightarrow}$, that is we shall say that the relationship of $X \overset{U,V}{\sim} Y$ is valid if such a sequence of elements $Z_1, Z_2 \ldots, Z_n$ exists (eventually void: $n=0$) such that:

$$X \overset{U,V}{\longleftrightarrow} Z_1 \overset{U,V}{\longleftrightarrow} Z_2 \overset{U,V}{\longleftrightarrow} \cdots \overset{U,V}{\longleftrightarrow} Z_n \overset{U,V}{\longleftrightarrow} Y$$

Result:

5) $X \overset{U,V}{\sim} X$

6) $X \overset{U,V}{\sim} Y \Longrightarrow Y \overset{U,V}{\sim} X$

7) $X \overset{U,V}{\sim} Y$, $Y \overset{U,V}{\sim} Z \Longrightarrow X \overset{U,V}{\sim} Z$

8) $X \overset{U,V}{\sim} Y$, $X' \overset{U,V}{\sim} Y' \Longrightarrow XX' \overset{U,V}{\sim} YY'$

9) $U \overset{U,V}{\sim} V$

5), 6) and 9) are evidently consequences of 1), 2) and 4); 7) is valid by definition; as to 8) it is a consequence of 3) which is immediately verifiable if $X \overset{U,V}{\sim} Y$ and $X' \overset{U,V}{\sim} Y'$ come from two sequences of the same length. In fact if

$$X' \overset{U,V}{\longleftrightarrow} Z'_1 \overset{U,V}{\longleftrightarrow} Z'_2 \overset{U,V}{\longleftrightarrow} \cdots \overset{U,V}{\longleftrightarrow} Z'_n \overset{U,V}{\longleftrightarrow} Y$$

one also has the following:

$$XX' \overset{U,V}{\longleftrightarrow} Z_1Z_1' \overset{U,V}{\longleftrightarrow} Z_2Z_2' \overset{U,V}{\longleftrightarrow} \cdots \overset{U,V}{\longleftrightarrow} Z_nZ_n' \overset{U,V}{\longleftrightarrow} YY'$$

However, in case the two sequences should not be of the same length, the shorter sequence can be lenthened by repeating the same element in consequence to 1).

It is also evident that

$$X \overset{U,V}{\sim} Y, \quad U = V \Longrightarrow X = Y$$

We shall finally say that $U \| V$ (U being separable from V) if $S_\Gamma \overset{U,V}{\sim} K_\Gamma$.

We can finally state our theorem:

The necessary and sufficient condition for such a non-trivial $\Gamma' < \Gamma$ to exist so that $P_1 \overset{\Gamma_1}{=} P_2$ is that, provided $P_1^\Gamma = U$ and $P_2^\Gamma = V$, $U \| V$ will not be.

Demonstration

In order to demonstrate the necessity, let us begin by establishing that, if $\Gamma' < \Gamma$ then $P_1^\Gamma \| P_2^\Gamma \Longrightarrow P_1^{\Gamma'} \| P_2^{\Gamma'}$.

(It is sufficient to observe that the first relationship depends on a chain of equalities which must be kept in Γ' since $\Gamma' < \Gamma$.)

However, if it were $U \| V$ in case such a model $\Gamma' < \Gamma$ existed so that $P_1 \overset{\Gamma}{=} P_2$ said $W = P_1^{\Gamma'} = P_2^{\Gamma'}$ one would have the following: $W \| W$ and therefore for 10) it would be: $S_{\Gamma'} = K_{\Gamma'}$, that is Γ' would be trivial

In order to prove sufficiently let us call Γ' the quotient model between the Γ model and the relationship $\overset{U,V}{\sim}$: that is the model the elements of which are the classes of elements of E_Γ which are equivalent among themselves (the relationship $\overset{U,V}{\sim}$ is in fact an equality for 5) 6) and 7) and it is a congruence for 8) as regards to τ_Γ). In other words Γ' is that model such that:

11) $$P \overset{\Gamma'}{=} P' \Longleftrightarrow P_\Gamma \overset{U,V}{\sim} P'_\Gamma$$

Consequently $\Gamma' < \Gamma$ for 5); and $P_1^\Gamma = P_2$ for 6).

If it is not $U \| V$, that is if S_Γ is not equal to K_Γ then for 11) it results that $S_{\Gamma'} \neq K_{\Gamma'}$; Γ' is non-trivial and it provides the required example.

Therefore the sufficiency of the condition is also shown.

An Application

From now on we shall consider the model Λ . Let us give an example to show how this condition can be used in practice.

We shall now assume that: $V = \lambda x(xx)\lambda x(xx)$ and that U is any λ-formula. Let us demonstrate that it is never: $U \| V$; that is that the formula V can be identified with any other closed formula without contradicting the axioms of combinatory logic (the same is true for all the zeroes which are reducible objects only for themselves, but not for all the zeroes which are objects in general).

Meanwhile we must make two observations on the formula V .

Obs. 1) - If $\mathfrak{F}[V]$ (formula containing at least one occurence of V) is reducible to normal form then $\mathfrak{F}[x]$ is also reducible to the same normal form.

In fact all the immediate reductions of the formula $\mathfrak{F}[V]$ can be simulated with analogous reductions of the formula $\mathfrak{F}[x]$ by ignoring only the useless reductions of the subformula V to itself. If the reduction reaches a normal form (thus not containing the formula V) it means that at a certain point of the reduction itself, all the occurences of V disappear as will also disappear the occurrences of x .

Obs. 2) - Let $\mathfrak{F}[V]$ and $\mathfrak{F}'[V]$ be two convertible formulae, then $\mathfrak{F}[x]$ and $\mathfrak{F}'[x]$ are reducible respectively to two formulae of the type:

$$\mathfrak{R}[x,x,V] \; , \quad \mathfrak{R}[x,V,x]$$

In fact, if $\mathfrak{F}[V]$ and $\mathfrak{F}'[V]$ are convertible they are both reducible to a single formula R . By simulating the same reduction with the formulae $\mathfrak{F}[x]$ and $\mathfrak{F}'[x]$ in place of R we shall have to obtain two formulae which become identical when x will be substituted by V ; therefore they can only be of the abovementioned type.

We must now introduce another relationship:

$$X \xrightarrow{U,V} Y$$

It is verified if and only if such a P exists that $PU = X$, $PV = Y$ and its symmetric: $X \xleftarrow{U,V} Y \Longleftrightarrow Y \xrightarrow{U,V} X$

It results:

12) $$S_\wedge \xleftarrow{U,V} Z \Longrightarrow S_\wedge = Z$$

In fact for Obs. 1):

$$PV = S_\wedge \Longrightarrow Px = S_\wedge \Longrightarrow PU = Z = S_\wedge$$

and analogously:

13) $$Z \xrightarrow{U,V} K_\wedge \Longrightarrow Z = K_\wedge$$

It is also evident that:

14) $$X \xleftrightarrow{U,V} Y \Longrightarrow X \xleftarrow{U,V} Z \xrightarrow{U,V} Y$$

where:

$$X = QUV, \quad Z = QUU \quad , \quad Y = QVU$$

and finally:

15) $$X \xrightarrow{U,V} Z \xleftarrow{U,V} Y \Longrightarrow X \xleftrightarrow{U,V} Y$$

in fact for Obs. 2):

$$Q_1U = X, \quad Q_2V = Z = Q_2V, \quad Q_2U = X \Longrightarrow$$

$$Q_1x = \mathcal{R}[x, x, V], \quad Q_2x = \mathcal{R}[x, V, x] \Longrightarrow$$

$$X = \mathcal{R}[U, U, V], \quad Y = \mathcal{R}[U, V, U] \Longrightarrow$$

$$X = (\lambda x \lambda y\, \mathcal{R}[U, X Y])UV, \quad Y = (\lambda x \lambda y\, \mathcal{R}[U, X, Y])VU$$

Let us now assume that:

$$S_\wedge \xleftrightarrow{U,V} * \xleftrightarrow{U,V} * \xleftrightarrow{U,V} * \xleftrightarrow{U,V} K_\wedge$$

(where $*$ indicates an opportune element). For 14) one would have the following:

$$S_\wedge \xleftarrow{U,V} * \xrightarrow{U,V} * \xleftarrow{U,V} * \xrightarrow{U,V} * \xleftarrow{U,V} * \xrightarrow{U,V} * \xleftarrow{U,V} * \xrightarrow{U,V} K_\wedge$$

For 12) and 13):

$$S_\wedge \xrightarrow{U,V} * \xleftarrow{U,V} * \xrightarrow{U,V} * \xleftarrow{U,V} * \xrightarrow{U,V} * \xleftarrow{U,V} K_\wedge$$

Finally for 15):

$$S_\wedge \xleftrightarrow{U,V} * \xleftrightarrow{U,V} * \xleftarrow{U,V} K_\wedge$$

The procedure could continue.

It is evident that $U \| V$ cannot be. In fact if it were

$S_\wedge \overset{U,V}{\sim} K_\wedge$ would imply $S_\wedge = K_\wedge$ which is obviously absurd (starting however from a long chain that links $S_\wedge$ and $K_\wedge$ by iteratively applying the above-shown procedure, the length of the chain would be progressively shortened until the length zero were obtained: $S_\wedge = K_\wedge$)

Conclusions

In general a more abbreviated way of expressing the condition of separability of two λ formulae U and V is the following: $U \parallel V$ if and only if a Δ and certain $Z_1, Z_2 \ldots Z_n$ exist such that:

$$\Delta UV = \langle S_\wedge, Z_1, Z_2, \ldots Z_{n-1}, Z_n \rangle$$
$$\Delta VU = \langle Z_1, Z_2, Z_3, \ldots Z_n, K_\wedge \rangle$$

where:

$$\langle t_0, t_1, t_2, \ldots, t_n \rangle = \lambda x (x t_0 t_1 t_2 \ldots t_n)$$

To lead this condition back to the preceeding one is a simple exercise.

Bohm (bibl.1)) has demonstrated that for any two normal forms there always exists such a Δ that:

$$\Delta N_1 = S_\wedge \quad , \quad \Delta N_2 = K_\wedge$$

This shows that two distinct normal formulae are always separable. Yet one must not think that for any two $U \parallel V$ there always exists such a Δ. In fact, for example, there does not exist such a Δ that:

$$\Delta(\theta K_\wedge) = S_\wedge \quad , \quad \Delta K = K_\wedge$$

although: $\theta K_\wedge \parallel K_\wedge$

BIBLIOGRAPHY

1) Böhm, C., Alcune proprietà delle forme β-normali del λ-k-calcolo. Pubbl. IAC n.696 (1968)

2) Curry, H.B., Feys R., Combinatory Logic, North Holland Publishing Company, Amsterdam (1968)

3) Rosenbloom, P.C., The Elements of Mathematical Logic, Dover Publication Inc., London (1950)

TYPED MEANING IN SCOTT'S λ-CALCULUS MODELS

Herbert Egli

Forschungsinstitut für Mathematik der ETHZ

CH-8006 Zürich

Abstract

A theorem is proved which relates the meaning of typed and typeless terms in corresponding λ-calculus models over complete partially ordered sets (cpo's). This theorem allows us for instance, to define the semantics of high-level programming languages using Scott's extensional λ-calculus models. This application is outlined briefly.

0. Motivation

A logic for computable functions, LCF, has been developed by Robin Milner [4], [5]. LCF is a formalization of typed λ-calculus models over cpo's (complete partially ordered sets) [6]. The inference rules of this logic are roughly those of the λ-calculus supplemented by rules for least elements, least fixed point operators and conditionals. Various methods for proving correctness of programs have been shown to be justifiable within LCF [15]. However, if we want to define the semantics of high level programming languages in the spirit of "mathematical semantics" [9], [11], the _typed_ models used for LCF are too restrictive. What we need for that purpose are _typeless_ semantic domains. A number of illustrations have already been given to show how recursively defined domains (which are typeless domains whose existence is guaranteed in Scott's theory [8]) can be used for programming language semantics, e.g. [12], [14]. Those domains used are non-extensional λ-calculus models. What we want to show now is that we can use Scott's extensional λ-calculus models (which we define over cpo's [2]) for programming language semantics as well. Their advantage over recursively defined domains is that their logical properties can be expressed concisely. In fact, the inference rules of LCF (with types disregarded) need only a few modifications in order to formalize properties of extensional models. In other words, we can use a type free version of LCF to prove properties about semantic definitions if we can write those typeless definitions using the extensional models. This, however can be done only if the semantics that those models provide can be understood intuitively. This is not obvious at all but is achieved by our main theorem which is the heart of this paper. In short, the theorem states that we do not lose the intended typed meaning in the typeless interpretation. From this we deduce that the semantics provided by extensional models for typeless definitions is natural, i.e. it gives us "what we expect". What we exactl mean by that is illustrated by a very simple example. We define an extension of the kind of languages defined by Milner in LCF [5], [7]. The term that defines the semantics of this language "mathematically" must be typeless, but we show that it expresses what we intended by comparing it with an appropriate typed term and applying our theorem. For a more elaborate example we refer to [3].

1. Typed and typeless languages

We summarize briefly the definitions and properties of typed and typeless languages and their semantics. For more details we refer to [2]

1.1 Complete partially ordered sets

Our semantic domains are complete partially ordered sets (cpo's), i.e. partially ordered sets with the property that

(i) each ascending chain has a least upper bound

(ii) there is a least element, denoted by $\perp$ (bottom).

- The appropriate functions between cpo's D and D' are the functions that respect least upper bounds of (nonempty) chains. They are called continuous functions. The set of all continuous functions from D to D' is itself a cpo under the pointwise induced partial ordering and is denoted by [D,D'].

- The product of any number of cpo's is obtained by taking the cartesian product of the underlying sets with the componentwise induced partial ordering.

- Continuous functions with several arguments can be viewed as functions on iterated function spaces since there is a continuous natural isomorphism $[D \times D', D''] \cong [D,[D',D'']]$ for any cpo's D, D' and D".

- Each continuous function $\Phi \in [D,D]$ has a least fixed point $f \in D$, given by $f = \bigsqcup \Phi^{(k)}(\perp)$, where $\Phi^{(k)}$ means k-times iterated application of Φ.

- The function that assigns to each $\Phi \in [D,D]$ its least fixedpoint in D is continuous and is called the least fixed point operator (for D).

1.2 Typed languages

Types are defined inductively by

- 0 is a type
- if τ,σ are types then $(\tau\rightarrow\sigma)$ is a type

These are all the types.

Notation: (i) $\tau \to \sigma \to \varrho$ means $(\tau\to(\sigma\to\varrho))$

(ii) A special selection of types are the integer types, defined by $n+1 = (n\to n)$.

By a typed language we mean a typed λ-language with some (non-standard) constants (denoted by a^τ, $b^\sigma,\ldots$), infinitely many variables for each type (denoted by x^τ, $y^\sigma,\ldots$) and for each type τ the standard constants L_τ, C_τ and U_τ (of type $(\tau\to\tau)\to\tau$, $0\to\tau\to\tau\to\tau$ and τ respectively). Terms are built up from these atoms by means of typed application and λ-abstraction, i.e.

- a^τ, x^τ are terms of type τ ;
- $t^{(\tau\to\sigma)}(s^\tau)$ is a term of type σ ;
- $[\lambda x^\tau \cdot t^\sigma]$ is a term of type $(\tau\to\sigma)$.

These are all the terms.

A (standard) model (structure) is determined by

(a) a cpo D^0 ,

(b) two disjoint sets TRUE, FALSE $\subseteq D^0$ s.t. the contraction to the flat 3-element cpo$\{\bot,t,f\}$ is continuous.

(c) the meaning of the (non-standard) constants (i.e. $\underline{a}^\tau \in D^\tau$, where $D^{\varrho\to\sigma} = [D^\varrho,D^\sigma]$) .

Notation: We use $\alpha^\tau,\beta^\sigma,\ldots$ as variables ranging over $D^\tau,D^\sigma\ldots$.

The meaning of the standard constants is given by:

- $\underline{U_\tau} = \bot^\tau \in D^\tau$
- $\underline{L_\tau} : [D^\tau,D^\tau] \to D^\tau$ is the least fixed point operator
- $\underline{C_\tau} : D^0 \times D^\tau \times D^\tau \to D^\tau$ is the conditional, i.e.

$$\underline{C}_{\tau}(\alpha^{o})(\beta^{\tau})(\gamma^{\tau}) = \begin{cases} \beta^{\tau} & \text{if } \alpha^{o} \in \text{TRUE} \\ \gamma^{\tau} & \text{if } \alpha^{o} \in \text{FALSE} \\ \bot^{\tau} & \text{ow.} \end{cases}$$

Each <u>interpretation</u> of the variables (i.e. $\underline{x}^{\tau} \in D^{\tau}$) extends in the obvious way to an interpretation of all terms ($\underline{t}^{\tau} \in D^{\tau}$).

<u>1.3 Extensional λ-calculus models</u>

The definition and the properties of the extensional λ-calculus model D over a cpo D^{o} have been discussed in detail in [2]. The construction is the same as in [10], except that we use cpo's rather than continuous lattices. All we need to know about these models in this report is summarized below. The name <u>extensional λ-calculus model</u> will be justified in the next subsection.

<u>Proposition</u>:

The extensional λ-calculus model D over a cpo D^{o} has the following properties:

(1) D is a cpo. For each type τ we have a pair of continuous functions

$$D^{\tau} \underset{\iota_{\tau}}{\overset{\pi_{\tau}}{\rightleftarrows}} D$$

such that $\pi_{\tau} \cdot \iota_{\tau} = Id_{D^{\tau}}$ and $\iota_{\tau} \cdot \pi_{\tau} \sqsubseteq Id_{D}$.

(2) D is isomorphic to its own function space ($D \cong [D,D]$, i.e. each element $\alpha \in D$ can be used as a continuous function from D to D and vica-versa).

(3) (i) $\iota_{\tau\to\sigma}(\alpha^{\tau\to\sigma}) = \iota_{\sigma}\cdot\alpha^{\tau\to\sigma}\cdot\pi_{\tau}$

(ii) $\pi_{\tau\to\sigma}(\alpha) = \pi_{\sigma}\cdot\alpha\cdot\iota_{\tau}$

These equations will be used over and over again.

1.4 Typeless languages

By a typeless language we mean a λ-language with constants. Terms are built up by means of application and λ-abstraction from non-standard constants (denoted by a,b,...), (infinitely many) variables (denoted by x,y,...) and the standard constants L,C,U .

A standard model (structure) is determined by

(a) a cpo D^0

(b) TRUE, FALSE as for typed models

(c) the meaning of non-standard constants (i.e. $\underline{a} \in D$, where D is the λ-calculus model over D^0).

Notation: We use $\alpha,\beta,\ldots$ as variables ranging over D .

The meaning of the standard constants is given by

$\underline{U} = \perp \in D$

$\underline{L} \in D \cong [[D,D],D]$ is the least fixed point operator

$\underline{C} \in D \cong [D\times D\times D,D]$ is the conditional, i.e.

$$\underline{C}(\alpha)(\beta)(\gamma) = \begin{cases} \beta & \text{if } \pi_0\alpha \in \text{TRUE} \\ \gamma & \text{if } \pi_0\alpha \in \text{FALSE} \\ \perp & \text{ow.} \end{cases}$$

As in the typed case, each interpretation of the variables (i.e. $\underline{x} \in D$) extends to an interpretation of all terms ($\underline{t} \in D \cong [D,D] \cong \ldots$). Before saying more about that we introduce some notation.

- $t|^x_s$, where t and s are terms and x is a variable, means substitution, i.e. "replace each free occurrence of x in t by s after suitable changes of bound variables in t such that no free variable in s becomes bound after substitution". Consequently, we identify a priori terms that differ only in the naming of their bound variables (α-converted terms).

- $\underline{t|^x_\alpha}$, where t is a term, x a variable and α an element of D , is used as an abbreviation for $\underline{\underline{t}}$, where $\underline{\underline{\ }}$ is the same interpretation as $\underline{\ }$ except that $\underline{\underline{x}} = \alpha$.

Proposition:

Each interpretation of the variables in D (i.e. $\underline{x} \in D$) extends to an interpretation of all terms such that

(i) $\underline{[\lambda x.t(x)]} = \underline{t}$ (provided that x is not free in t)

(ii) $\underline{[\lambda x.t](s)} = \underline{t|^x_s}$.

Proof: Given the meaning of the constants and an interpretation of the variables, the interpretation of the other terms is defined inductively by

- $\underline{t(s)} = \underline{t}(\underline{s})$
- $\underline{[\lambda x.t]}$ = element in D corresponding to the function

$$\alpha \mapsto \underline{t|^x_\alpha} .$$

(i) and (ii) are obviously satisfied.

Remark: This proposition states that the interpretation (which is invariant under α-conversion by definition) is also invariant under

η-conversion (i) and β-conversion (ii) of λ-terms, thus D is a model for the λ-calculus. It is extensional since unrestricted η-conversion can be applied, as opposed to other models as given for instance in [16].

2. Relation between typed and typeless terms

We now want to "make typed terms typeless" (we call it transformation) and compare the typed meaning of a term with the meaning of the transformed term in a "corresponding" typeless model.

2.1 Derived languages

We start out with a typed language and derive from it a typeless language as follows.

The derived language is a typeless language s.t. the non-standard constants are in 1-1 correspondence with the non-standard constants of the typed language. We will write $\{a^{\tau}\}$ to denote the typeless constant that corresponds to the typed constant denoted by a^{τ} .

The semantic domain for the derived language is clearly given by the D^{O} and the subsets TRUE, FALSE which determine the typed structure. The meaning of the non-standard constants is given by $\underline{\{a^{\tau}\}} = \iota_{\tau} \underline{a}^{\tau}$.

Even if we are interested only in closed terms, free variables are needed over and over again in inductive proofs. If we transform a typed term with free variables then the transformed term will contain free variables, so its meaning depends on an interpretation of the typeless variables. What we clearly want to do is to interpret a free typeless variable as the image of the interpretation of the typed variable from which it was obtained. To keep track in our meta language of the origin of free variables in transformed terms we extend the notion of typeless terms by adding atoms of the form $\{x^\tau\}$, where x^τ is a typed variable.

We want to interpret an $\{x^\tau\}$ with respect to a typed interpretation ($\underline{\{x^\tau\}} = \iota_\tau \underline{x^\tau}$), so the meaning of a typeless term now may depend on a typeless and a typed interpretation.

A term is semi-closed if it contains no free (typeless) variables.

Examples:

$\{x^\tau\}$, $[\lambda x.\ a(x(\{x^\tau\}))]$,... are semi-closed.

x , $[\lambda y.\ a(x(\{x^\tau\}))]$,... are not semi-closed.

The meaning of a semi-closed term is clearly independent of any interpretation of the typeless variables. In other words, each typed interpretation extends to an interpretation of semi-closed terms.

2.2 Transformation

We now can make precise what we mean by transforming a typed term into a typeless term.

Tr: {typed terms} → {semi-closed terms}

is defined by

- $\text{Tr } L_\tau = L$; $\text{Tr } C_\tau = C$; $\text{Tr } U_\tau = U$
- $\text{Tr } a^\tau = \{a^\tau\}$ for all non-standard constants a^τ
- $\text{Tr } x^\tau = \{x^\tau\}$
- $\text{Tr}(t^{\tau\to\sigma}(s^\tau)) = (\text{Tr } t^{\tau\to\sigma})(\text{Tr } s^\tau)$
- $\text{Tr}[\lambda x^\tau.t^\sigma] = [\lambda x.(\text{Tr } t^\sigma)\,|\,{}^{\{x^\tau\}}_{x}]$ for a new variable x .

In general we cannot expect that $\underline{\text{Tr } t^\sigma} = \iota_\sigma \underline{t^\sigma}$. A simple example is $\underline{\text{Tr}[\lambda x^0.x^0]} = \underline{[\lambda x.x]} \neq \iota_1 \underline{[\lambda x^0.x^0]}$. Another example is the factorial on the positive integers which can be defined by either

$$t_1 = L_1[\lambda y^1.[\lambda x^0.C_0(x^0{=}1)(1)(x^0{*}y^1(x^0{-}1))]]$$

or $$t_2 = L_1[\lambda y^1.[\lambda x^0.C_0(x^0{=}1)(x^0)(x^0{*}y^1(x^0{-}1))]]$$

It is easy to see that

$$\underline{Tr\ t_1} = \iota_1\underline{t_1} = \iota_1\underline{t_2} \neq \underline{Tr\ t_2}$$, where for instance

$$\underline{Tr\ t_2} = L[\lambda y.[\lambda x.\ C(x \circleddash ①)\ (x)\ (x \circledast y(x \ominus ①))]]\ .$$

We used ○ here to indicate that the respective names now are names for typeless objects. Once we have decided on the typeless semantics we will clearly prefer a simpler notation, for instance

$$L\ \lambda fx.(x = 1 \rightarrow x,\ x * f(x-1))$$

or even $$f(x) \leftarrow (x = 1 \rightarrow x,\ x * f(x-1))\ .$$

But first we have to show that we do not lose the intended typed meaning when we use the typeless semantics.

2.3 Main theorem

THEOREM:	(I)	$\underline{t}^\tau = \pi_\tau(\underline{Tr\ t^\tau})$
	(II)	$\iota_0\underline{t}^0 = \underline{Tr\ t^0}$

We prove this theorem in several steps. A straightforward induction proof does not work because

in general $\pi_{\tau\rightarrow\sigma}(\alpha)(\pi_\tau\beta) \neq \pi_\sigma(\alpha(\beta))$

and also $\underline{L_\tau}(\pi_{\tau\rightarrow\tau}\alpha) \neq \pi_\tau(\underline{L}(\alpha))$.

Lemma 1: Each typed term has a (β-η-) normal form.

Actually, each typed term is strongly normalizable. A proof of this fact can be found in [13].

Remark: A normal form (NF) has one of the following shapes:

$$\left.\begin{array}{l} - \; a^{\tau}(NF_1)\ldots(NF_k) \\ - \; x^{\tau}(NF_1)\ldots(NF_k) \end{array}\right\} k \geq 0$$

- $[\lambda x^{\tau}.NF_1]$

Lemma 2: The theorem is true for L_{τ}-free terms with no C_{τ}'s for $\tau \neq 0$.

Lemma 2.1: Transformation commutes with conversion. (Proof: immediate)

Proof of lemma 2: Due to lemma 1 and 2.1 it is sufficient to show lemma 2 for terms in normal form. We do this by induction on the length of normal forms, i.e. we show that the theorem is true for

(1) $t = \xi^{\tau}(NF_1)\ldots(NF_k)$, provided it is true for $NF_1,\ldots,NF_k$ and ξ^{τ} is either a variable, a non-standard constant or C_0 .

(2) $t = [\lambda x^{\tau}.NF_1]$, provided it is true for NF_1.

The induction hypothesis is included in (1) for $k = 0$.

Case (1a): $\xi^{\tau} = C_0$.

- Part (I) of the theorem follows from the equations

$$\begin{aligned} &(\pi_{0\to0\to0\to0}\underline{C})\;(\alpha^0)\;(\beta^0)\;(\gamma^0) \\ =\;&(\pi_{0\to0\to0}\{\underline{C}(\iota_0\alpha^0)\})\;(\beta^0)\;(\gamma^0) \\ =\;&(\pi_{0\to0}\{\underline{C}(\iota_0\alpha^0)(\iota_0\beta^0)\})\;(\gamma^0) \\ =\;&\pi_0\{\underline{C}(\iota_0\alpha^0)\;(\iota_0\beta^0)\;(\iota_0\gamma^0)\} \\ =\;&\underline{C_0}\;(\alpha^0)\;(\beta^0)\;(\gamma^0) \end{aligned}$$

- Part (II) of the theorem (for k=3) follows from the equation

$$\underline{C}\;(\iota_0\alpha^0)(\iota_0\beta^0)(\iota_0\gamma^0) = \iota_0\{\underline{C_0}(\alpha^0)(\beta^0)(\gamma^0)\}\;.$$

<u>Case (1b)</u>: $\xi^{\tau} \neq C_0$.

We show by induction on k that

$$\underline{Tr\{\xi^{\tau}(NF_1)\ldots(NF_k)\}} = \iota_{\sigma}\underline{\{\xi^{\tau}(NF_1)\ldots(NF_k)\}}$$

where NF_i has type τ_i and $\tau = \tau_1 \to \tau_2 \to \ldots \to \tau_k \to \sigma$

(i) $k = 0$: immediate

(ii) $\underline{Tr\{\xi^{\tau}(NF_1)\ldots(NF_{k-1})(NF_k)\}}$

$$= \underline{Tr\{\xi^{\tau}(NF_1)\ldots(NF_{k-1})\}}\ \underline{(Tr\ NF_k)}$$

$$= \iota_{\tau_k \to \sigma}\underline{\{\xi^{\tau}(NF_1)\ldots(NF_{k-1})\}}\ \underline{(Tr\ NF_k)}$$

$$= \iota_{\sigma}\underline{\{\xi^{\tau}(NF_1)\ldots(NF_{k-1})}(\pi_{\tau_k}\ \underline{Tr\ NF_k})\}$$

$$= \iota_{\sigma}\underline{\{\xi^{\tau}(NF_1)\ldots(NF_{k-1})(NF_k)\}}$$

<u>Case (2)</u>: Since $[\lambda x^{\tau}.NF^{\sigma}]$ can not be of type 0 , we have to prove only part (I) of the theorem.

$$\underline{[\lambda x^{\tau}.NF^{\sigma}](\alpha^{\tau})} = \underline{NF^{\sigma}|^{x^{\tau}}_{\alpha^{\tau}}} = \pi_{\sigma}\underline{\{(Tr\ NF^{\sigma})|^{\{x^{\tau}\}}_{\iota_{\tau}\alpha^{\tau}}\}}$$

$$= \pi_{\sigma}\underline{\{[\lambda x.(Tr\ NF^{\sigma})|^{\{x^{\tau}\}}_{x}}](\iota_{\tau}\alpha^{\tau})\} = \{\pi_{\tau\to\sigma}\ \underline{Tr[\lambda x^{\tau}.NF^{\sigma}]}\}\ (\alpha^{\tau})$$

<u>Lemma 3</u>: The theorem is true for L_{τ} - free terms.

<u>Proof</u>: This is an easy generalization from lemma 2. Each C_{τ} can be replaced by a typed λ-term containing only C_0 as a constant, and this replacement is compatible with transformation. These replacements are given inductively by

$$C_{\sigma\to\varrho} \sim [\lambda x^0.\lambda y^{\sigma\to\varrho}.\lambda z^{\sigma\to\varrho}.\lambda w^{\sigma}.C_{\varrho}(x^0)(y^{\sigma\to\varrho}(w^{\sigma}))(z^{\sigma\to\varrho}(w^{\sigma}))]$$

The corresponding transformations are

C and $[\lambda x.\lambda y.\lambda z.\lambda w.\ C(x)(y(w))(z(w))]$.

They have the same meaning in the typeless model.

We now want to include also least fixed point operators. For this we define for each k a type-preserving construction.

S_k: {typed terms} → {L_τ - free typed terms}

by $S_k L_\tau = [\lambda x^{\tau\to\tau}.(x^{\tau\to\tau})^{(k)}(U_\tau)]$; $[(x^{\tau\to\tau})^{(k)}(U_\tau)$ means $\underbrace{x^{\tau\to\tau}(\ldots x^{\tau\to\tau}}_{k\text{-times}}(U_\tau)\ldots)]$

$S_k a^\tau = a^\tau$ for all constants $\neq L_\tau$;

$S_k x^\tau = x^\tau$

$S_k(t^{\tau\to\sigma}(s^\tau)) = (S_k t^{\tau\to\sigma})(S_k s^\tau)$;

$S_k[\lambda x^\tau.t^\sigma] = [\lambda x^\tau.S_k t^\sigma]$.

<u>Lemma 4</u>: (i) $\underline{t^\tau} = \bigsqcup_k \underline{S_k t^\tau}$

(ii) $\underline{\mathrm{Tr}\ t^\tau} = \bigsqcup_k \underline{\mathrm{Tr}\ S_k t^\tau}$

<u>Proof</u>: (i) Straightforward induction.

(ii) To prove this we need a stronger version:

For all typeless interpretations,

$$\underline{\mathrm{Tr}\ t^\tau|^{\{x^\sigma\},\ldots}_{x\ \ ,\ldots}} = \bigsqcup_k \underline{\mathrm{Tr}\ S_k t^\tau|^{\{x^\sigma\},\ldots}_{x\ \ ,\ldots}} ,$$

where all subterms of the form $\{y^\varrho\}$ are replaced by typeless variables.

We prove this by induction on the length of t .

- $t = L_\tau$: $\underline{\mathrm{Tr}\ L_\tau} = \underline{L} = \bigsqcup_k \underline{[\lambda x.x^{(k)}(U)]} = \bigsqcup_k \underline{\mathrm{Tr}\ S_k L_\tau}$.

- $t = a^\tau$, x^τ : immediate

- $t = s^{\tau\to\sigma}(r^\tau)$:

$$\underline{\mathrm{Tr}\ s^{\tau\to\sigma}(r^\tau)|^{\dots}_{\dots}} = \underline{(\mathrm{Tr}\ s^{\tau\to\sigma}|^{\dots}_{\dots})}\,\underline{(\mathrm{Tr}\ r^\tau|^{\dots}_{\dots})}$$

$$= (\bigsqcup_k \underline{\mathrm{Tr}\ S_k s^{\tau\to\sigma}|^{\dots}_{\dots}})(\bigsqcup_e \underline{\mathrm{Tr}\ S_e r^\tau|^{\dots}_{\dots}})$$

$$= \bigsqcup_k \underline{(\mathrm{Tr}\ S_k s^{\tau\to\sigma})(\mathrm{Tr}\ S_k r^\tau)|^{\dots}_{\dots}} = \bigsqcup_k \underline{\mathrm{Tr}\ S_k(s^{\tau\to\sigma}(r^\tau))|^{\dots}_{\dots}} .$$

- $t = [\lambda x^\tau.s^\sigma]$:

$$\underline{\mathrm{Tr}[\lambda x^\tau.s^\sigma]|^{\{y^\rho\},\dots}_{y\ \ ,\dots}}(\alpha) = \underline{[\lambda x.\ \mathrm{Tr}\ s^\sigma|^{\{x^\tau\},\{y^\rho\},\dots}_{x\ \ ,\ y\ \ ,\dots}]}(\alpha)$$

$$= \underline{\mathrm{Tr}\ s^\sigma|^{\{x^\tau\},\{y^\rho\},\dots}_{x\ \ ,\ y\ \ ,\dots}|^x_\alpha} = \bigsqcup_k \underline{(\mathrm{Tr}\ S_k s^\sigma|^{\{x^\tau\},\{y^\rho\},\dots}_{x\ \ ,\ y\ \ ,\dots})|^x_\alpha}$$

$$= \bigsqcup_k \underline{[\lambda x.\ \mathrm{Tr}\ S_k s^\sigma|^{\{x^\tau\},\{y^\rho\},\dots}_{x\ \ ,\ y\ \ ,\dots}]}(\alpha) = \bigsqcup_k \underline{\mathrm{Tr}[\lambda x^\tau.S_k s^\sigma|^{\{y^\rho\},\dots}_{y\ \ ,\dots}]}(\alpha)$$

$$= \bigsqcup_k \underline{\mathrm{Tr}\ S_k[\lambda x^\tau.s^\sigma]|^{\{y^\rho\},\dots}_{y\ \ ,\dots}}(\alpha)$$

Proof of the theorem:

$$\pi_\tau\ \underline{\mathrm{Tr}\ t^\tau} = \pi_\tau \bigsqcup_k \underline{\mathrm{Tr}\ S_k t^\tau} = \bigsqcup_k \pi_\tau\ \underline{\mathrm{Tr}\ S_k t^\tau} = \bigsqcup_k \underline{S_k t^\tau} = \underline{t}^\tau$$

Lemma 4   Lemma 3   Lemma 4

$$\underline{\mathrm{Tr}\ t^O} = \underline{\mathrm{Tr}\ S_k t^O} = \bigsqcup_k \iota_O\ \underline{S_k t^O} = \iota_O \bigsqcup_k \underline{S_k t^O} = \iota_O \underline{t}^O$$

3. Application

We now want to illustrate how the theorem that we just proved allows us to use typeless models for the definition of programming languages. Space here allows only to give a very simple example of a programming language. We have to refer to [3] for the more elaborate example of a language which includes blocks and arbitrary procedures. Milner defined in LCF Algol-like languages with assignment, conditional, while and compound statements [5], [7]. All variables are global, so that the state of the system can be characterized by a function which assigns to each identifies its current value. This simple semantic concept needs no change when we add parameter free recursive procedures to such a language. For simplicity, we also want to allow only procedures used as statements in order to avoid that the evaluation of an expression might change the state.

The model in which we will interpret the semantic definitions is built up over the following basic sets.

SYN: It contains all syntactic entities (related by the abstract syntax). In particular, it contains ID = VAR $\dot{\cup}$ PRN (identifies which are either a variable or a procedure name), EXPR (expressions) and STM (statements = programs).

PRIM: primitive values, consisting of INT and BOOL. D^O then is the cpo obtained from the disjoint union of the basic sets by adding a bottom The truth values of BOOL are used to denote true and false respectively also in D^O. The state of the system is characterized by a function which assigns to each indentifier its value in the extensional model D over D^O. We write STATE = ID $\rightarrow$ D to indicate this. What we actually mean, of course, is the least extension of such a function to $[D,D] \cong D$.

We do not go into details of what an expression and its meaning in a given state is.

A straightforward definition gives us the meaning-function E for expressions. E is thought of as a function E : EXPR → STATE → PRIM. Using this function we can define the meaning-function for statements M : STM → STATE → STATE by

$$
\begin{aligned}
M \equiv\ & L\lambda M.\lambda q.\lambda\sigma. \\
& \quad Isass(q) \rightarrow [\sigma | lhsof(q) | E(rhsof(q),\sigma)]\ , \\
& \quad Iscond(q) \rightarrow \{E(ifof(q),\sigma) \rightarrow M(thenof(q),\sigma), M(elseof(q),\sigma)\}, \\
& \quad Iswhile(q) \rightarrow \{E(testof(q),\sigma) \rightarrow M(q,M(stmof(q),\sigma)),\sigma\}, \\
& \quad Iscompnd(q) \rightarrow M(secndof(q),M(firstof(q),\sigma))\ , \\
& \quad Ispdef(q) \rightarrow [\sigma | idof(q) | M(bodyof(q))], \\
& \quad Ispcall(q) \rightarrow \sigma(nameof(q))(\sigma), U\ .
\end{aligned}
$$

(<u>Notation</u>: $[\alpha|\beta|\gamma]$ is used as an abbreviation for $\lambda k.k = \beta \rightarrow \gamma,\alpha(k)$.) Notice that this definition is "mathematical" since the value associated with a procedure name is the statefunction that a call of p causes. This makes it impossible to view M as a typed term. However, if we avoid storing the mathematical meaning of procedures by simply storing their syntactic definition, we get a meaning-function M' which is the transformation of a typed term whose typed meaning can be considered to be the correct intended meaning. M' looks as follows

$$
\begin{aligned}
M' \equiv\ & L\lambda M.\lambda q.\lambda\sigma. \\
& \quad \vdots \\
& \quad Ispdef(q) \rightarrow [\sigma | idof(q) | bodyof(q)], \\
& \quad Ispcall(q) \rightarrow M(\sigma(nameof(q)),\sigma),\ U\ .
\end{aligned}
$$

Obviously, M and M' are not equivalent terms (they handle procedures differently). However, they yield equal output values for any program q, i.e. for the initial state $\sigma_o = U$ and for any variable x we have $M(q,\sigma_o,x) \equiv M'(q,\sigma_o,x)$. Now applying our theorem tells us that $M'(q,\sigma_o,x)$ is equivalent to its typed meaning which is the correct one by definition. Thus $M(q,\sigma_o,x)$ has the correct meaning too.

Once we know that M is correct, we prefer M over M' because M gives us more flexibility. For instance, we can handle easily library procedures. It is sufficient to know their semantics in order to preset the initial state for the meaning of a program using such procedures. As an example, let us look at a procedure p which assigns the value a ! (for a fixed variable a) to the variable fact. This procedure may be given in any language. For presetting the state with respect to the meaning-function M it is sufficient to know the statefunction that a call of p causes, whereas for M' we have to find a semantically equivalent procedure definition $q = (\underline{\text{proc}}\ p \leftarrow \pi)$ in the specific Algol-language at hand, e.g.

```
proc p ← If a = 0 then  fact := 1 else
                  begin
                        a := a-1; call p; a := a+1; fact := fact*a
                  end
```

The preset initial state σ_o' for M' would be $\sigma_o' \equiv [U|p|\pi]$. If q happened to be the actual definition of this procedure then we could take $\sigma_o \equiv M(q,U)$ as the initial state when we use M . If not, we would adopt the semantics of the actual definition in whatever language it was given. For instance, this might give us $\bar{\sigma}_o \equiv [U|p|\lambda\sigma.[\sigma|fact|\sigma(a)!]]$ (with $\sigma(a)! \equiv (L\lambda f.\lambda x.x = 0 \rightarrow 1\ ,\ x{*}f(x-1))(\sigma(a))$ if you like). σ_o and $\bar{\sigma}_o$ are essentially equivalent,

i.e. they yield equal output values; in other words,
$M(q,\sigma_o,x) \equiv M(q,\bar{\sigma}_o,x)$ for any program q and variable x .

Conclusion

We have shown that the typeless semantics given by extensional λ-calculus models preserves the intended typed meaning and that it can therefore be used in a natural way for programming language definitions. We can view this approach either as a generalization of LCF to a typeless system which allows us to define the semantics of programming languages in the spirit of "mathematical semantics". On the other hand, we can view it as a unification of semantic definitions using recursively defined domains, thus providing a general logical system for such definitions. With minor modifications we could use the transformed inference rules of LCF for a typeless logic (i.e. to formalize properties of the extensional λ-calculus models). It is another question however, precisely how the typeless logic should be formulated in order to be appropriate for mechanical proofs. Analyzing first the kind of things that we want to prove about language definitions should help to answer this question.

References

[1] Constable, R.L. and Egli, H., "Computability on Continuous Higher Types and its role in the semantics of programming languages", TR 74-209, Computer Science Department, Cornell University (1974).

[2] Egli, H., "An Analysis of Scott's λ-Calculus Models", TR 73-191, Computer Science Department. Cornell University (1973).

[3] Egli, H., "Programming Language Semantics using Extensional λ-Calculus Models", TR 74-206, Computer Science Department, Cornell University (1974).

[4] Milner, R., "Logic for Computable Functions, Description of a Machine Implementation", AIM-169/CS-288, Computer Science Department, Stanford University (1972).

[5] Milner, R., "Implementation and Applications of Scott's Logic for Computable Functions", Proc. of an ACM Conference on Proving Assertions about Programs, Las Cruces, New Mexico (1972).

[6] Milner, R., "Models of LCF", AIM-186/CS-332, Computer Science Department, Stanford University (1973).

[7] Milner, R. and Weyhrauch, R., "Proving Compiler Correctness in a Mechanized Logic", Machine Intelligence 7, ed. D. Michie, Edinburgh University Press (1972).

[8] Reynolds, J.C., "Notes on a Lattice-Theoretic Approach to the Theory of Computation", Systems and Information Science, Syracuse University (1972).

[9] Scott, D., "Outline of a Mathematical Theory of Computation", Oxford University Computing Laboratory, Technical Monograph PRG-2 (1970).

[10] Scott, D., "Continuous Lattices", Proc. Dalhousie Conference on Toposes, Algebraic Geometry and Logic, Springer Lecture Notes in Mathematics # 274, (1972).

[11] Scott, D. and Strachey, C., "Toward a Mathematical Semantics for Computer Languages", Proc. of a Symposium on Computer and Automata, New York, Polytechnic Institute of Brooklyn (1971).

[12] Strachey, C. and Wadsworth, C.P., "Continuations. A Mathematical Semantics for Handling Full Jumps", Oxford University Computing Laboratory, Technical Monograph PRG-11 (1974).

[13] Stenlund, S., "Combinators, λ-Terms and Proof Theory", D. Reidel Publishing Company/Dordrecht-Holland (1972).

[14] Tennent, R.D., "Mathematical Semantics and Design of Programming Languages", Ph.D. Thesis, Department of Computer Science, University of Toronto (Sept. 1973).

[15] Vuillemin, J., "Proof Techniques for Recursive Programs", Ph.D. Thesis, Computer Science Department, Stanford University (1973).

[16] Wadsworth, C.P., "Semantics and Pragmatics of the Lambda-Calculus", Ph.D. Thesis, University of Oxford (1971).

* * *
*

PROGRAMMING LANGUAGE SEMANTICS IN A TYPED LAMBDA-CALCULUS

by

Luigia Aiello
Istituto di Elaborazione dell'Informazione-CNR Pisa

and

Mario Aiello
Istituto di Scienze dell'Informazione, Università di Pisa

1. Introduction.

Various methods have been proposed to formally define the semantics of programming languages. They may be classified into: operational methods [3,7,8], axiomatic methods [6] (see [7] for an application), and mathematical methods [11](see also [5,14]).

In this paper we are concerned with the mathematical methods for defining the semantics of programming languages. To provide a mathematical semantics for a programming language means to associate with the constructs of the language certain mathematical entities. Programming language constructs are represented in an "abstract syntactic" form. All the problems concerning the parsing are ignored. A semantic function is associated with each construct, which is a mapping from stores into stores. Stores are a mean to keep the information relevant to the "computation". However, the mathematical semantics is extensional. This means that what is relevant for defining the meaning of a particular syntactic construct is not the history of the computation, i.e. the successive modifications of the stores, but the function (from stores into stores) which is associated to that construct.

In 1969 Dana Scott proposed, in an unpublished memo, an equation calculus based on terms in the typed lambda-calculus whose most powerful rule of inference is Kleene's first recursion theorem stated as a rule. This calculus has been implemented in the form of an interactive proof checker (LCF) by Robin Milner in 1972 [10].

Dana Scott discovered a model also for the type free lambda-calculus [12] and suggested it as a framework in which to define a mathematical semantics for programming languages [11] and discuss properties of programs. The semantics he suggests for a programming language consists in interpreting each programming language construct into a term of the typed or untyped lambda-calculus language. This term represents the computed function. Lambda-calculus terms can also be viewed as computation procedures. For this reason the Scott axiomatic approach to the definition of the semantics of programming languages seems natural: it is extensional and, at the same time, it clearly exhibits the operational nature of the meaning of programming language constructs.

In 1972 Milner and Weyhrauch [16] defined in LCF the semantics of a simple but complete[4], programming language. Later on this language was extended [1] and finally the entire (arithmetic part) of PASCAL was treated [2].

A type free logic seems the most natural environment for discussing properties of programming languages, since we have not to worry about type conflicts arising when defining the semantics of some programming language feature. For instance, it is general opinion that a type free logic is necessary for dealing with full jumps in a natural way.

In [2] a semantics for the programming language PASCAL [15 ,7] is defined in LCF. In this semantics many theorems have been proved regarding properties of particular PASCAL programs and properties of the language itself.

The aim of this paper is to show how a very reasonable semantics can be given to a programming language even in a typed lambda-calculus environment. We want to point out, using PASCAL as example, that such semantics not only looks "natural" but is usable in practice. In fact, theorems of a certain complexity and of a practical interest have been checked in LCF with a reasonable proof.

We do not want to claim, in this paper, that the typed logic is better or more natural than the type free one. When working with Richard W. Weyhrauch at the axiomatization of PASCAL we chose the typed logic since no type free proof checker was available. At the moment there is no experience of machine-checking proofs in the type free logic, so no comparison is possible.

After a brief discussion on the definition of the semantics for a programming language in LCF, the paper contains the detailed description of the semantics of goto's and of procedure activations. In fact, these are the crucial points when representing the semantics of a programming language in a typed logic. We have avoided type conflicts by acting at a "syntactic level", instead of "semantic level". This will be clarified in the sequel: it amounts to dealing with type zero objects, instead of dealing directly with functions.

2. Programming language semantics in LCF.

As already said, LCF is an implementation, in the form of an interactive proof checker, of a logic using an equation calculus between lambda-calculus expressions. LCF has been used to define the semantics of programming languages and to prove the correctness of various programs and of a simple compiler.

In [2] the interpretation of PASCAL in LCF and many of the theorems proved about that interpretation are presented. We now shortly describe that semantic definition and show, in more details, how the jump and the procedure activation are handled. The solution we propose for PASCAL is not ad hoc for this language. It is valid for any ALGOL-like language and it can be generalized to deal with other control disciplines.

The semantics provided to PASCAL is in the style of Scott and Strachey [11]. Statements are represented in terms of their abstract syntax. With each program a function is associated which is its meaning. The association is made by a functional named FUNCT:

$$\text{FUNCT} \equiv \{\lambda\ p\ o\ .\{\lambda\ i\ .\ \text{INPUT} \otimes \text{PASCAL}\ (p,o) \otimes \text{OUTPUT}(i)\}\}$$

that returns a function from input data into output data. In the above formula $\otimes$ denotes the functional composition: $\lambda f\ g\ x.\ g(f(x))$. PASCAL is the LCF combinator that, applied to a program p and an inizialization of the output buffer o, returns a function from stores into stores. It thus represents an explicit extensional definition of the behavior of PASCAL statements. Stores consist of frames, each one identified by a frame pointer:

stores: framepointers $\rightarrow$ frames
frames: locations $\rightarrow$ values

Frames have been introduced to take into account the change of environment due to procedure and function activations.The combinator PASCAL defines the meaning of a PASCAL program as the composition of MD and MS. MD gives a meaning to the declaration part of a program by creating an appropriate location in the store for each declared item. MS has the form of a conditional. Each branch corresponds to a different statement of the language. For any statement st and frame pointer f specifying the environment in which the statement must be interpreted, MS(st,f) is a function from stores into stores.

The semantic functions MS defined in[16] and [1] are homorphisms with respect to "mkcmpnd", the syntactic constructor for compound statements, i.e. for every statement st1 and st2:

(+) MS(mkcmpnd(st1, st2)) $\equiv$ MS(st1) $\otimes$ MS(st2)

This relation does not hold if the language contains a goto statement. In fact, during the evalutation of st1, the control can be moved to a different place of the text. In such case st2 is never evaluated.

The notion of <u>continuation</u> has been introduced for dealing with the semantics of jumps in programming languages. It has been suggested by L. Morris and C. Wadsworth, indipendently. It is reported by Strachey and Wadsworth in [13] and can be shortly described as fol lows: a semantic function is not associated to a single statement, but to an entire text,by specifying the meaning of its first statement and a continuation function. The continuation represents the place where the "computation" goes on if no jump occurs to move the control to another point of the text. So the meaning of the jump is just to change the continuation.

In the definition of PASCAL in LCF the semantics of statements is given by a function MS for which the above relation (+) in general does not hold. If st1 is a simple statement different from a goto then its meaning is composed by the functional composition with the meaning of st2. If st1 is a structured statement (conditional, repetition or concatenation)then an append combinator for pieces of syntax is used in the recursive definition of MS. For instance, if we want to specify the meaning of mkcmpnd (st1,st2) where st1 is:

<u>if</u> test <u>then</u> truecase <u>else</u> falsecase;

and if MBEXPR (<u>M</u>eaning of <u>B</u>oolean <u>EXPR</u>essions) is the evaluation function for boolean expressions, then:

MS(mkcmpnd (st1, st2)) ≡
  COND(MBEXPR(test), MS(append(truecase, st2)),
                     MS(append(falsecase, st2)))

where COND≡ λ p q1 q2 s.p(s) → q1(s), q2(s) . To describe it in a very informal way, we may say that the decision on whether evaluating st2 or not is "delayed" until truecase or falsecase are evaluated. st2 will be evaluated only if no jump is encountered first. Obviously the semantics of repetitive statements, for instance the while, cannot be defined by a minimal fixpoint definition of this kind:

WHILE ≡ { μ F . { λ t b . COND(t,b ⊗ F(t,b), ID) } }

where μ is the minimal fixpoint operator and ID the identity function

In fact a jump can stop the execution of the body b, hence of the repetitive statement itself.

3. Semantics of jumps and procedure activations.

In the previous section we have shortly explained how the presence of goto's influences the definition of the semantics for other statements of the language. We now describe how the semantics of the goto is defined. To this purpose we first shortly describe the structure of a store. A store is a collection of frames. Each procedure (function) activation produces a new frame where all the information relevant to that procedure (function) activation is kept. A store, expressed as a conditional, has the following form:

$$\{\lambda f.f=0\rightarrow\{\lambda\ n.\ n=n1\ \rightarrow\ v1,\ n = n2\rightarrow v2,\ldots\ \},$$
$$f=1\rightarrow\{\lambda\ n.\ n=m1\ \rightarrow\ w1,\ n = m2\rightarrow w2,\ldots\ \},$$
$$\ldots\ \}$$

Since the ALGOL-like discipline of procedure activations is a strict stack discipline, frames are pointed by integers.

In each frame a text-location is created, where the text to be evaluated in that frame is saved.

$\{\lambda n.n=textloc\rightarrow$ text-to-be-evalutaed, ...$\}$

The combinator TEXT applied to the appropriate framepointer returns the content of such text-location. If a jump to a label lb is evaluated in the frame pointed by f, its meaning is defined as:

GOTO(lb,f) = MS(segm(lb,TEXT(f)),f)

This means that MS applies (recursively) to the text returned by the combinator segm.segm is a generalized selector function which finds the first occurrence of lb in the text returned by TEXT(f) and gives, as result, the "tail" of such text, starting from the first occurrence of the label. Multiple occurrences of the label lb are ignored. On the other hand, if no occurrence of lb is found, the result of segm is undefined.

No restriction is imposed on the place where the label may appear in the text. This means that jumps into (or out of) the body of repetitive statements are allowed. In this interpretation jumps out of a procedure or function activation are not allowed. A generalization to include this case causes no problem.

When defining the mathematical semantics of procedures with procedural arguments in a lambda-calculus environment, type conflicts may arise from the self application of functions. This happens if the meaning of a procedure is defined as a function applied to the meaning of its actual arguments. Type conflicts are avoided by associating with each formal argument of a procedure the text of the actual argument, expressed in abstract syntactic form. This text is evaluated each time the corresponding formal argument appears in the body of the activated procedure.

As already said, when a procedure is activated, a new frame is set up in the store to keep record of all the information local to that activation. In such frame the text of the activated procedure is saved, as well as the bindings between formal and actual parameters and a pointer to the frame where such procedure has been declared. This pointer represents the access link for that procedure activation, i.e. the specification of the environment where the free variables of the procedure have to be bound.

In order to store the name parameters a "binding list" is created (each formal name corresponds to the actual name). Value parameters are evalutated and stored into locations whose names are those of the corresponding formal parameters. Finally, if the procedure has a procedural parameter pp, the text of the actual parameter is passed (not the corresponding meaning function). Such text is stored into a procedure location pp and evaluated whenever pp is activated.

When defining the semantics of a real language the mechanism for binding parameters in procedure activation is more complex than described above (see, for instance, the function MKBINDING in [2]). In fact, subscripts have to be evaluated, if any, and a type checking must be done. (Note that by "type" here we mean a type of the programming language, not of LCF!). To better explain the modifications of the store caused by a procedure activation,consider the following situation: a program p declares a and b integer variables and assignes 3 and 4 to them.It also declares a procedure p1 with three parameters: an integer name parameter f1, an integer value parameter f2, and f3 which is a procedure parameter. Let txt be the text of the procedure p1. STORE1 represents the situation of the store before the activation of the procedure p1, while STORE2 represents the situation after p1 has been activated with actual arguments (a,a+b,p1).

```
STORE1  ≡ {λf.f=0 →
          {λn.n=textloc    →  statmof(p),
               n=typeloc(a) → INT,
               n=typeloc(b) → INT,
               n=procloc(p1) → procedure-p1-specification,
               n=acclink(p1) → 0,
               n=a   →   3,
               n=b   →   4,      UU}, UU    }
```

Note that statmof is the selector that gives the statement part of a program. A meaning is given to the declaration part of the program by creating a location for the type of the variables and two locations for each procedure: one for the procedure specification and the other one for its access link.

```
STORE2  = { λf.f=0 → STORE1(0)
             f=1→{λn.
             n=textloc → statmof (txt)
             n=alink   → 0,
             n=bindloc (f1) → a,
```

```
n=typeloc(f2) → INT,
n=f2    →    7,
n=procloc(f3) → procedure-p1-specification,
n=acclink(f3) → 0, UU},    UU }
```

In practice, the fact that f3 is a parameter is ignored: it is as if a procedure named f3 is declared whose specification coincides with that of p1.

4. Remarks and Conclusions.

We have interpreted a programming language with jumps and procedures in LCF. The interpretation we have given to jumps and procedure activations directly reflects the operational ideas we have about these notions. For this reason the semantics defined for this language looks "natural". More important, this semantic definition has been used for proving properties of programs and equivalences between various syntactic constructs. Among others, all the equivalences reported in [15] have been checked in LCF, very often with a one-step proof.

Up to now there is no experience of proving properties of programs or classes of programs in a type free logic, since no proof checker is available for the type free lambda-calculus.

F.W. Henke is working, at Stanford, at the implementation of an interactive proof checker based on the type free logic proposed by Scott. Even though this logic seems "cleaner" for defining the mathematical semantics for programming languages,experiments on how actually carrying out proofs about programs in that environment have still to be done. It is not yet evident whether or not the "cleaness" of the axiomatization results in more straightforward proofs.

Finally, we want to point out that the interpretation given to a programming language in a LCF-like environment can constitute a fruitful framework in which to discuss and design the features of a programming language. Probably the LCF combinators allow to represent a programming language at the right level of abstraction, without entering into the problems faced by an implementor, but with every construct completely specified by the associate meaning function.

Acknowledgements.

The research that led to this paper has been done while the authors were visitors at the Artificial Intelligence Project of the Stanford University. The axiomatization of PASCAL in LCF has been done in cooperation with Richard W. Weyhrauch. Our gratitude to him for many stimulating discussions and for the fun we had when working together.

Many thanks to Mrs. Patrizia Grassi for typing this paper.

References.

1 Aiello, L. and Aiello, M., Proving Program Correctness in LCF, Presented at the Colloquium on Programming, Paris, 9-11 April 1974.

2 Aiello, L., Aiello, M. and Weyhrauch,R., W., The semantics of PASCAL in LCF, AI Memo n.221, Stanford University, August 1974.

3 Böhm, C., The CUCH as a formal and description language,in Formal language description languages for computer programming (T.B.Steel ed.) North Holland 1966.

4 Böhm, C. and Jacopini G., Flow diagrams, Turing machines and languages with only two formation rules. Comm.ACM, 9, (1966) 366-371.

5 Egli, H., Programming language semantics using extensional λ-calculus models. TR 74-206, Dept. of Computer Science, Cornell University, April 1974.

6 Hoare, C.A.R., An Axiomatic Basis for Computer Programming, Comm. ACM, 12, (1969) 576-580,583.

7 Hoare, C.A.R. and Wirth, N., An Axiomatic Definition of the Programming Language PASCAL, Acta Informatica, 2 (1973) 335-355.

8 Landin, P.J., A correspondence between ALGOL 60 and Church's lambda notation, Comm. ACM 8 (1965) 89-101,158-165.

9 Lucas, P. and Walk, K., On the Formal Description of PL/1, Ann.rev. in Automatic Programming, 6, Part 3 (1969).

10 Milner, R., Logic for computable functions, description of a machine implementation. Artificial Intelligence Memo n.169, Stanford University (1972).

11 Scott, D.S. and Strachey, C., Towards a Mathematical Semantics for Computer Languages, Proc. of the symposium on Computers and Automata, Microwave Research Institute Symposia Series, Vol. 12, Polytechnic Institute of Brooklyn (1971).

12 Scott, D.S. Continous Lattices, Proc. of 1971 Dalhousie Conf., Springer Lecture Notes Series, Springer-Verlag, Heidelberg (1971).

13 Strackey, C. Wadsworth C.P., Continuations,A Mathematical semantics for handling full jumps Oxford University Computing Laboratory Technival Manograph PRG-11 January 1974.

14 Tennent, R.D., Mathematical semantics and Design of programming languages. Ph.D. Thesis Dept of Computer Science, University of Toronto, Sept. 1973.

15 Wirth, N., The programming Language PASCAL, Acta Informatica, 1, (1971), 35-63.

16 Weyhrauch, R.W. and Milner R., Program Semantics and Correctness in a Mechanized Logic, Proc. 1st USA-Japan Computer Conf. Tokyo (1972).

BIG TREES IN A λ-CALCULUS WITH λ-EXPRESSIONS AS TYPES

Roel de Vrijer

Department of Mathematics, Technological University Eindhoven,
The Netherlands. *)

0. Outline.

The abstract term system λλ studied in this paper is a close relative of the Automath family of languages. In the investigation of normalization and decidability properties of these languages, λλ came up as a natural generalization of AUT-QE, the language currently in use for mechanical proof checking at the Automath project in Eindhoven. For introductory reference, see Van Daalen [4].
The introduction, section 1, is an informal account of the system λλ and its relation to other systems.
The formal description of λλ is given in the sections 2 and 3.
In 4 the main results are stated, mostly without proof.
Section 5 is devoted to proving that the big trees are well founded (BT).

1. Introduction.

1.1. Heuristic description.

Before describing the main results of the paper we make a few heuristic comments, especially on the generalized typestructure involved. Here we use the "formulas-as-types" notion for interpreting mathematical statements and proofs, originated independently by De Bruijn [3] and Howard [6] (the term comes from Howard). Further references are given in 1.4.

1.1.1. Type structure.

To illustrate the transition from the type structure of traditional type theory, e.g. the typed λ-calculus exhibited in Hindley et al. [5], to the types we have here, we consider constructive versions of propositional and predicate logic respectively. If we identify a proposition α with the type of its constructions (or proofs), then the implication $\alpha \to \beta$ will be the type of constructions that map constructions of α to constructions of β. That is, $\alpha \to \beta$ corresponds essentially to the cartesian power β^{α}.
In predicate logic a construction c of $\forall x.P(x)$ will map any object t from the domain of quantification α to a construction of the proposition $P(t)$. Hence the type of c(t) depends on the choice of t. The notion of power doesn't suffice any longer; we need that of cartesian product: $\prod_{x\in\alpha} P(x)$.

*) Part of this research was done, while the author was supported by the Netherlands Organization for the Advancement of Pure Science (Z.W.O.).

1.1.2. Abstraction and application, two interpretations.

Automath exploits the formal similarity between two kinds of abstraction: functional abstraction to form the functionlike construction $\lambda x \varepsilon \alpha . c(x)$ and the product construction $\prod_{x \varepsilon \alpha} P(x)$. It is convenient to unify these principles in the notations $[x,\alpha]c(x)$ and $[x,\alpha]P(x)$, respectively. Observe that now functional application in the former case corresponds to specification of coordinate axis in the latter. Also here we use the same notations: $\langle t\rangle[x,\alpha]c(x)$ and $\langle t\rangle[x,\alpha]P(x)$, which reduce to $c(t)$ and $P(t)$, respectively. Now this uniform syntactical treatment of both kinds of abstraction, very convenient for our purposes, may cause some confusion in interpretation. For example, vis à vis of the formula-type analogy it amounts to using the same notation for both the predicate, i.e. "propositional function", $\lambda x \varepsilon \alpha . P(x)$ and its universal quantification $\forall x \varepsilon \alpha . P(x)$.

1.1.3. Supertypes, type inclusion.

We further introduce the constant type as a "supertype" of types. Then, e.g. $[x,\alpha]$type will be the supertype of all those types β, such that whenever $t \varepsilon \alpha$, $\langle t\rangle\beta$ is a meaningful type. Hence, carrying on the example from 1.1.2, we have $[x,\alpha]P(x) \varepsilon [x,\alpha]$type. Moreover, because of the possibility of interpreting $[x,\alpha]P(x)$ as a proposition $\forall x \varepsilon \alpha . P(x)$, we require that $[x,\alpha]P(x) \varepsilon$ type. This motivates the facility in $\lambda\lambda$ (and in AUT-QE) to pass here from $[x,\alpha]$type to type, known as the principle of type inclusion: $[x,\alpha]$type $\subseteq$ type (cf. [3], [4] and 3.5.2 below).

In order to clarify this slightly ambiguous situation one could for the product construction introduce the Π's again, and obtain $\Pi[x,\alpha]P(x) \varepsilon$ type for the product and $[x,\alpha]P(x) \varepsilon \Pi[x,\alpha]$type for the type-valued function, respectively.

1.1.4. $\lambda\lambda$-theories.

Expressions are built up by using the principles of abstraction and application mentioned above, starting from variables, parameters and constants. A particular choice of the constants and their (super) type assignments will depend on the interpretation one has in mind. Such a choice is formally fixed by a base (cf. 3.1). Each base determines a specific $\lambda\lambda$-theory.

In informal mathematics new notions are always introduced in a context, possibly indicated by the presence of certain parameters and assumptions. This observation is reflected in $\lambda\lambda$ by the fact that constants are allowed to depend on parameters. We now illustrate the treatment of constants in $\lambda\lambda$ and the parameter mechanism involved.

Let $C_1(\alpha,\beta)$ be a type constant, to be interpreted as the proposition $\exists x \varepsilon \alpha . \langle x\rangle\beta$, where β is supposed to represent a predicate on the type α. Informally introducing $C_1(\alpha,\beta)$ one might stipulate:

(1) "Let P be a type, Q be a predicate on P. Then we will consider $C_1(P,Q)$ as a proposition."

In $\lambda\lambda$ the (super) types of parameters are indicated by superscripts and hence the corresponding axiom reads:

(2) $C_1(P^{\underline{type}}, Q^{[x,P]\underline{type}}) \in \underline{type}$.

The rule of existence introduction can now be formalized by adding another constant $C_2(\alpha,\beta)$ together with the axiom

(3) $C_2(P^{\underline{type}}, Q^{[x,P]\underline{type}}) \in [x,P][y,\langle x\rangle Q]C_1(P,Q)$.

When actually given $\alpha \in \underline{type}$ and $\beta \in [x,\alpha]\underline{type}$, $C_1(\alpha,\beta) \in \underline{type}$ and $C_2(\alpha,\beta) \in [x,\alpha][y,\langle x\rangle\beta]C_1(\alpha,\beta)$ can now be obtained as instances of (2) and (3), respectively. Moreover, for objects $t \in \alpha$ and $s \in \langle t\rangle\beta$ application and β-reduction yield: $\langle s\rangle\langle t\rangle C_2(\alpha,\beta) \in C_1(\alpha,\beta)$.
For further explanation on the subject of interpretation we refer to the treatment of AUT-QE in [4] and to Van Benthem Jutting [2].

1.2. Applicability.

Usually, in type theory as in the Automath languages, term application is subjected to the applicability condition: $\langle t\rangle f$ is a term iff there are types α and β such that $t \in \alpha$ and $f \in [x,\alpha]\beta$. Now in typed λ-calculus this condition is easy to formulate. The type structure and the assignments of types to terms are given in advance, i.e. all of the syntax precedes the generation of theorems. In our case however, types depend on objects and the type assignments are themselves treated as theorems in $\lambda\lambda$. Hence here the applicability condition would make derivability interfere with term formation. A common way of dealing with this complication (cf. Automath, Martin-Löf, etc.) is to generate the terms (including the types) simultaneously with the theorems.
By contrast we take the approach of allowing unrestricted application in $\lambda\lambda$, but instead now subjecting the rule of β-reduction

(4) $\langle t\rangle[x,\alpha]\Sigma >_1 [\![x/t]\!]\Sigma$

to the condition $t \in \alpha$. We can then formulate an applicability condition by referring to derivability in $\lambda\lambda$ and so define the set of legitimate terms. The legitimate fragment $\lambda\lambda - \ell$ of $\lambda\lambda$ is the system one obtains by restricting $\lambda\lambda$ to the language containing only legitimate terms. Hence $\lambda\lambda - \ell$ may be considered as the part of $\lambda\lambda$ that is significant for interpretation. (Though, of course, the illegitimate terms do have a computational interpretation in the term model.) The justification for the above sketched procedure lies in the following result:

(5) $\lambda\lambda$ is a conservative extension of $\lambda\lambda - \ell$.

This property may be regarded as a soundness criterion for the notion of legitimacy as defined above, and hence for $\lambda\lambda$: if the equality of two significant (read: legitimate) terms can be proved in $\lambda\lambda$, it can be done using only significant terms. The proof of (5) uses the result on "big trees" described below.

1.3. Decidability, big trees.

We now turn to a second desirable property of the systems:

(6) $\lambda\lambda$ and hence $\lambda\lambda - \ell$ are decidable.

Decidability of the typed λ-calculus is an easy corollary of the strong normalizability property (SN) and the Church-Rosser property (CR). Every term reduces effectively to its normal form (nf) and two terms are equal iff their nf's are identical. However, although both SN and CR go through for $\lambda\lambda$, they are not sufficient for the decidability, as we will now explain.
In the discussion we make use of an effective function τ, which assigns canonically to every object a type such that $t \in \tau(t)$. Then, since we have uniqueness of types (cf. 3.3.5):

(7) $t \in \alpha \Leftrightarrow \tau(t) = \alpha$

(where by CR, = is equivalent to having the same nf).
So, in order to see if (4) holds, we must first determine if $\tau(t)$ and α have the same nf (by (7)). Then in the process of reducing these terms questions of the form (4) may arise again, and so on.
To deal with this problem, we proceed as follows. Let $\rightarrow$ be the improper reduction relation generated by

(i) usual $\beta\eta$-reduction,
(ii) applying τ,
(iii) taking proper subterms.

Call the tree of $\rightarrow$-reduction sequences of a term Σ the "big tree" of Σ. Then we prove instead of SN the stronger property:

(BT) big trees of terms in $\lambda\lambda$ are well founded.

Together with CR this result easily implies the decidability. Further, as mentioned above, it is also used in the proof of (5).
In his thesis Nederpelt [8] stated as a conjecture for his system Λ, the closure property:

(8) Legitimate terms reduce to legitimate terms.

It turns out that BT (for Λ) implies (8). Further it seems that BT can be proved for Λ by a method, similar to the one used here. (Note that by contrast (8) for λλ is a simple consequence of the formulation of the system and its proof does not require BT.)

We feel that, apart from the applications described, BT may have some interest on its own.

1.4. Historical remarks.

The first proof of normalization of an Automath system was given in Van Benthem Jutting [1]. Nederpelt [8] proved strong normalization for his system Λ. He made two conjectures: the above mentioned closure property for Λ and CR for the system with η-reduction. The latter conjecture was proved by Van Daalen (to appear in his thesis). The result is assumed in this paper.

Scott [10] suggested to use the ideas of De Bruijn [3] for the formalization of an intuitionistic theory of constructions. At about the same time Howard [6] came up with similar ideas. The line is pursued by Martin-Löf [7]. His theory of types is claimed to be a natural formal framework for intuitionistic mathematics. The different accents in motivation - Automath more practical, Martin-Löf more philosophical - might be responsible for some of the differences in the investigated systems.

2. The language, expressions.

In this paragraph we specify the language of a λλ-theory. This language is affected by the choice of a base (cf. 3.1). A similarity type (defined below) codes the information, which is relevant for the formation of expressions. Hence for each similarity type s we define the language $\mathcal{L}_s$.

2.1. Alphabet.

All formal symbols used are from the *alphabet* consisting of the symbols for

variables	$x,y,z,\dots$
parameters	$P,Q,R,\dots$
constants	$C_1,C_2,C_3,\dots$ and <u>type</u>
binary relations	$=,\geq,>,>_1,\varepsilon$,

and the auxiliary symbols [,], <,>, (,), ,.

Variable symbols will be indexed by types to become (object-) variables, parameter symbols by types and supertypes to become object- and type-parameters, respectively. The set of variables is assumed to be such that whenever needed, we are able to choose uniquely a "new" variable of the desired type, not yet occurring in the context. The enumeration of the constant symbols is meant to show the order in which they can be introduced in a particular interpretation (cf. 1.1.4 and the notions of date and base). In Automath this would be the order in which they appear in a "book" (cf. [4]).

2.2. Similarity type.

A *similarity type* s is a triple $<S_0,S_1,\sigma>$, where S_0 and S_1 are disjoint sets of natural numbers and σ is a function from $S_0 \cup S_1$ to $\{0,1\}^*$, the set of finite (possibly empty) sequences of zeros and ones.

Here S_0 indicates the set of constant symbols used for object-constants, S_1 the set of constant symbols used for type-constants and if $i \in S_0 \cup S_1$, then $\sigma(i)$ determines the positions of object- and type-parameters of C_i (cf. 2.3.1 (ii)).

2.3. Expressions.

The expressions fall into three *sorts*: objects, types and supertypes. These are simultaneously defined in 2.3.1. In the definition we use already the notion of *closed* expression, to be defined in 2.3.4. However, it is clear that the definitions could have been given simultaneously.

2.3.1. Definition. Given a similarity type s, the sets of *variables*, *parameters*, *constants*, *objects*, *types* and *supertypes*, building together the set E_s of *expressions* is defined by simultaneous induction.

(i) If x is a variable symbol, P a parameter symbol, α a type, β a closed (cf. 2.3.4) type and β^* a closed supertype, then x^α is a variable, P^β is an object-parameter and P^{β^*} is a type-parameter.

(ii) Let $\sigma(i) = \delta_1,\ldots,\delta_n$ and $\Sigma_1,\ldots,\Sigma_n$ be expressions such that Σ_j is an object if $\delta_j = 0$ and Σ_j is a type if $\delta_j = 1$, then $C_i(\Sigma_1,\ldots,\Sigma_n)$ is an object-constant if $i \in S_0$ and a type-constant if $i \in S_1$.

(iii) Variables, object-parameters and object-constants are *atomic* objects. Type-parameters and type-constants are atomic types and type is the only atomic supertype.

(iv) If f and t are objects, α and β are types, α^* is a supertype and x^α a variable, then $<t>f$ and $[x^\alpha,\alpha]t$ are objects, $<t>\beta$ and $[x^\alpha,\alpha]\beta$ are types and $<t>\alpha^*$ and $[x^\alpha,\alpha]\alpha^*$ are supertypes.

2.3.2. Conventions. As syntactical variables we use $\Sigma,\Gamma,\ldots$ for expressions in general, $f,g,t,s,\ldots$ for objects, $\alpha,\beta,\ldots$ for types and $\alpha^*,\beta^*,\ldots$ for supertypes. The symbols for variables, parameters and constants are used themselves as syntactical variables for their respective categories as well.

As long as no confusion arises we will freely add and omit indexes. In particular the superscripts of variables and parameters are suppressed where possible, e.g. we write $[x,\alpha]x$ instead of $[x^\alpha,\alpha]x^\alpha$.

Vectorial notation is introduced for sequences of expressions; e.g. $\vec{\alpha}$ is short for the sequence $\alpha_1,\ldots,\alpha_n$, where the number n is either known or not essential.

As = is a symbol of the language we use $\equiv$ for syntactic equality between expressions.

Now follow some more technical and notational definitions concerning expressions.

2.3.3. Complexity, length and date.

According to definition 2.3.1 each expression has a construction, easily seen to be unique, consisting of a finite number of applications of the rules (i) to (iv).

The *complexity* $c(\Sigma)$ of an expression Σ is the number of steps in its construction.

By induction on $c(\Sigma)$ we define two more measures on Σ: its *length* $\ell(\Sigma)$ and *date* $d(\Sigma)$.

$$\ell(\Sigma) = 1 \text{ if } \Sigma \text{ is either a variable, a parameter or } \underline{\text{type}};$$

$$\ell(C(\Sigma_1,\ldots,\Sigma_n)) = \max(\ell(\Sigma_1),\ldots,\ell(\Sigma_n)) + 1;\ \ell(<t>\Gamma) = \max(\ell(t),\ell(\Gamma)) + 1;$$

$$\ell([x,\alpha]\Gamma) = \max(\ell(\alpha),\ell(\Gamma)) + 1.$$

$$d(\underline{\text{type}}) = 0;\ d(x^\alpha) = d(\alpha);\ d(P^\Gamma) = d(\Gamma);\ d(C_i(\Sigma_1,\ldots,\Sigma_n)) =$$

$$= \max(i,d(\Sigma_1),\ldots,d(\Sigma_n));\ d(<t>\Gamma) = \max(d(t),d(\Gamma));\ d([x,\alpha]\Gamma) =$$

$$= \max(d(\alpha),d(\Gamma)).$$

Notice that $d(\Sigma)$ is the greatest natural number i, such that C_i appears in the construction of Σ.

2.3.4. Free variables, parameters, special variables.

By induction of $\ell(\Sigma)$ we define the sets $FV(\Sigma)$ of *free variables* and $Par(\Sigma)$ of *parameters* of Σ.

$FV(\underline{\text{type}}) = \emptyset$	; $Par(\underline{\text{type}}) = \emptyset$
$FV(x) = \{x\}$	; $Par(x) = \emptyset$
$FV(P) = \emptyset$	; $Par(P) = \{P\}$
$FV(C(\Sigma_1,\ldots,\Sigma_n)) = \bigcup_{i\le n} FV(\Sigma_i)$	; $Par(C(\Sigma_1,\ldots,\Sigma_n)) = \bigcup_{i\le n} Par(\Sigma_i)$
$FV(<t>\Gamma) = FV(t) \cup FV(\Gamma)$	; $Par(<t>\Gamma) = Par(t) \cup Par(\Gamma)$
$FV([x,\alpha]\Gamma) = FV(\alpha) \cup (FV(\Gamma) \setminus \{x\})$	; $Par([x,\alpha]\Gamma) = Par(\alpha) \cup Par(\Gamma)$.

The set $SV(\Sigma)$ of *special variables* of Σ is defined as

$$SV(\Sigma) = \bigcup_{x^\alpha \in FV(\Sigma)} FV(\alpha) .$$

An expression Σ is called *closed* if $FV(\Sigma) = \emptyset$.

For a sequence $\Sigma_1,\ldots,\Sigma_n$ we introduce the notation $F(\vec{\Sigma}) = \bigcup_{i\le n} F(\Sigma_i)$, F is FV, SV or Par.

2.3.5. Proper subexpressions.

The relation $\succ$ (contains as a proper subexpression) between expressions is the smallest transitive relation such that $C(\Sigma_1,\ldots,\Sigma_n) \succ \Sigma_i$; $<t>\Gamma \succ t$; $<t>\Gamma \succ \Gamma$; $[x,\alpha]\Gamma \succ \alpha$ and $[x,\alpha]\Gamma \succ \Gamma$.

2.3.6. Simultaneous substitution.

Let $P_1,\dots,P_m$ and $x_1,\dots,x_n$ be sequences of distinct parameters and variables. And let $t_1,\dots,t_n$ be a sequence of objects and $\Sigma_1,\dots,\Sigma_m$ a sequence of expressions, such that Σ_i is of the same sort (i.e. object or type) as P_i. Then the result $[\![\vec{P},\vec{x}/\vec{\Sigma},\vec{t}]\!]\Sigma$ of *simultaneous substitution* of $\vec{\Sigma}, \vec{t}$ for $\vec{P}, \vec{x}$ is defined by induction on $\ell(\Sigma)$. In the definition we abbreviate $[\![\vec{P},\vec{x}/\vec{\Sigma},\vec{t}]\!]\Gamma$ by Γ'.

$x_i' \equiv t_i$ $(1 \le i \le n)$ and $x' \equiv x$ if $x \notin \{x_1,\dots,x_n\}$.

$P_i' \equiv \Sigma_i$ $(1 \le i \le m)$ and $P' \equiv P$ if $P \notin \{P_1,\dots,P_m\}$.

$(C(\Sigma_1,\dots,\Sigma_k))' \equiv C(\Sigma_1',\dots,\Sigma_k')$, $\underline{type}' \equiv \underline{type}$.

$(\langle t\rangle\Gamma)' \equiv \langle t'\rangle\Gamma'$.

$([x,\alpha]\Gamma)' \equiv [y,\alpha']([\![x/y]\!]\Gamma)'$, where y is a new variable.

By $[\![\vec{P},\vec{x}/\vec{\Sigma},\vec{t}]\!]\vec{\Gamma}$ we denote the sequence $\Gamma_1',\dots,\Gamma_k'$.

2.3.7. α-equivalence.

The relation $\approx$ of *α-equivalence* between expressions is the smallest equivalence relation such that, if $\Sigma \approx \Gamma$, $t \approx s$ and $\alpha \approx \beta$, then also $x^\alpha \approx x^\beta$, $P^\Sigma \approx P^\Gamma$, $C(\Lambda_1,\dots,\Sigma,\dots,\Lambda_n) \approx C(\Lambda_1,\dots,\Gamma,\dots,\Lambda_n)$, $\langle t\rangle\Sigma \approx \langle s\rangle\Gamma$ and if $y \notin FV(\Gamma)$ also $[x,\alpha]\Sigma \approx [y,\alpha][\![x/y]\!]\Gamma$.

In the sequel we shall simply identify α-equivalent expressions. Formally one might pass to α-equivalence classes, considering an expression as merely a name, denoting the class it belongs to, and show that the preceding definitions behave well with respect to $\approx$.

In some places names of bound variables will be tacitly assumed to be chosen such that no "clashes" arise.

2.3.8. Lemma. Let $\{Q_1,\dots,Q_m\} \cap Par(\vec{\Sigma},\vec{t}) = \{y_1,\dots,y_n\} \cap FV(\vec{\Sigma},\vec{t}) = \emptyset$.
Then $[\![\vec{P},\vec{x}/\vec{\Sigma},\vec{t}]\!][\![\vec{Q},\vec{y}/\vec{\Gamma},\vec{s}]\!]\Sigma \equiv [\![\vec{Q},\vec{y}/[\![\vec{P},\vec{x}/\vec{\Sigma},\vec{t}]\!](\vec{\Gamma},\vec{s})]\!][\![\vec{P},\vec{x}/\vec{\Sigma},\vec{t}]\!]\Sigma$.

2.3.9. Lemma. Let Σ be closed and $Par(\Sigma) \subseteq \{Q_1,\dots,Q_n\}$.
Then $[\![\vec{P},\vec{x}/\vec{\Sigma},\vec{t}]\!][\![\vec{Q}/\vec{\Gamma}]\!]\Sigma \equiv [\![\vec{Q}/[\![\vec{P},\vec{x}/\vec{\Sigma},\vec{t}]\!]\vec{\Gamma}]\!]\Sigma$.

2.4. Formulas, language.

Let s be a similarity type. Then the *language* $\mathcal{L}_s$ consists of the *formulas*: $\Sigma = \Gamma$ (equals), $\Sigma \ge \Gamma$ (reduces to), $\Sigma > \Gamma$ (properly reduces to), $\Sigma >_1 \Gamma$ (reduces in one step to) and $\Sigma \in \Gamma$ (has type or has supertype), where Σ and Γ are expressions in E_s.

If R is a relation symbol we write $\Sigma_1,\dots,\Sigma_n \; R \; \Gamma_1,\dots,\Gamma_n$ $(\vec{\Sigma} \; R \; \vec{\Gamma})$ for the sequence of formulas $\Sigma_1 \; R \; \Gamma_1,\dots,\Sigma_n \; R \; \Gamma_n$.

3. $\lambda\lambda$-theories.

3.1. Base.

According to what has been said in 1.1.4, the set of axioms and rules of a $\lambda\lambda$-theory can be divided into two parts.

(i) A fixed set, characterizing the basic system (the same for any $\lambda\lambda$-theory).

(ii) In addition the assignments of types and supertypes to the relevant constants, determined by a base (defined below).

The situation may be compared to e.g. predicate logic, where one adds for each particular theory a set of mathematical axioms to the fixed framework of logical axioms and rules.

Now recall the example in 1.1.4. It involves an instance (1) of the general assumption scheme:

(*) "Let P_1 be a Σ_1, let P_2 be a Σ_2, ... and P_n be a Σ_n. Then ... ."

In such a scheme it is assumed that the Σ_i's are well defined in the given context, which leads to the requirement that Σ_{i+1} should not contain free variables or parameters other than $P_1,\ldots,P_i$. This observation motivates the following definition.

3.1.1. <u>Definition</u>. A *regular sequence of parameters* (rsop) is a sequence $P_1^{\Sigma_1},\ldots,P_n^{\Sigma_n}$ of distinct parameters, where the Σ_i's are closed types or supertypes and for $0 \le i < n$, $\mathrm{Par}(\Sigma_{i+1}) \subseteq \{P_1,\ldots,P_i\}$.

We now proceed to the definition of a base. Notice the requirements on dates, motivated by the remark made in 2.1.

3.1.2. <u>Definition</u>. A base Ω is a triple $<s,\rho,\tau'>$, where

(i) s is a similarity type $<S_0,S_1,\sigma>$.

(ii) ρ is an effective function from $S_0 \cup S_1$ to rsop's, such that for all $i \in S_0 \cup S_1$, if $\rho(i) = P_1^{\Sigma_1},\ldots,P_n^{\Sigma_n}$, then $C_i(P_1^{\Sigma_1},\ldots,P_n^{\Sigma_n}) \in E_s$ and $\max(d(\Sigma_1),\ldots,d(\Sigma_n)) < i$.

(iii) τ' is an effective function from S_0 to closed types in E_s and from S_1 to closed supertypes in E_s, such that for $i \in S_0 \cup S_1$, if $\rho(i) = P_1,\ldots,P_n$, then $\mathrm{Par}(\tau'(i)) \subseteq \{P_1,\ldots,P_n\}$ and $d(\tau'(i)) < i$.

3.2. Axioms and rules of $\lambda\lambda[\Omega]$.

3.2.1. Given a base Ω, the $\lambda\lambda$-theory $\lambda\lambda[\Omega]$ is formulated in the language $\mathcal{L}_s$. The axioms and rules of $\lambda\lambda[\Omega]$ are the following.

I type assignment.

a) $x^\alpha \in \alpha$; $P^\Sigma \in \Sigma$.

b) $C_i(\Sigma_1,\ldots,\Sigma_n) \in [\![P_1,\ldots,P_n/\Sigma_1,\ldots,\Sigma_n]\!]\tau'(i)$ if $i \in S_0 \cup S_1$ and $\rho(i) = P_1,\ldots,P_n$.

c) $\alpha \in \underline{type}$ (type inclusion).

d) $\Sigma \in \Gamma \vdash [x,\alpha]\Sigma \in [x,\alpha]\Gamma$, provided $x \notin SV(\Sigma)$.

e) $\Sigma \in \Gamma \vdash <t>\Sigma \in <t>\Gamma$.

II one step reduction.

β-reduction: $t \in \alpha \vdash \langle t\rangle[x,\alpha]\Sigma >_1 [\![x/t]\!]\Sigma$.

η-reduction: $f \in \beta,\ \beta \in [x,\alpha]\alpha^* \vdash [x,\alpha]\langle x\rangle f >_1 f$, provided $x \notin FV(f)$,

$\beta \in [x,\alpha]\alpha^* \vdash [x,\alpha]\langle x\rangle\beta >_1 \beta$, provided $x \notin FV(\beta)$.

monotonicity rules:

a) $\Sigma >_1 \Gamma \vdash C(\Sigma_1,\ldots,\Sigma,\ldots,\Sigma_n) >_1 C(\Sigma_1,\ldots,\Gamma,\ldots,\Sigma_n)$.

b) $\Sigma >_1 \Gamma \vdash \langle t\rangle\Sigma >_1 \langle t\rangle\Gamma$; $t >_1 s \vdash \langle t\rangle\Sigma >_1 \langle s\rangle\Sigma$.

c) $\Sigma >_1 \Gamma \vdash [x,\alpha]\Sigma >_1 [x,\alpha]\Gamma$, provided $x \notin SV(\Sigma)$.

d) $\alpha >_1 \beta \vdash [x,\alpha]\Sigma >_1 [y,\beta][\![x/y]\!]\Sigma$, provided $y \notin FV(\Sigma)$.

III proper reduction, reduction and equality.

a) $\Sigma >_1 \Gamma \vdash \Sigma > \Gamma$; $\Sigma > \Gamma,\ \Gamma > \Lambda \vdash \Sigma > \Lambda$.

b) $\Sigma > \Gamma \vdash \Sigma \geq \Gamma$; $\Sigma \geq \Sigma$.

c) $\Sigma \geq \Gamma \vdash \Sigma = \Gamma$; $\Sigma = \Gamma \vdash \Gamma = \Sigma$; $\Sigma = \Gamma,\ \Gamma = \Lambda \vdash \Sigma = \Lambda$.

d) $\Sigma = \Gamma,\ \Lambda \in \Sigma \vdash \Lambda \in \Gamma$; $\Sigma = \Gamma,\ \Sigma \in \Lambda \vdash \Gamma \in \Lambda$.

3.2.2. Remarks.

(i) Ic) amounts to the principle of type inclusion (cf. 1.1.3 and 3.5).

(ii) The motivation of the restriction in Id) is clear from following example. Suppose one had $[x,\alpha]y^{C(x)} \in [x,\alpha]C(x)$. Then for arbitrary $t \in \alpha$ by application and β-reduction $y^{C(x)} \in C(t)$, which is obviously not intended.

(iii) The restriction in IIc) excludes the possibility of both
$\langle t\rangle[x,\alpha]\langle y^{C(x)}\rangle[z,C(x)]z >_1 \langle t\rangle[x,\alpha]y^{C(x)} >_1 y^{C(x)}$ and
$\langle t\rangle[x,\alpha]\langle y^{C(x)}\rangle[z,C(x)]z >_1 \langle y^{C(x)}\rangle[z,C(t)]z$, both in nf, violating CR.

In the sequel we assume an arbitrary base Ω to be fixed. By just stating a formula we mean that it is derivable in $\lambda\lambda[\Omega]$, for convenience further referred to as $\lambda\lambda$.

Syntactical variables for expressions are supposed to range over E_s.

3.2.3. <u>Lemma</u>. The monotonicity rules IIa)-d) hold also with $>_1$ replaced by $>$, $\geq$ or $=$.

3.2.4. Now follows a, rather technical, definition, auxiliary to the important substitution lemma 3.2.5. Compare also definition 3.1.1 (rsop's).

<u>Definition</u>. Given an expression Σ, a sequence $P_1^{\Sigma_1},\ldots,P_m^{\Sigma_m},x_1^{\alpha_1},\ldots,x_n^{\alpha_n}/\Gamma_1,\ldots,\Gamma_m,t_1,\ldots,t_n$ is called a *regular substitution sequence* (rss) *for* Σ, if the following conditions are satisfied:

(i) $\Gamma_i \in [\![\vec{P}/\vec{\Gamma}]\!]\Sigma_i \quad (1 \leq i \leq m)$.

(ii) $t_i \in [\![\vec{P},\vec{x}/\vec{\Gamma},\vec{t}]\!]\alpha_i \quad (1 \leq i \leq n)$.

(iii) If $Q^\Lambda \in Par(\Sigma) \setminus \{P_1,\ldots,P_m\}$, then $Par(\Lambda) \cap \{P_1,\ldots,P_m\} = \emptyset$.

(iv) If $y^\beta \in FV(\Sigma) \setminus \{x_1,\ldots,x_n\}$, then $FV(\beta) \cap \{x_1,\ldots,x_n\} =$
$= Par(\beta) \cap \{P_1,\ldots,P_m\} = \emptyset$.

It is easily verified that the conditions (iii) and (iv) are fulfilled if in particular:

. $m = 0$ and $\{x_1,\ldots,x_n\} \cap SV(\Sigma) = \emptyset$, or

. $Par(\Sigma) \subseteq \{P_1,\ldots,P_m\}$ and $FV(\Sigma) \subseteq \{x_1,\ldots,x_n\}$, and hence if

. $Par(\Sigma) \subseteq \{P_1,\ldots,P_m\}$ and Σ is closed.

3.2.5. Lemma. Let $\vec{P},\vec{x}/\vec{\Sigma},\vec{t}$ be both an rss for Σ and for Γ, and let $\Sigma\ R\ \Gamma$, where R is $>_1$, $>$, $\geq$, $=$ or ε. Then also $[\![\vec{P},\vec{x}/\vec{\Sigma},\vec{t}]\!]\Sigma\ R\ [\![\vec{P},\vec{x}/\vec{\Sigma},\vec{t}]\!]\Gamma$.

Proof. Simultaneous induction on the length of deduction of $\Sigma\ R\ \Gamma$.

3.3. Canonical type assignment, uniqueness of types.

The assignment function τ' generates a function τ, which assigns canonically to each object a type and to each type a supertype, such that always $\Sigma\ \varepsilon\ \tau(\Sigma)$.

3.3.1. Definition. $\tau(\Sigma)$ is defined by induction on $\ell(\Sigma)$.

$\tau(x^{\alpha}) \equiv \alpha$; $\tau(P^{\Gamma}) \equiv \Gamma$.

$\tau(C_i(\Sigma_1,\ldots,\Sigma_n)) \equiv [\![\vec{P}/\vec{\Sigma}]\!]\tau'(i)$ for $i \in S_0 \cup S_1$, where $\rho(i) = P_1,\ldots,P_n$.

$\tau(<t>\Gamma) \equiv <t>\tau(\Gamma)$; $\tau([x,\alpha]\Gamma) \equiv [x,\alpha]\tau(\Gamma)$, where x is chosen such that $x \notin SV(\Gamma)$.

3.3.2. Lemma. $\Sigma\ \varepsilon\ \tau(\Sigma)$ holds for any object or type Σ.

Proof. Induction on $\ell(\Sigma)$.

3.3.3. Lemma. $[\![\vec{x}/\vec{t}]\!]\tau(C(\vec{\Sigma})) \equiv \tau(C([\![\vec{x}/\vec{t}]\!]\vec{\Sigma}))$.

Proof. Immediate by lemma 2.3.9.

3.3.4. Lemma. Let $\vec{x}/\vec{t}$ be an rss for Σ, then $\tau([\![\vec{x}/\vec{t}]\!]\Sigma) = [\![\vec{x}/\vec{t}]\!]\tau(\Sigma)$.

Proof. Induction on $\ell(\Sigma)$. Use lemma 3.3.3 in case Σ is a constant.

3.3.5. Theorem (uniqueness of types). $t\ \varepsilon\ \alpha \Leftrightarrow \alpha = \tau(t)$.

Proof. One side is implied by lemma 3.3.2. For the other side, prove by simultaneous induction on the length of deduction of $t\ \varepsilon\ \alpha$ and $t = s$, respectively, the two statements $t\ \varepsilon\ \alpha \Rightarrow \alpha = \tau(t)$ and $t = s \Rightarrow \tau(t) = \tau(s)$. The proof makes use of the previous lemma.

3.3.6. Remark. The analogous result for supertypes does not hold (cf. 3.5). However, in $\lambda\lambda$ without rule Ic) one would obtain theorem 3.3.5 for supertypes as well.

3.4. Legitimacy.

In this section we define the set L of legitimate expressions. Then the legitimate fragment $\lambda\lambda - \ell$ of $\lambda\lambda$ is the theory obtained by restricting the axioms and rules of $\lambda\lambda$, to use only expressions from L.

3.4.1. Remark that L depends on the choice of Ω. We might call Ω a *legitimate base* if $\{C_i(\rho(i)) \mid i \in S_0 \cup S_1\} \cup \{\tau'(i) \mid i \in S_0 \cup S_1\} \subseteq L$.

3.4.2. For the sake of the characterization of the legitimate expressions we now introduce a function τ^*, assigning canonically to each expression a supertype.

Definition.

$\tau^*(\alpha^*) \equiv \alpha^*$ for supertypes α^*.

$\tau^*(\alpha) \equiv \tau(\alpha)$ for types α.

$\tau^*(t) \equiv \tau(\tau(t))$ for objects t.

Remark. τ^* may be compared to Typ^* in Nederpelt [8].

3.4.3. Definition. The set L of *legitimate expressions* is specified by defining by induction on $(d(\Sigma),c(\Sigma))$ (i.e. $\omega\cdot d(\Sigma) + c(\Sigma)$, cf. 5.2), what it means for an expression Σ to be *legitimate*.

$x^\alpha \in L$ iff $\alpha \in L$; $P^\Sigma \in L$ iff $\Sigma \in L$.

$C_i(\vec{\Sigma}) \in L$ iff $\Sigma_1,\ldots,\Sigma_n$, $\tau'(i) \in L$ and $\rho(i) / \vec{\Sigma}$ is an rss for $\tau'(i)$.

$\langle t\rangle\Gamma \in L$ iff $t,\Gamma \in L$ and for some α, α^*: $t \in \alpha$ and $\tau^*(\Gamma) = [x,\alpha]\alpha^*$.

$[x,\alpha]\Gamma \in L$ iff $\alpha,\Gamma \in L$, provided $x \notin SV(\Gamma)$.

3.4.4. Lemma. Let $\vec{P},\vec{x}/\vec{\Gamma},\vec{t}$ be an rss for $\Sigma_1,\ldots,\Sigma_n$, respectively, and let $\vec{Q}/\vec{\Sigma}$ be an rss for the closed expression Σ. Then also $\vec{Q}/[\![\vec{P},\vec{x}/\vec{\Gamma},\vec{t}]\!]\vec{\Sigma}$ is an rss for Σ.

Proof. Apply lemma 3.2.5.

3.4.5. Lemma. Let $\Sigma,\Gamma_1,\ldots,\Gamma_m,t_1,\ldots,t_n \in L$ and let $\vec{P},\vec{x}/\vec{\Gamma},\vec{t}$ be an rss for Σ, then $[\![\vec{P},\vec{x}/\vec{\Gamma},\vec{t}]\!]\Sigma \in L$.

Proof. Induction on $\ell(\Sigma)$. Use lemmas 3.2.5 and 3.4.4.

3.4.6. Theorem (Extended Closure). Let $\Sigma \in L$ and let either $\Sigma \geq \Gamma$, or $\Sigma \not> \Gamma$ or $\tau(\Sigma) \equiv \Gamma$. Then also $\Gamma \in L$ (i.e. $\Sigma \in L$ and $\Sigma \to \Gamma \Rightarrow \Gamma \in L$).

3.5. Type inclusion, uniqueness of domains.

The analogue of the uniqueness of types theorem for supertypes does not hold. E.g. we have both $[x,\alpha]\beta \in [x,\alpha]\underline{type}$ and $[x,\alpha]\beta \in \underline{type}$ (cf. 1.1.3). However, one does obtain a weaker result, viz. uniqueness of domains:

$$\alpha \in [x,\beta]\beta^* \text{ and } \alpha \in [x,\gamma]\gamma^* \Rightarrow \beta = \gamma .$$

This property is important as a justification for the above characterization of legitimate expressions.

We state here without proof:

3.5.1. Theorem. $\alpha \in [x,\beta]\beta^*$ iff for some supertype γ^*, $\tau(\alpha) = [x,\beta]\gamma^*$.

In order to say something more on the structure of supertypes in $\lambda\lambda - \ell$, we define the relation $\subseteq$ of *type inclusion* between supertypes in L.

3.5.2. Definition. First define the relation $\subset$ between supertypes in L inductively by

(i) $\alpha^* \subset \underline{type}$ for any supertype α^*.

(ii) If $\alpha^* \subset \beta^*$, then also $[x,\alpha]\alpha^* \subset [x,\alpha]\beta^*$ and $<t>\alpha^* \subset <t>\beta^*$.

Then $\subseteq$ is the smallest transitive relation in L extending = and $\subset$.

3.5.3. Theorem. Let $\alpha,\beta,\alpha^*,\beta^* \in L$. Then

(i) $\alpha \varepsilon \alpha^*$ and $\alpha \varepsilon \beta^* \Rightarrow \alpha^* \subseteq \beta^*$ or $\beta^* \subseteq \alpha^*$.

(ii) $\alpha \varepsilon \alpha^* \Rightarrow \tau(\alpha) \subseteq \alpha^*$.

Hence τ assigns to a legitimate type its minimal supertype. Remark that a supertype in L, which is in nf, is always of the form $[x_1,\alpha_1] \ldots [x_k,\alpha_k]\underline{type}$.

4. Decidability and conservativity.

4.1. Sequences, trees.

We use $\sigma,\rho,\ldots$ to range over, finite or infinite, sequences of expressions. We define $\ell h(\sigma)$ to be the length of σ if σ is finite, $\ell h(\sigma) = \infty$ if σ is infinite. Σ will also stand for the sequence of length one, consisting of Σ only. If $\ell h(\sigma) < \infty$, then σ,ρ stands for the concatenation of σ and ρ. We define: $\sigma < \rho$ (ρ extends σ) iff there exists a sequence τ, such that $\sigma,\tau = \rho$.

4.1.1. Definition. A sequence $\Sigma_0,\Sigma_1,\ldots$ is called a

(i) *reduction sequence* of Σ_0 iff $\Sigma_i >_1 \Sigma_{i+1}$,

(ii) *rs-sequence* of Σ_0 iff either $\Sigma_i >_1 \Sigma_{i+1}$ or $\Sigma_i \not> \Sigma_{i+1}$,

(iii) $\twoheadrightarrow$*-sequence* of Σ_0 iff either $\Sigma_i >_1 \Sigma_{i+1}$ or $\Sigma_i \not> \Sigma_{i+1}$ or $\tau(\Sigma_i) \equiv \Sigma_{i+1}$.

4.1.2. Definition. The finite reduction sequences of a term Σ form under the partial order < a tree, the *reduction tree* of Σ. Analogously we have the *rs-tree* and the $\twoheadrightarrow$-tree of Σ. The latter is called the *big tree* of Σ. The set of $\twoheadrightarrow$-sequences of Σ is denoted by $S(\Sigma)$. $B(\Sigma) = \{\Gamma \mid \Sigma \twoheadrightarrow \Gamma\}$.

4.1.3. Definition. $h(\Sigma)$ will be the *height* of the reduction tree of Σ:
$h(\Sigma) = \max(\{\ell h(\sigma) \mid \sigma \text{ is a reduction sequence of } \Sigma\})$.
Analogously, $b(\Sigma) = \max(\{\ell h(\sigma) \mid \sigma \in S(\Sigma)\})$ is the height of the big tree of Σ.

4.2. Normal forms, strong normalization.

An expression Σ is in *normal form* (nf) if there does not exist an expression Γ such that $\Sigma >_1 \Gamma$.

An expression Σ is called *strongly normalizable* if $h(\Sigma) < \infty$, i.e., if the reduction tree of Σ is well founded.

4.3. Results.

We now state the main results of the paper. The details of proofs are generally omitted. However, section 5 will be devoted to sketching the proof of BT (theorem 4.3.2).

4.3.1. Theorem (CR). If $\Sigma = \Gamma$, then there exists an expression Λ, such that $\Sigma \geq \Lambda$ and $\Gamma \geq \Lambda$.

A proof shall not be given here. Let it suffice to remark that in $\lambda\lambda$ without the rule of η-reduction the property follows easily from the strong normalizability of $\lambda\lambda$. In the present situation, where η-reduction is included, the proof is more complicated. It was proved by Van Daalen (cf. 1.4).

4.3.2. Theorem (BT). For every expression Σ, $b(\Sigma) < \infty$. I.e., big trees in $\lambda\lambda$ are well founded.

This result implies that every expression is strongly normalizable (SN). Moreover, by CR one obtains that for each Σ, there exists a unique nf Γ, such that $\Sigma = \Gamma$. (In contrast to its use in "uniqueness of types", uniqueness is here to be understood with respect to $\equiv$.) This unique expression will be denoted by $nf(\Sigma)$.

4.3.3. Corollary. Given an expression Σ, its big tree can be effectively constructed.
Proof. Given the big trees of an object t and a type α, one can decide if $t \in \alpha$; viz. by merely checking if $nf(\tau(t)) \equiv nf(\alpha)$. By this observation it is easy to devise an algorithm, which, when applied to an expression Σ, constructs the big tree of Σ, and which can be proved to be correct by induction on $b(\Sigma)$.

4.3.4. Corollary. $\lambda\lambda$ is decidable.

4.3.5. Let $(\Sigma,\Gamma) \vdash \Lambda \; R \; \Lambda'$ assert the existence of a deduction of $\Lambda \; R \; \Lambda'$ in $\lambda\lambda$, in which occur only expressions from $B(\Sigma) \cup B(\Gamma)$.

Lemma (transitivity). If $\Sigma',\Gamma' \in B(\Sigma) \cup B(\Gamma)$ and $(\Sigma',\Gamma') \vdash \Lambda = \Lambda'$, then $(\Sigma,\Gamma) \vdash \Lambda = \Lambda'$.

4.3.6. Definition. A new measure $n(\Gamma)$ is defined by induction on $b(\Gamma)$:
$$n(\Gamma) = \Big(\sum_{(\sigma,\Lambda)\in S'(\Gamma)} n(\Lambda)\Big) + 1, \text{ where } S'(\Gamma) = \{\rho \in S(\Gamma) \mid \ell h(\rho) > 1\}.$$

4.3.7. Theorem. Let $\Sigma \; R \; \Gamma$, where R is $=$, $\geq$, $>$, $>_1$ or ε. Then $(\Sigma,\Gamma) \vdash \Sigma \; R \; \Gamma$.
Proof. Induction on $n(\Sigma) + n(\Gamma)$. Let us restrict attention to equalities. If Σ and Γ are both in nf, then by CR, $\Sigma \equiv \Gamma$ and we are done. So assume that $\Sigma >_1 \Sigma'$. Then by the induction hypothesis and transitivity, $(\Sigma,\Gamma) \vdash \Sigma' = \Gamma$. Hence it is enough to show that $(\Sigma,\Gamma) \vdash \Sigma = \Sigma'$. Now distinguish cases as to the last rule applied in a deduction of $\Sigma >_1 \Sigma'$. We treat only one case.
Let $\Sigma \equiv [x,\alpha]\langle x\rangle f >_1 f \equiv \Sigma'$ and $\tau^*(f) = [x,\alpha]\alpha^*$. It must be shown that $(\Sigma,\Gamma) \vdash \tau^*(f) = [x,\alpha]\beta^*$ for some β^* (cf. the rule of η-reduction and theorem 3.5.1). By CR, $\tau^*(f)$ and $[x,\alpha]\alpha^*$ have a common reduct $[y,\gamma]\gamma^*$. Now $n(\alpha) + n(\gamma) < n(\Sigma)$ and $n(\tau^*(f)) + n([y,\gamma]\gamma^*) < n(\Sigma)$ imply that $(\Sigma,\Gamma) \vdash \alpha = \gamma$ and $(\Sigma,\Gamma) \vdash \tau^*(f) = [y,\gamma]\gamma^*$, respectively, and consequently $(\Sigma,\Gamma) \vdash \tau^*(f) = {}$ $= [x,\alpha]\gamma^*$.

4.3.8. Corollary. $\lambda\lambda$ is a conservative extension of $\lambda\lambda - \ell$.

Proof. By theorem 4.3.7 and the closure theorem 3.4.6.

5. Proof of the big tree theorem.

The strategy of the proof of BT (theorem 4.3.2) will be to define an extension $\lambda\lambda - p$ of $\lambda\lambda$, by adding an extra rule of term formation for ordered pairs: if $\tau(\Sigma) = \Gamma$, then $\ulcorner\Sigma,\Gamma\urcorner$ is an expression. A pair $\ulcorner\Sigma,\Gamma\urcorner$ may be considered as just a copy of Σ, Γ being present only for bookkeeping reasons. The reduction relation is extended to include the projections $\ulcorner\Sigma,\Gamma\urcorner >_1 \Sigma$ and $\ulcorner\Sigma,\Gamma\urcorner >_1 \Gamma$. Strong normalization of expressions in $\lambda\lambda - p$ is proved by using a computability argument. Subsequently a map φ is defined, embedding $\lambda\lambda$ in $\lambda\lambda - p$ such that $\rightarrow$-sequences in $\lambda\lambda$ give rise to longer rs-sequences in $\lambda\lambda - p$. Termination of rs-sequences is an easy corollary of SN. Hence we may conclude that $\rightarrow$-sequences in $\lambda\lambda$ do terminate.

5.1. Introduction of $\lambda\lambda - p$.

The base Ω, which was fixed under 3.2.2, is still assumed here. So $\lambda\lambda - p$ will be in fact an extension of $\lambda\lambda[\Omega]$.

The definition of the set $E - p$ of expressions of $\lambda\lambda - p$ involves a "forget function" p from expressions of $\lambda\lambda - p$ to expressions of $\lambda\lambda$, consistently deleting the second coordinates of pairs. (Hence p acts as the identity on expressions of $\lambda\lambda$.) The next two definitions should be taken as simultaneously defining the set $E - p$ and the function p.

5.1.1. Definition. For the definition of $E - p$ take clauses (i) to (iv) of the inductive definition of E (2.3.1) and add a fifth clause:

(v) If Σ and Γ are in $E - p$ and $\tau(p(\Sigma)) = p(\Gamma)$ is deducible in $\lambda\lambda$, then $\ulcorner\Sigma,\Gamma\urcorner$ is an object if Σ is an object and a type if Σ is a type, respectively.

5.1.2. Definition. The function p: $E - p \rightarrow E$ is defined inductively.

$p(\underline{type}) \equiv \underline{type}$; $p(P^{\Sigma}) \equiv P^{p(\Sigma)}$; $p(x^{\alpha}) \equiv x^{p(\alpha)}$;

$P(C(\Sigma_1,\dots,\Sigma_n)) \equiv C(p(\Sigma_1),\dots,p(\Sigma_n))$.

$p(\langle t\rangle\Sigma) \equiv \langle p(t)\rangle p(\Sigma)$; $p([x,\alpha]\Sigma) \equiv [x,p(\alpha)]p(\Sigma)$.

$p(\ulcorner\Sigma,\Gamma\urcorner) \equiv p(\Sigma)$.

5.1.3. The definitions, notations and conventions from section 2.3 are generalized to $E - p$. In particular, $\ell(\ulcorner\Sigma,\Gamma\urcorner) = \max(\ell(\Sigma),\ell(\Gamma)) + 1$; $d(\ulcorner\Sigma,\Gamma\urcorner) = \max(d(\Sigma),d(\Gamma))$; $Par(\ulcorner\Sigma,\Gamma\urcorner) = Par(\Sigma) \cup Par(\Gamma)$; $FV(\ulcorner\Sigma,\Gamma\urcorner) = FV(\Sigma) \cup FV(\Gamma)$; $\ulcorner\Sigma,\Gamma\urcorner \triangleright \Sigma$, $\ulcorner\Sigma,\Gamma\urcorner \triangleright \Gamma$. $[\![\vec{P},\vec{x}/\vec{\Lambda},\vec{t}]\!]\ulcorner\Sigma,\Gamma\urcorner \equiv \ulcorner[\![\vec{P},\vec{x}/\vec{\Lambda},\vec{t}]\!]\Sigma, [\![\vec{P},\vec{x}/\vec{\Lambda},\vec{t}]\!]\Gamma\urcorner$. Substitution is only admitted if the substitution result is in $E - p$ again, i.e., if the substitution does not violate the restriction in 5.1.1 (v). A sufficient condition for this requirement is given in 5.1.6 below.

5.1.4. The formulas of $\lambda\lambda - p$ are defined as in 2.4.

5.1.5. The axioms and rules of $\lambda\lambda - p$ are those of $\lambda\lambda$ (cf. 3.2.1) and additionally

II projection: $\ulcorner\Sigma,\Gamma\urcorner >_1 \Sigma$; $\ulcorner\Sigma,\Gamma\urcorner >_1 \Gamma$.
e) $\Sigma >_1 \Lambda \vdash \ulcorner\Sigma,\Gamma\urcorner >_1 \ulcorner\Lambda,\Gamma\urcorner$; $\Gamma >_1 \Lambda \vdash \ulcorner\Sigma,\Gamma\urcorner >_1 \ulcorner\Sigma,\Lambda\urcorner$.

Remark that now, by projection, an expression may reduce to an expression of a different sort, i.e. an object to a type and a type to a supertype, respectively. For that reason a few obvious restrictions are to be made in some of the rules. In IIa) and IIId) we require Σ and Γ to be of the same sort. In IIb), t and s have to be both objects; in IId), α and β have to be both types.

5.1.6. The definitions and results of sections 3.2 and 3.3 are generalized to $\lambda\lambda - p$. Remark in particular that by lemma 3.2.5 we obtain: If $\vec{P},\vec{x}/\vec{\Gamma},\vec{t}$ is an rss for Σ in $E - p$, then $[\![\vec{P},\vec{x}/\vec{\Gamma},\vec{t}]\!]\Sigma$ is in $E - p$ again, and hence an admitted substitution. Add to definition 3.3.1 the clause: $\tau(\ulcorner\Sigma,\Gamma\urcorner) \equiv \tau(\Sigma)$.

5.2. Norms.

The proof of SN for $\lambda\lambda - p$ is essentially based on the method of proof orginated by Tait [11], and used e.g. by Prawitz [9, Appendix A] for a system of natural deduction. The key notion of this method, computability (alternative terminologies: convertability, validity, reductibilité), could be defined by induction on the length of type in [11] and on the length of the end formula of a deduction in [9]. Here it is essential that the type of a term and the end formula of a deduction do not change under reduction of the term and the deduction, respectively. In our proof their task will be fulfilled by a norm on expressions $\gamma(\Sigma)$. Auxiliary to its definition we first introduce the measure $m(\Sigma)$.

Note. Pairs of natural numbers are supposed to be ordered lexicographically.

5.2.1. Definition. $m(\Sigma)$ is defined by induction on $(d(\Sigma),c(\Sigma))$.
$m(\underline{type}) = 0$; $m(P^\Gamma) = m(\Gamma) + 1$; $m(x^\alpha) = m(\alpha) + 1$;
$m(C_i(\Sigma_1,\ldots,\Sigma_n)) = \max(m(\Sigma_1),\ldots,m(\Sigma_n)) + m(\tau'(i)) + 1$;
$m(\langle t\rangle\Gamma) = \max(m(t),m(\Gamma))$; $m([x,\alpha]\Gamma) = \max(m(\alpha),m(\Gamma))$ and
$m(\ulcorner\Gamma,\Lambda\urcorner) = \max(m(\Gamma),m(\Lambda))$.

5.2.2. Lemma.

(i) If Σ is an atomic expression (not $\underline{type}$), then $m(\tau(\Sigma)) < m(\Sigma)$.

(ii) For all objects and types Σ, $m(\tau(\Sigma)) \leq m(\Sigma)$.

(iii) If $\Sigma \geq \Gamma$, then $m(\Gamma) \leq m(\Sigma)$.

5.2.3. The norm $\gamma(\Sigma)$ is going to be a, possibly empty, string of the brackets [and]. Let G,H,... range over such strings. They are well ordered by <: G < H iff the number of brackets in G is less then the number of brackets in H. λ denotes the empty string.

Definition. $\gamma(\Sigma)$ is defined by induction on $(m(\Sigma),\ell(\Sigma))$.

$\gamma(\underline{type}) = \lambda$; $\gamma(\Sigma) = \gamma(\tau(\Sigma))$ for other atomic Σ's;

$\gamma([x,\alpha]\Gamma) = [\gamma(\alpha)]\gamma(\Gamma)$; $\gamma(\ulcorner\Gamma,\Lambda\urcorner) = \gamma(\Gamma)$;

$$\gamma(\langle t\rangle\Gamma) = \begin{cases} G \text{ if } \gamma(\Gamma) = [\gamma(t)]G, \\ \lambda \text{ otherwise.} \end{cases}$$

5.2.4. Lemma. If $\gamma(t_i) = \gamma(\alpha_i)$ $(1 \le i \le n)$, then $\gamma([\![x_1^{\alpha_1},\ldots,x_n^{\alpha_n}/t_1,\ldots,t_n]\!]\Sigma = \gamma(\Sigma)$.

5.2.5. Lemma.

(i) If $t \in \alpha$, then $\gamma(t) = \gamma(\alpha)$.

(ii) If $\Sigma = \Gamma$, then $\gamma(\Sigma) = \gamma(\Gamma)$.

Proof. Prove (i) and (ii) simultaneously by induction on the length of deduction in $\lambda\lambda - p$. Use lemma 5.2.4.

5.3. Computability.

The notion of computability can now be defined by induction on $\gamma(\Sigma)$.

5.3.1. Definition. An expression Σ is *computable* (comp) if both

(i) Σ is strongly normalizable;

(ii) whenever $\Sigma \ge [x,\alpha]\Gamma$ and $t \in \alpha$ and t is comp, then also $[\![x/t]\!]\Gamma$ is comp.

The definition is correct. For if $\Sigma \ge [x,\alpha]\Gamma$ and $t \in \alpha$, then $\gamma(t) = \gamma(\alpha) < [\gamma(\alpha)]\gamma(\Gamma) = \gamma(\Sigma)$ and $\gamma([\![x/t]\!]\Gamma) = \gamma(\Gamma) < \gamma(\Sigma)$.

5.3.2. Lemma.

(i) If $\Sigma \ge \Gamma$ and Σ is comp, then so is Γ.

(ii) Let Σ not have the form $[x,\alpha]\Gamma$. Then Σ is comp iff all Σ_1 such that $\Sigma >_1 \Sigma_1$ are comp.

Proof. Immediate by inspection of the definition.

5.3.3. Lemma. If $\Sigma_1,\ldots,\Sigma_n$ are comp, then so is $C(\vec{\Sigma})$.

Proof. Induction on $h(\Sigma_1) + \ldots + h(\Sigma_n)$.

5.3.4. Lemma. If Σ and t are comp, then so is $\langle t\rangle\Sigma$.

Proof. Induction on $h(\Sigma) + h(t)$. Assuming that $\langle t\rangle\Sigma >_1 \Gamma$, prove that Γ is comp, and apply lemma 5.3.2 (ii). Distinguish two cases:

(i) Either $t >_1 t_1$ and $\Gamma \equiv \langle t_1\rangle\Sigma$ or $\Sigma >_1 \Sigma_1$ and $\Gamma \equiv \langle t\rangle\Sigma_1$. Then Γ is comp by the induction hypothesis.

(ii) $\Sigma \equiv [x,\alpha]\Lambda$ and $\Gamma \equiv [\![x/t]\!]\Lambda$. Then Γ is comp by clause (ii) of the computability definition 5.3.1.

5.3.5. Lemma. If Σ and Γ are comp, then so is $\ulcorner\Sigma,\Gamma\urcorner$.

Proof. Again prove by induction on $h(\Sigma) + h(\Gamma)$, that $\ulcorner\Sigma,\Gamma\urcorner >_1 \Lambda$ implies that Λ is comp.

5.3.6. Definition. Σ is called *computable under substitution* (cus), if for all comp expressions $t_1,\dots,t_n$ and variables $x_1,\dots,x_n$, such that $\vec{x}/\vec{t}$ is an rss for Σ, $[\![\vec{x}/\vec{t}]\!]\Sigma$ is comp.

5.3.7. Theorem. All expressions Σ of $E-p$ are cus.

Proof. Induction on $\ell(\Sigma)$. Let $\vec{x}/\vec{t}$ be an arbitrary rss for Σ, such that $t_1,\dots,t_n$ are comp. Throughout the proof we abbreviate $\Lambda' \equiv [\![\vec{x}/\vec{t}]\!]\Lambda$.

The only case which is not immediate by the lemmas 5.3.3-5 and the induction hypothesis is $\Sigma \equiv [x,\alpha]\Gamma$, $\Sigma' \equiv [x,\alpha']\Gamma'$, where α' and Γ' are comp by the induction hypothesis. We check (i) and (ii) of definition 5.3.1.

(i) Suppose $\sigma = \Sigma_0,\Sigma_1,\dots$ is a nonterminating reduction sequence of Σ'. Distinguish two cases:

a) There exist finite reduction sequences $\sigma_0,[x,\alpha_1]\langle x\rangle f_0$ of Σ', and σ_1,α_1 of α', and $\sigma_2,\langle x\rangle f_0$ of Γ', such that $\sigma = \sigma_0,[x,\alpha_1]\langle x\rangle f_0,f_0,f_1,\dots$. Then $\sigma_2,\langle x\rangle f_0,\langle x\rangle f_1,\dots$ would be a nonterminating reduction sequence of Γ', contradicting the computability of Γ'.

b) Case a) does not apply, i.e., no outer η-reductions are performed in σ. Then σ would induce reduction sequences σ_0 of α' and σ_1 of Γ', such that either σ_0 or σ_1 or both are nonterminating, contradicting the fact that α' and Γ' are both comp.

(ii) Suppose $\Sigma' \geq [x,\alpha_1]\Gamma_1$. Again distinguish two cases:

a) $\alpha' \geq \alpha_2$, $\Gamma' \geq \langle x\rangle f$, $x \notin FV(f)$, and so $\Sigma' \geq [x,\alpha_2]\langle x\rangle f >_1 f$ and $f \geq [x,\alpha_1]\Gamma_1$. Let $t \in \alpha_1$ be comp. Then $[\![x/t]\!]\Gamma' \geq [\![x/t]\!](\langle x\rangle f) \equiv \langle t\rangle f \geq \langle t\rangle[x,\alpha_1]\Gamma_1 >_1 [\![x/t]\!]\Gamma_1$. Further $\vec{x},x/\vec{t},t$ is an rss for Γ and $[\![x/t]\!]\Gamma' \equiv [\![\vec{x},x/\vec{t},t]\!]\Gamma$. Hence by the induction hypothesis $[\![x/t]\!]\Gamma'$ is comp and by lemma 5.3.2 (i) so is $[\![x/t]\!]\Gamma_1$.

b) Case a) does not apply. Then $\alpha' \geq \alpha_1$ and $\Gamma' \geq [\![x^{\alpha_1}/x^{\alpha'}]\!]\Gamma_1$. Hence, if $t \in \alpha_1$, also $[\![x/t]\!]\Gamma' \geq [\![x/t]\!]\Gamma_1$ and repeating the argument in a) we find that for comp $t \in \alpha_1$, $[\![x/t]\!]\Gamma_1$ is comp.

5.3.8. Corollary. All expressions of $E-p$ are strongly normalizable.

5.3.9. Corollary. If Σ is an expression in $E-p$, then every rs-sequence of Σ terminates.

Proof. Induction on $(h(\Sigma),\ell(\Sigma))$, observing that if $\Sigma \succ \Gamma$, then $h(\Gamma) \leq h(\Sigma)$.

5.4. Embedding $\lambda\lambda$ in $\lambda\lambda - p$.

We now define a map φ: $E \to E-p$, such that to each $\to$-sequence of an expression Σ in $\lambda\lambda$ corresponds a longer rs-sequence of $\varphi(\Sigma)$ in $\lambda\lambda - p$. Then corollary 5.3.9 guarantees the well foundedness of big trees in $\lambda\lambda$.

5.4.1. Definition. $\varphi(\Sigma)$ is defined by induction on $(m(\Sigma),\ell(\Sigma))$.
$\varphi(x^\alpha) \equiv \ulcorner x^\alpha,\varphi(\alpha)\urcorner$; $\varphi(P^\Sigma) \equiv \ulcorner P^\Sigma,\varphi(\Sigma)\urcorner$;
$\varphi(C(\Sigma_1,\ldots,\Sigma_n)) \equiv \ulcorner C(\varphi(\Sigma_1),\ldots,\varphi(\Sigma_n)),\varphi(\tau(C(\vec{\Sigma})))\urcorner$;
$\varphi(\langle t\rangle\Gamma) \equiv \langle\varphi(t)\rangle\varphi(\Gamma)$ and $\varphi([x,\alpha]\Gamma) \equiv [y,\varphi(\alpha)][\![x/y]\!]\varphi(\Gamma)$.

5.4.2. Lemma. If $\Sigma \in E$, then $\Sigma = \varphi(\Sigma)$ (in $\lambda\lambda - p$).
Proof. By induction on $\ell(\Sigma)$, check that $\varphi(\Sigma) \geq \Sigma$.

5.4.3. Corollary. If $\Sigma \,\varepsilon\, \Gamma$ in $\lambda\lambda$, then $\varphi(\Sigma) \,\varepsilon\, \varphi(\Gamma)$.

5.4.4. Lemma. If $t \,\varepsilon\, \alpha$ in $\lambda\lambda$, $\Sigma \in E$, then $[\![x^\alpha/\varphi\cdot t]\!]\varphi(\Sigma) \geq \varphi([\![x/t]\!]\Sigma)$.
Proof. Induction on $(m(\Sigma),\ell(\Sigma))$. We show only three cases.

(i) $[\![x/\varphi(t)]\!]\varphi(x) \equiv [\![x/\varphi(t)]\!](\ulcorner x,\varphi(\alpha)\urcorner) \equiv \ulcorner\varphi(t),\varphi(\alpha)\urcorner >_1 \varphi(t)$.

(ii) $[\![x/\varphi(t)]\!]\ \varphi(C(\vec{\Gamma})) \equiv \ulcorner C([\![x/\varphi.t]\!]\varphi(\vec{\Gamma})),[\![x/\varphi(t)]\!]\varphi(\tau(C(\vec{\Gamma})))\urcorner \geq$
$\geq \ulcorner C(\varphi([\![x/t]\!]\vec{\Gamma})),\varphi([\![x/t]\!]\tau(C(\vec{\Gamma}))\urcorner \equiv \varphi([\![x/t]\!]C(\vec{\Gamma}))$. Here we applied the induction hypothesis on $\Gamma_1,\ldots,\Gamma_n$ and $\tau(C(\vec{\Gamma}))$ and we used lemma 3.3.3.

(iii) $[\![x/\varphi(t)]\!]\varphi([y,\beta]\Gamma) \equiv [z,[\![x/\varphi(t)]\!]\varphi(\beta)][\![y/z]\!][\![x/\varphi(t)]\!]\varphi(\Gamma) \geq$
$\geq [u,\varphi([\![x/t]\!]\beta)][\![y/u]\!]\varphi([\![x/t]\!]\Gamma) \equiv \varphi([\![x/t]\!][y,\beta]\Gamma)$. (Apply induction hypothesis on β and Γ.)

5.4.5. Lemma. If $\Sigma >_1 \Gamma$ in $\lambda\lambda$, then $\varphi(\Sigma) > \varphi(\Gamma)$.
Proof. Induction on the length of deduction of $\Sigma >_1 \Gamma$. We show only one case. Let $\Sigma \equiv \langle t\rangle[x,\alpha]\Lambda >_1 [\![x/t]\!]\Lambda \equiv \Gamma$ and $t \,\varepsilon\, \alpha$ (β-reduction). Then
$\varphi(\Sigma) \equiv \langle\varphi(t)\rangle[y,\varphi(\alpha)][\![x/y]\!]\varphi(\Lambda) >_1 [\![x/\varphi(t)]\!]\varphi(\Lambda) \geq \varphi(\Gamma)$, by lemmas 5.4.3 and 5.4.4.

5.4.6. Lemma. If $\Sigma \in E$, then $\varphi(\Sigma) > \varphi(\tau(\Sigma))$ (Σ either object or type).
Proof. Induction on $\ell(\Sigma)$. Two examples are:

(i) $\varphi(x^\alpha) \equiv \ulcorner x^\alpha,\varphi(\alpha)\urcorner >_1 \varphi(\alpha) \equiv \varphi(\tau(x^\alpha))$;

(ii) $\varphi(\langle t\rangle\Gamma) \equiv \langle\varphi(t)\rangle\varphi(\Gamma) > \langle\varphi(t)\rangle\varphi(\tau(\Gamma)) \equiv \varphi(\tau(\langle t\rangle\Gamma))$, by the induction hypothesis for Γ.

5.4.7. Lemma. If $\Sigma \not> \Gamma$ in $\lambda\lambda$, then $\varphi(\Sigma) \not> \Gamma$ in $\lambda\lambda - p$.

5.4.8. Corollary. If $\Sigma_0,\ldots,\Sigma_n$ is a $\not\to$-sequence in $\lambda\lambda$, then there exists an rs-sequence from $\varphi(\Sigma_0)$ to $\varphi(\Sigma_n)$ in $\lambda\lambda - p$ of equal or greater length.
Proof. Induction on n, using the lemmas 5.4.5-7.

5.4.9. Theorem. If $\Sigma \in E$, then every $\not\to$-sequence of Σ terminates.
Proof. Immediate from the corollaries 5.3.9 and 5.4.8.

References.

[1] Benthem Jutting, L.S. van, On normal forms in AUTOMATH; Unpublished, 1971.

[2] Benthem Jutting, L.S. van, The development of a text in AUT-QE; proceedings of the Symposium APLASM (Orsay, December 1973), ed. P. Braffort (to appear).

[3] Bruijn, N.G. de, The mathematical language AUTOMATH, its usage, and some of its extensions; Symposium on Automatic Demonstration (Versailles, December 1968), Springer Lecture Notes in Mathematics, Vol. 125 (1970), 29-61.

[4] Daalen, D.T. van, A description of AUTOMATH and some aspects of its language theory; proceedings of the Symposium APLASM (Orsay, December 1973), ed. P. Brafford (to appear).

[5] Hindley, J.R., B. Lercher and J.P. Seldin, Introduction to Combinatory Logic, Cambridge University Press (1972).

[6] Howard, W.A., The formulae-as-types notion of construction; Unpublished (1969).

[7] Martin-Löf, P., An intuitionistic theory of types; Unpublished (1972).

[8] Nederpelt, R.P., Strong normalization in a typed lambda calculus with lambda structured types; Doctoral dissertation, Technological University Eindhoven (1973).

[9] Prawitz, D., Ideas and results in proof theory; Proc. of the second Scandinavian Logic Symposium, ed. J.E. Fenstad, North Holland 1971.

[10] Scott, D., Constructive validity; Symposium on Automatic Demonstration (Versailles, December 1968), Springer Lecture Notes in Mathematics, vol. 125 (1970), 237-275.

[11] Tait, W.W., Intensional interpretations of functionals of finite type I. J. of Symbolic Logic 32 (1967), 198-212.

Eindhoven, January 1975.

NORMED UNIFORMLY REFLEXIVE STRUCTURES

Henk Barendregt
Mathematisch Instituut
Boedapestlaan, Utrecht
The Netherlands

§0. Introduction. The theory of Uniformly Reflexive Structures (URS) studied by Wagner and Strong ([8],[6],[1]), is an elegant axiomatization of parts of recursion theory. The theory abstracts some properties of the function $\{n\}(m)$ (i.e. the nth partial recursive function applied to m) by considering arbitrary domains with a binary operation application. The standard URS is $\mathfrak{R}$ with domain $\omega \cup \{*\}$ and application $n.m = \{n\}(m)$ if defined $*$ else.

However the URS are not completely adequate for the description of recursion theory. Real computations do have a length, a feature which is missing in the URS. In fact there are sentences in the language of URS undecided by the axioms. E.g. let $e = \lambda x.xx$, i.e. $ex = xx$ for all x, then $ee = *$ is such a sentence. But this sentence holds in the intended interpretation $\mathfrak{R}$ as follows from an argument using length of computation.

Moreover in a URS it is not always possible to represent the partial recursive functions.

To overcome these defects we introduce a concept of a norm.

A Normed Uniformly Reflexive Structure (NURS) is a URS which a norm $|..;..|$ can be defined satisfying:

1. $|x;y| \in \omega \cup \{\infty\}$
2. $|x;y| = \infty \Longleftrightarrow x.y = *$
3. $|s.x.y;z| > |x.z;y.z| + |x;z| + |y;z|$, if $|s.x.y;z| \neq \infty$

The intended interpretation of $|x;y|$ is "the length of computation of x.y".

The following facts motivate the introduction of NURS. As was intended $\mathfrak{R}$ is a NURS. Wagners (highly) constructible URS are NURS. In every NURS $ee = *$ holds. More generally, for a NURS $\mathfrak{U}$ and a term M of the theory, M has no normal form $\Longleftrightarrow \mathfrak{U} \models M = *$. In a NURS all splinters are semi-computable, and hence can be used to represent the partial recursive functions.

The use of length of computation in recursion theory has also been stressed by Y.Moschovakis [3]. In fact the axioms of the norm in a URS imply Moschovakis' condition on the length of computation.
Familiarity with URS is assumed. See e.g. Wagner [8] and Strong [6].
In §1 the defects of URS mentioned above are shown. A formal theory WS, convenient for the study of URS, is introduced in §2. The term model of an extension of WS provides some counter examples for the relation between semi-computable and recursively enumerable. The results about the NURS are proved in §3.

§1. The definition of a URS given below is not exactly the same as those of Wagner and Strong. The axioms are written down in a way showing the correspondence with combinatory logic. Axiom 7 is added; it implies that we may assume that terms with different normal forms are unequal in a URS (2.10).

1.1. Def. A URS is a structure $\mathfrak{U} = \langle U,*,i,k,s,\delta,\cdot\rangle$ such that the following holds where a,b,c are variables ranging over $U - \{*\}$:

1. $*.a = a.* = *.* = *$
2. $i.a = a$
3. $k.a.b = a$
4. $s.a.b.c = (a.c).(b.c) \quad ; \quad s.a.b \neq *$
5. $a = b \;\rightarrow\; \delta.a.b = k \quad ; \quad a \neq b \;\rightarrow\; \delta.a.b = k.i$
6. $i \neq k$
7. $s.a.b = s.a'.b' \;\rightarrow\; a = a' \wedge b = b'$.

1.2. Def. Kleenes URS, $\mathfrak{K}$, is the structure $\langle \omega^*,*,i,k,s,\delta,\cdot\rangle$ such that $\omega^* = \omega \cup \{*\}$ with, $* \notin \omega$, $n.m = \begin{cases}\{n\}(m) & \text{if defined}\\ * & \text{else}\end{cases}$

$*.n = n.* = *.* = *$, and i,k,s,δ are to be found by the s-m-n theorem such that axioms 2,...,7 hold. As an example we construct k. Let $\psi(x,y) = x$. Then ψ is partial recursive. Hence

$$\begin{aligned} x &= \psi(x,y) \\ &= \{e\}(x,y) && \text{for some index } e \text{ of } \psi. \\ &= \{s^1_1\,(e,x)\}(y) \\ &= \{\{k\}(x)\}(y) && k \text{ index of } \lambda x.s^1_1(e,x). \\ &= k.x.y. \end{aligned}$$

By pumping up the indices, cf. Rogers [4], p.83, we can assure that axiom 7 holds.

1.3. Theorem. Let e = s.i.i. Then e.e = * is independent in the theory of the URS. [1)]

Proof. It will be shown that e.e = * is true in $\mathfrak{R}$ but false in a modification $\mathfrak{R}^\circ$.

We have $\mathfrak{R} \models$ e.a = (i.a)(i.a) = a.a, i.e. {e}(a) = {a}(a).

The computation of {e}(a) runs as follows:

Read a; compute {a}(a). Hence the computation of {e}(e) is:

Read e; compute {e}(e); Read e; compute {e}(e); ...

Therefore {e}(e) is undefined. Hence $\mathfrak{R} \models$ e.e = *.

Let $\mathfrak{R}^\circ = \langle \omega^*, *, i, k, s^\circ, \delta, \circ \rangle$ be the following modification of $\mathfrak{R}$.

$$a \circ b = a.b \quad \text{if } a \neq e \text{ or } b \neq e$$
$$ = 0 \quad \text{else}$$

Then $\circ$ is partial recursive. Let $s^\circ.a.b.c = (a \circ c) \circ (b \circ c)$. Again by pumping up the indices we may assume that $s^\circ \neq e$, $s^\circ.a \neq e$ for all a and $s^\circ.a.b = e$ iff a = b = i. Hence $s^\circ \circ a \circ b \circ c = s^\circ.a.b.c = (a \circ c) \circ (b \circ c)$, unless perhaps $s^\circ.a.b. = c = e$. But then a = b = i and $(i \circ e) \circ (i \circ e) = e \bullet e$. It is clear that $i, k, \delta \neq e$ and the axioms 2,3 and 5 follow. Axiom 7 can be assured as in 1.2. Clearly $\mathfrak{R}^\circ \not\models$ e.e = *. ⊠

Another defect of the URS is the following. The partial recursive functions can be represented in a URS provided one has an infinite semi-computable (SC) splinter, Strong [6],3.2. However, H.Friedman has shown that there is a URS without infinite SC splinter.

1.4. Def. Let $\mathfrak{A}$ be a non-standard model of Peano arithmetic with universe A. Let $\mathfrak{R}_{\mathfrak{A}}$ be the structure $\langle A^*, *, i, k, s, \delta, \circledast \rangle$ where $* \notin A$, i, k, s, δ are as in 1.2 and $\circledast$ is defined by

$$* \circledast a = a \circledast * = * \circledast * = *$$
$$a \circledast b = c \quad \text{if} \quad \mathfrak{A} \models \{\underline{a}\}(\underline{b}) = \underline{c} \quad \text{i.e.} \quad \mathfrak{A} \models \exists z[T(\underline{a},\underline{b},z) \wedge U(z) = \underline{c}]$$
$$ = * \quad \text{else.}$$

U and T are the components of Kleene's normal form theorem. Then $\mathfrak{R}_{\mathfrak{A}}$ is a URS; e.g. $\mathfrak{R}_{\mathfrak{A}} \models$ k.a.b = a holds since {{k}(a)}(b) = a is provable in Peano arithmetic, hence $\mathfrak{A} \models$ {{k}(a)}(b) = a.

---

1) Compare this with the following : Let E = {x | x ∈ x}. Then E ∈ E is independent in ZF without foundation, but refusable in ZF itself.

1.5. Theorem (H.Friedman). $\mathfrak{R}_{\mathfrak{A}}$ is a URS without infinite SC splinter.

Proof. If $\mathfrak{R}_{\mathfrak{A}}$ would contain an infinite SC splinter, each splinter would be SC, Strong [6] 3.11. Therefore the set of standard numbers would be SC. But this is absurd since SC sets are definable ($x \in A \iff f(x) \neq *$), and the set of standard numbers is not. ☒

1.6. Cor. There exists a URS with an infinite non SC splinter on which the partial recursive functions can be represented.

Proof. Let $\mathfrak{N}$ be the standard model of Peano arithmetic. Let $\mathfrak{A} \equiv \mathfrak{N}$ be a non-standard model. For each partial recursive function ψ with index e we have

$$\mathfrak{N} \models \{\underline{e}\}(\underline{n}) = \underline{m} \iff \psi(n) = m$$

$$\mathfrak{N} \models \neg\exists z\, T(\underline{e},\underline{n},z) \iff \psi(n) \text{ is undefined.}$$

Therefore, since $\mathfrak{A} \equiv \mathfrak{N}$, $\mathfrak{R}_{\mathfrak{A}} \models \underline{e}\,\underline{n} = \underline{m} \iff \psi(n) = m$

$$\mathfrak{R}_{\mathfrak{A}} \models \underline{e}\,\underline{n} = \underline{*} \iff \psi(n) \text{ is undefined.}$$ ☒

However, there exists a URS such that only partial recursive functions with recursive domain can be represented on any of the infinite splinters.

1.7. Theorem. There exists a URS such that for no infinite splinter X the partial recursive functions can be represented on X.

Proof. Let $\mathfrak{A}$ be a non-standard model of Peano arithmetic in which only the recursive r.e. sets are definable on ω, see [2], Exc.7, p123. Let ψ be a partial recursive function with non recursive domain A. Then ψ is not representable on the splinter of standard integers for otherwise A would be definable on ω. But then ψ is not representable on any infinite splinter X, since all infinite splinters are in bijective computable correspondence, [6], 3-7. ☒

§2. The following theory WS is convenient for the study of URS.

2.1. Def. WS has the following language.

Alphabet:	$x_0, x_1, \ldots$	variables
	$I, K, S, \Delta, \underline{*}$	constants
	$\geqslant, =$	reduction, equality
	(,)	brackets

Terms are inductively defined by

1. A variable or constant is a term
2. If M,N are terms, so is (MN).

Formulas are $M \geqslant N$ and $M = N$ where M,N are terms.

Notation: x,y,z,... denote arbitrary variables

M,N,L denote arbitrary terms

$M_1M_2...M_n$ stands for $(..(M_1M_2)...M_n)$

$M \subset M'$ if M is a subterm of M'

$x \in M$ if x occurs in M

M is closed if for no x $x \in M$

$\equiv$ denotes syntactic equality.

If M is a closed WS term and $\mathfrak{U} = \langle U,*,i,k,s,\delta,\cdot\rangle$ is a URS, then $M^{\mathfrak{U}}$ is the obvious interpretation of M in $\mathfrak{U}$: $\underline{*}^{\mathfrak{U}} = *$, $I^{\mathfrak{U}} = i$, etc, $(MN)^{\mathfrak{U}} = M^{\mathfrak{U}}.N^{\mathfrak{U}}$; $\mathfrak{U} \models M = N$ iff $M^{\mathfrak{U}} = N^{\mathfrak{U}}$.

A term M is <u>in</u> normal form (nf) if it has no subterms of the form $\underline{*}$, IA, KAB, SABC or ΔAB.

WS is defined by the following axioms and rules:

I 0. $\underline{*}M \geqslant \underline{*}$ $\quad M\underline{*} \geqslant \underline{*}$

1. $IM \geqslant M$
2. $KMN \geqslant M$ if N is in nf
3. $SMNL \geqslant ML(NL)$
4. a $\Delta MM \geqslant K$ if M is in nf
   b $\Delta MN \geqslant KI$ if M,N are in nf and $M \not\equiv N$

II 1. $M \geqslant M$

2. $M \geqslant M' \Rightarrow ZM \geqslant ZM'$, $MZ \geqslant M'Z$
3. $M \geqslant N,\ N \geqslant L \Rightarrow M \geqslant L$

III 1. $M \geqslant N \Rightarrow M = N$

2. $M = N \Rightarrow N = M$
3. $M = N,\ N = L \Rightarrow M = L$

2.2. <u>(Church-Rosser theorem)</u> If WS $\vdash M = N$, then for some term Z WS $\vdash M \geqslant Z$ and WS $\vdash N \geqslant Z$.

Proof. Well-known. See e.g. [5], T. 12, p. 144. ⊠

2.3. <u>Def</u>. A WS-term M <u>has</u> a nf if WS $\vdash M = M'$ and M' is in nf.

By 2.2 the normal form of a term is unique if it exists. If M has a nf, all its reduction sequences terminate, by the restriction in axioms I2, 4.

2.4. Def. Let $\mathcal{U}$ be a URS with domain U. WS($\mathcal{U}$) is the theory WS modified as follows. For each $a \in U$, $\underline{a}$ is an additional constant. A term of WS($\mathcal{U}$) is in nf, if it does not contain a subterm $\underline{*}$, IA, etc. or $\underline{a}$M. WS($\mathcal{U}$) has the additional axioms $\underline{a}M \geq \underline{a.M}$. Axiom I4.b should be replaced by

$\Delta MN \geq KI$ if M,N are nf's and $\mathcal{U} \models M \neq N$.

Clearly $\mathcal{U} \models$ WS($\mathcal{U}$).

2.2 and 2.3 apply also to WS($\mathcal{U}$).

2.5. (Abstraction) Let M be a WS($\mathcal{U}$) term not containing $*$. Then there exists a WS($\mathcal{U}$) term $\lambda x.M$ such that

1. $\lambda x.M$ is in nf; $x \notin \lambda x.M$
2. WS($\mathcal{U}$) $\vdash (\lambda x.M)N = [x/N]M$ for N in nf.

Proof. As in combinatory logic. ☒

Note, however, that also there exists a WS term $\lambda x.\underline{*}$ in nf such that $\mathcal{U} \models (\lambda x.\underline{*})a = \underline{*}$ for all $\mathcal{U}$.
Take e.g. $\lambda x.\underline{*} = S(K\omega)(K\omega)$ with $\omega = \lambda x.\Delta(KI)(xx)$.

2.6. Def. Let $M \sim M'$ denote $Mx = M'x$ for $x \notin MM'$.

2.7. (Fixed Point Theorem) There exists a WS term FP such that

1. WS $\vdash$ FP f $\sim$ f(FP f)
2. FP f is in nf.

Proof. Let $\omega_f = \lambda xz.f(xx)z$ and FP f $= \omega_f\omega_f$. ☒

2.8. Lemma. Let M be a WS($\mathcal{U}$) term. Then M is a nf $\Rightarrow$ $\Rightarrow \mathcal{U} \models M \neq *$.

Proof. The set of normal forms NF can be defined inductively by 1. $\underline{a}$,I,K,S,$\Delta \in$ NF. 2. AB $\in$ NF $\Rightarrow$ KA,SA,ΔA and SAB $\in$ NF. Then the result follows inductively realizing that in a URS k.a, s.a, δ.a, s.a.b $\neq *$. ☒

The pumping up of indices used in 1.2 and 1.3 can be done in each URS due to axiom 7.

2.9. Lemma.
Then there exists a term P such that for all $\mathcal{U}$

1. $\mathcal{U} \models Pab \neq *$
2. $\mathcal{U} \models Pab \sim a$
3. $\mathcal{U} \models Pab = Pa'b' \rightarrow a = a' \wedge b = b'$.

Proof. Let $P \equiv \lambda abx.\ K(ax)b$. Clearly P satisfies 1 and 2. By writing out P in terms of I, K and S, one sees that P satisfies 3 due to axiom 7. ⊠

2.10. Cor. Let $M \not\equiv M'$ be WS terms in nf. Then we may assume $\mathfrak{U} \models M \neq M'$ for all $\mathfrak{U}$.
Proof. By changing if necessary the basic constants i,k,s, and δ, using P. See e.g. [9], p. 133 bottom. ⊠

What we may we will.

2.11. Cor. WS($\mathfrak{A}$) is a conservative extension of WS.
Proof. The only axiom of WS not in WS($\mathfrak{A}$) is I4b. However, this follows from the modified axiom by 2.10. Hence WS($\mathfrak{A}$) is an extension of WS. If M,N are WS terms and WS($\mathfrak{U}$) $\vdash M = N$ (or $\vdash M \geqslant N$), then the proof involves only WS terms (unless WS $\vdash M = N = \underline{*}$). The WS($\mathfrak{U}$) axioms only can hold for $A \not\equiv B$, by 2.10. Hence WS $\vdash M = N$ ($\vdash M \geqslant N$). ⊠

2.12. Theorem 1. WS($\mathfrak{U}$) $\vdash M = N \Rightarrow \mathfrak{U} \models M = N$
2. M has a nf $\Rightarrow \mathfrak{U} \models M \neq *$
Proof. 1. Induction on the length of proof of M = N using 2.10.
2. By 1. and 2.6. ⊠

The converse of 2.12. 1,2 are false. E.g. in $\mathcal{K}^{\circ} \models EE \neq *$ where E = SII. But EE has no nf. However, if $\mathfrak{U}$ is a NURS the converse of 2.12.2 is true. See 3.3.

2.13. Def. Let WS* be WS augmented by the axioms:
$M \geqslant \underline{*}$ if M has no nf.

For each NURS $\mathfrak{U}$ we will have the completeness result:
WS* $\vdash M = N \iff \mathfrak{U} \models M = N$, for closed M,N;
see 3.5.

2.14. Def. $\mathfrak{U}(WS^*_o)$ (respectively $\mathfrak{U}(WS^*_c)$) is the term model consisting of arbitrary (respectively closed) WS terms modulo provable equality in WS*. Clearly they are URS.
Similarly we define $\mathfrak{U}(WS^*_{o,c}(\mathfrak{U}))$.

These term models can be used for some counter-examples

2.15. <u>Def</u>. A subset X of a URS $\mathfrak{U}$ is RE if $X = \emptyset$ or $X = \mathrm{Ra}\, f =$ $= \{a \mid \mathfrak{U} \models \exists x\, (\underline{fx} = \underline{a})\}$ for some total f in $\mathfrak{U}$ (i.e. $\forall a\ fa \neq *$).
In $\mathfrak{K}$, X is RE $\Leftrightarrow$ X is SC.

2.16. <u>Theorem</u> 1). For $\mathfrak{U}(WS_0^*)$ we have
1. X is SC $\not\Rightarrow$ X is RE
2. X is RE $\not\Rightarrow$ X is SC
3. X is computable $\Rightarrow$ X is finite or cofinite.
Proof.
2.16.1 <u>Def</u>. The family of F, $\mathcal{F}(F)$, is the set $\{N \mid \exists F' \vdash F \geqslant F' \wedge N \subset F'\}$. If F has a nf, $\mathcal{F}(F)$ is finite.
Each reduction of FA to a nf can be written in the form

$$FA \geqslant_\beta M_0[A] \geqslant_\delta M_0'[A] \geqslant_\beta M_1[A] \geqslant_\delta M_1'[A] \geqslant \ldots \geqslant M[A] \qquad (*)$$

where $\geqslant_\beta$ is axiomatized leaving out the Δ reduction axioms and $\geqslant_\delta$ is axiomatized leaving out the $\underline{*}$,I,K,S axioms. A may not actually occur in $M[A]$. Referring to the sequence (*) we define:

2.16.2 <u>Def</u>. $\mathrm{Diag}_n(F,A) = \{\Delta C_1[A]\, C_2[A] \mid \Delta C_1[A]\, C_2[A] \subset M_n\}$.
B satisfies $\mathrm{Diag}_n(F,A) \Leftrightarrow \Delta C_1[A]\, C_2[A] = \Delta C_1[B]\, C_2[B]$, for all members of $\mathrm{Diag}_n(FA)$.

2.16.3 <u>Lemma</u>. Let FA have a nf for all A. Let $xa \not\subset F$. Consider the sequence (*) for F(xa). Then
0. B satisfies $\mathrm{Diag}_n(F,xa) \Rightarrow M_n[B] \geqslant_\delta M_n'[B]$.
1_n. xa is never "active" (i.e. in a subterm of the form ((xa)P)) in $M_n[xa]$, $M_n'[xa]$.
2_n. For almost all, i.e. all except finitely many, B satisfies $\mathrm{Diag}_n(F,xa)$.

Proof. 0 is obvious.
1_0 follows by substituting for xa a nf ω such that ωP has no nf for all P.
$1_n \Rightarrow 2_n$ by realizing that the only possible exceptions are in $\mathcal{F}(F)$.
$2_n \Rightarrow 1_{n+1}$ follows as 1_0 with ω satisfying $\bigcup_0^n \mathrm{Diag}_n(F,xa)$ and using 0. ☒

---

1) A different example of 1. was given in Wagner [8], 6.13.
3. was proved by Strong [7] for the URS $\mathfrak{U}(WS_c^*)$.

2.16.4 Cor. Let FA have a nf for all A. Let $xa \not\subset F$ and $xa \not\subset M$, the nf of F(xa). Then for almost all B F(B) = F(xa).

Proof. Let $\mathrm{Diag}(F,xa) = \cup\, \mathrm{Diag}_n(F,xa)$ which is finite. This is satisfied by almost all B (2.16.3.2). Thus (2.16.3.0) $FB \geqslant M[B]$. Also $F(xa) \geqslant M[xa]$. But then, since $xa \not\subset M[xa]$, FB = F(xa). ☒

More easily one can prove the following.

2.16.5 Cor. Let F(xa) have a nf, where $xa \not\subset F$, $xa \not\subset$ the nf of F(xa). Then for $x' \not\subset F$ $F(x'a) = F(xa)$.

Proof. Since x'a is a non-active term, it does not matter if it occurs in an active place. ☒

2.16.6 Cor. Suppose RA $F \subset$ closed normal forms. Then Ra F is finite.

Proof. Take $xa \not\subset F$. By the assumption, never $xa \subset M$, the nf of FA. Hence for almost all B, FB = F(xa). ☒

Now we can prove 2.16.

1. Take $X = \{K^nI \mid n \in \omega\}$. Then X is an infinite splinter hence SC (since $\mathcal{U}(WS_0^*)$ is a NURS, see §3). Suppose X were RE, say X = Ra F. Then F satisfies the assumption of 2.16.6, but Ra F = X is not finite. Contradiction.

2. Take X = Ra F, with Fa = xa. Suppose X were SC, i.e.

GM = I if $M \in X$

$\quad$ * else

for some G. Take $a \notin G$. Then $xa \not\subset G$. Also $xa \not\subset I$ which is the nf of G(xa). Hence for $x' \not\subset G$ it follows by 2.16.5 that G(x'a) = = G(xa) = I, i.e. $x'a \in X$, a contradiction.

3. Let $X = \emptyset$ be computable. Define

GM = M if $M \in X$

$\quad M_0$ else for some $M_0 \in X$.

Then X = Ra G. Suppose the complement of X is not finite. Then there is a variable $x \notin \mathrm{Ra}\, G \cup \mathcal{F}(G)$. Then $xa \not\subset G$, $xa \not\subset$ the nf of G(xa). Hence by 2.16.4 GB = G(xa) for almost all B, i.e. X = Ra G is finite. ☒

§3. For NURS it is convenient to define for elements of $\omega \cup \{\infty\}$: $p \gtrsim q$ iff $p = \infty \vee p > q$. Then $\gtrsim$ is transitive and axiom 3 for a norm can be stated as

$|s.a.b;c| \gtrsim |a.c;b.c| + |a;c| + |b;c|$.

3.1. Examples of NURS.

1. $\mathfrak{K}$ becomes a NURS by defining

$$|e;x| = \begin{cases} \mu z \ T(e,x,z) & \text{if defined} \\ \infty & \text{else} \end{cases}$$

Then an examination of the properties of the T predicate shows that this defines a norm on $\mathfrak{K}$.

2. $\mathcal{U}(WS^*_{\hat{0},c})$ are NURS by defining

$$|F;X| = \begin{cases} \text{the length of the inside out reduction of FX to nf} \\ \infty \quad \text{if FX has no nf.} \end{cases}$$

The inside out reduction only reduces redeces SABC,etc. when A, B and C are normal forms.

3. Let $\mathcal{U}$ be a (highly) constructible URS in the sense of [8]. Then $\mathcal{U}$ is a NURS:

Let $f(e;x) = \mu n[\langle e,x\rangle \in \Lambda_n]$ if defined
$\qquad\qquad \infty$ else.

Take $|e;x| = 4^{f(e;n)}$. This is a norm on $\mathcal{U}$, for let $f(sxy,z) = n$, then $sxy = \phi_6(x,y)$, $n > 0$ and $\langle x,z\rangle,\langle y,z\rangle,\langle xz,yz\rangle \in \Lambda_{n-1}$ (see [8],p.20-21 for the notation). Then $f(x,z),f(y,z),f(xz,yz) \leqslant n-1$, and $|sxy;z| = 4^n > 3.4^{n-1} \geqslant |x;z| + |y;z| + |xz;yz|$.

4. Let $\mathfrak{A}$ be a non-standard model of Peano arithmetic. Then $\mathfrak{K}_{\mathfrak{A}}$ is not a NURS. This follows from 1.5 and 3.4. Similarly it follows from 1.3 and 3.2 that $\mathfrak{K}^\circ$ is not a NURS.

The sentence EE = *, with E = SII, which was independent in the theory of URS becomes true in all NURS.

3.2. Let E = SII and $\mathcal{U}$ be a NURS. Then
$\mathcal{U} \models EE = *$.

Proof. Suppose $EE \neq *$. Then $|E;E| \neq \infty$. But then $|E;E| = |SII;E| > |IE;IE| = |E;E|$, a contradiction. ⊠

More general

3.3. <u>Theorem</u>. Let $\mathcal{U}$ be a NURS and M a WS($\mathcal{U}$) term. Then
M has no nf $\Longleftrightarrow$ $\mathcal{U} \models M = *$.

Proof. ⇐ By 2.12.2.

⇒ This will be proved in a number of steps.

3.3.1 Def. SC(M), the set of subcomputations of M, is defined inductively by:

If M is in normal form SC(M) = ∅; else M ≡ AB and SC(AB) = SC(A) ∪ SC(B) ∪ $\{|A^{\mathfrak{U}};B^{\mathfrak{U}}|\}$. Below we often omit the superscript $\mathfrak{U}$. Clearly SC(M) is a finite set ⊂ ω ∪ {∞} and if M ⊃ M', then SC(M) ⊃ SC(M').

3.3.2 Def. ‖M‖ = Max{SC(M)}. If SC(M) contains ∞, ‖M‖ = ∞.

3.3.3 Lemma. If M ⊃ M', then ‖M‖ ≥ ‖M'‖.

3.3.4 Lemma. ‖M‖ = ∞ ⟺ $\mathfrak{U} \models$ M = *.

Proof. ‖M‖ = ∞ ⟺ ∞ ∈ SC(M)

⟺ for some AB ⊂ M |A;B| = ∞

⟺ for some AB ⊂ M $\mathfrak{U} \models$ AB = *

⟺ $\mathfrak{U} \models$ M = *. ☒

3.3.5 Lemma. Let M ≥ M' be an axiom of WS($\mathfrak{U}$). Then ‖M‖ ≥ ‖M'‖ .

Proof. Let M ≡ SABC and M' ≡ AC(BC).

Then SC(M) = {|S;A|,|SA;B|,|SAB;C|} ∪ SC(A) ∪ SC(B) ∪ SC(C).

SC(M') = {|A;C|,|B;C| ,|AC;BC|} ∪ SC(A) ∪ SC(B) ∪ SC(C).

Since |SAB;C| ≥ Max{|A;C|,|B;C|,|AC;BC|}

‖M‖ ≥ ‖M'‖. Equality may occur, e.g. if SC(C) contains the largest subcomputation.

If M ≡ KAB, M ≡ IA or M ≡ M' , then M' ≡ A or M' ≡ M, hence M ⊃ M' and the result follows by 3.3.3.

If M ≡ ΔAB, then M' ≡ K or ≡ KI, so SC(M) ⊃ SC(M') = ∅, hence ‖M‖ ≥ ‖M'‖. **Similarly if M ≡ aN.** ☒

3.3.6 Cor. If WS($\mathfrak{U}$) ⊢ M ≥ M', then ‖M‖ ≥ ‖M'‖.

Proof. Induction on the length of proof of M ≥ M'.

Let us consider only the case that M ≥ M' is ZA ≥ ZA' and is a direct consequence of A ≥ A'. Then SC(ZA) = SC(Z) ∪ SC(A) ∪ {|Z;A|} and similarly for SC(ZA'). Now $\mathfrak{U} \models$ A = A', hence |Z;A| = |Z;A'|. Hence ‖ZA‖ ≥ ‖ZA'‖ by the induction hypothesis ‖A‖ ≥ ‖A'‖. ☒

3.3.7 Def. A special redex is a WS($\mathfrak{U}$)-term SABC, where A, B and C are in normal form.

3.3.8 Lemma. If SABC is a special redex, then
$\| SABC \| \underset{\sim}{>} \| AC(BC) \|$.
Proof. Since $\underline{SC}(A) = \underline{SC}(B) = \underline{SC}(C) = \emptyset$

$$\| SABC \| = \mathrm{Max}\{|S;A|, |SA;B|, |SAB;C|\} \geqslant |SAB;C| \underset{\sim}{>}$$
$$\underset{\sim}{>} \mathrm{Max}\{|A;C|, |B;C|, |AC;BC|\} = \| AC(BC) \| .$$ ⊠

3.3.9 Lemma. Let M be a WS($\mathfrak{M}$)term without normal form. Then there exists a special redex N without normal form in the family (see 2.16.1) of M, or else $\mathfrak{M} \models M = *$.
Proof. Consider the finite set T of subterms of M partially ordered by $\subset$. Let N be a minimal element of T without a normal form. Then all subterms of N have a normal form. Checking all possibilities it follows that N is of the form SABC. Let A^*, B^* and C^* be the normal forms of A, B and C. Now we have
$M \equiv$ ——(SABC) —— $\geqslant$ ——($SA^*B^*C^*$) —— and $SA^*B^*C^*$ is a special redex without normal form. ⊠

3.3.10 Cor. If M has no normal form, then there exists a term M' without normal form and $\| M \| \underset{\sim}{>} \| M' \|$.
Proof. Let N be as in 3.3.9, then $\| M \| \geqslant \| N \|$ by 3.3.6 and 3.3.3. Let $N \geqslant M'$. Then $\| N \| \underset{\sim}{>} \| M' \|$ by 3.3.8. Since N has no normal form, neither has M'. ⊠

Now the proof of 3.3.⇒ can be given.
Let M be a term without normal form. Suppose $\mathfrak{M} \models M \neq *$. Then $\| M \| \neq \infty$ by 3.3.4. Hence by 3.3.10 there exists a sequence $M, M', M'', \ldots$ such that $\| M \| > \| M' \| > \| M'' \| > \ldots$ is an infinite descending chain of integers. ⊠

3.4. Theorem. In a NURS $\mathfrak{M}$ all infinite splinters are SC.
Proof. Let $X = \{f^n o\}$ be an infinite splinter. Define by the fixed point lemma a WS($\mathfrak{M}$) term H such that

$$Hyx = I \quad \text{if } y = x$$
$$\quad\quad H(fy)x \text{ else.}$$

Then $h = (H\underline{o})^{\mathfrak{M}}$ is a semi-characteristic function of X:
If $a \in X$, clearly $H\,\underline{o}\,\underline{a} = I$, hence $ha \neq *$.
If $a \notin X$, then $H\,\underline{o}\,\underline{a} \geqslant H\,\underline{f(o)a} \geqslant \ldots$, i.e.
$H\,\underline{o}\,\underline{a}$ has no nf. Hence $ha = *$ by 3.3. ⊠

WS* is a complete axiomatization for the equations true in all NURS.

3.5. <u>Theorem</u>. Let $\mathfrak{U}$ be a NURS. Then for closed WS terms:
WS* $\vdash M = N \iff \mathfrak{U} \models M = N$.
Proof. $\Rightarrow$ By 2.12.1, 3.3. $\Leftarrow$ By 2.10,3.3. ⊠

3.6. <u>Theorem</u>. Each URS can be embedded in a NURS (cf.Wagner [8], p.31, 6.2), if the similarity type has no constants.
Proof. Clearly $\mathfrak{U} \hookrightarrow \mathfrak{U}(WS^*_{o,c}(\mathfrak{U}))$ which is a NURS by 3.1.2. ⊠

<u>Concluding remarks</u>.
A URS is almost a precomputation theory in the sense of Moschovakis [3] 1). Restricting the attention to single-valued functions, his computation theories have an additional length of computation $|e;\vec{x}|$ satisfying

(+) $|S^n_m(e,\vec{x});\vec{y}| > |e;\vec{x},\vec{y}|$, if defined.

Define in a NURS $|e;\vec{x}| = |e;x_1| + |e.x_1;x_2| + \ldots + |e.x_1\ldots x_{n-1};x_n|$.

Then it follows readily from the definition of S^n_m in a URS ([8],2.6) that this norm satisfies Moschovakis' axiom (+).

As suggested in [6], there is another way of extending a URS. A <u>selection</u> 2) URS is an URS containing a "selection operator" c such that

$$\exists a[f.a \neq *] \Rightarrow f.(c.f) \neq * .$$

1) Not quite, because a URS does not need to contain a computable successor set.

2) In [6] such a URS is called "well-ordered". This name is a little absurd as can be argued as follows. Let $\mathfrak{A}$ be a model of Peano arithmetic of power continuum. Then $\mathfrak{N}_{\mathfrak{A}}$ is a selection URS but cannot be well-ordered in ZF. On the other hand $\mathfrak{U}(WS^*_o)$ is countable and hence well-ordered, but has no selection operator.

In a selection URS a set is computable iff it is SC and co SC, [6],3.4. This is not true in a general URS, [8],p.39 bottom.

Having a norm or a selection operator are independent of each other. $\mathfrak{K}$ has a selection operator $\{c\}(e) = (\mu x\, T(e,(x)_0,(x)_1))_0$. Since this is provably in arithmetic a selection operator, $\mathfrak{K}_{\mathfrak{A}}$ is a selection URS but not a NURS. Conversely, it is not difficult to show that $\mathcal{U}(WS_0^*)$ is not a selection URS, although it is a NURS,

In a NURS it would be natural to require for a selection operator c

$$|c;a| \gtrsim |a;c.a|$$

cf.[3],p.225,(6-4).

Acknowledgement. The paper is an elaboration of part II of the author's dissertation. He wishes to thank his supervisor professor G.Kreisel for his stimulating personality.

References.

[1] Friedman,H. Axiomatic recursive function theory, in: R.Gandy and M.Yates (eds), Logic Colloquium '69, North Holland, Amsterdam (1971), 113-137.

[2] Kreisel,G., J. Krivine, Elements of Matnematical Logic, North-Holland, Amsterdam (1967).

[3] Moschovakis,Y. Axioms for computation theories - first draft, in: R.Gandy and M.Yates (eds), Logic Colloquium '69, North Holland, Amsterdam (1971), 199-255.

[4] Rogers,H. Theory of recursive functions and effective operations, McGrawHill (1967).

[5] Rosser,J. A mathematical logic without variables, Ann. of Math. ser.2, 36 (1936), 127-150.

[6] Strong,H. Algebraically generalized recursive function theory, IBM J.Research and Development (1968), 465-475.

[7] - . Construction of models for algebraically generalized recursive function theory, J.Symbolic Logic 35 (1970), 401-409.

[8] Wagner,E. Uniform reflexive structures: on the nature of Gödelizations and relative computability, Trans.Amer. Math.Soc.144 (1969), 1-41.

[9] Troelstra,A, et al. Metamathematical investigation of intuitionistic arithmetic and analysis, Lecture Notes in Mathematics 344, Springer (1973).

A MODEL WITH NONDETERMINISTIC COMPUTATION

Marisa Venturini Zilli

Istituto per le Applicazioni del Calcolo
C.N.R. - Roma

Abstract. In paragraph 2 of this paper an algorithmic model is described having a set R of rules. In paragraph 3 two new rules are added to the rules of R. Such new rules may change a nonterminating computation into a terminating one. In paragraph 4 some families of expressions to which the rules added to R can be applied are defined. In paragraph 5 it is shown that the considered model is a uniformly reflexive structure.

1. Introduction. It is well known that in programming it is possible to avoid a set of instructions whose execution compels the program to run perpetually. It is also well known that there is no such way of handling the indefinite in recursive function theory and in those computing models where the computation rules (which transform expressions of the model into expressions of the model) are such that computation must always start from the innermost parts of any expression (in other words, where computations are carried out in a way in some sense similar to what is known as "call by value" in programming jargon).

Now, by adding suitable rules, it is possible also in such computing models to skip some indefinite expressions, provided that these can be considered equal to other expressions having a definite meaning.

This is analogue to what is done informally, for example, in recursive function theory when the range of a function is extended by adding a new value in correspondence to an argument where we know the function to be undefined.

Such a point of view is opposite to that which considers "unnatural" the fact that an expression where computations never stop turns out to be equal to another expression having a definite meaning.

Moreover, in computing models of the above kind there are also other families of indefinite expressions for whose elements it is not possible to do what we said, but where there is sufficient information for handling them in some special way. For an analogue situation we may think of what is informally done in calculus when ∞ is introduced.

In the light of such intuitive considerations, from a formal point of view the problem reduces to distinguishing different kinds of indefinite within certain computing models.

In this paper we consider an algorithmic model having a set R of rules and two "special" rules that change, in infinitely many cases, a nonterminating computation into a terminating one. Nondeterminism characterizes this model, because every time a special rule is applicable rules of R are applicable too.

2. Algorithmic model. Let D be the domain generated by α, ψ, $*_1$, $*_2$, and let variables u,v,w,...,x,y,z ... vary on D.

Definition of D :

1) $\alpha, \psi, *_1, *_2 \in D$

2) if $u,v \in D$ then $uv \in D$

3) every expression in D is obtained by means of 1) and 2) only.

The unique operation inside D is the binary operation (uv), simply written xy, neither commutative nor associative. Moreover, association to the left is to be assumed, that is uvw stands for (uv)w.

Let $V \subsetneq D$ be the smallest set of expressions such that:

a) $\alpha, \psi \in V$

b) if $x,y \in V \cup \{*_1\}$ then $\alpha x,\ \psi x,\ \alpha xy,\ \psi xy \in V$.

We define the following predicates within D:

A) "$u \longrightarrow v$" ("u immediately reduces to v") as the predicate holding for two expressions u and v when v is obtained from u by applying one of the following rules (where $\equiv$ has the intended meaning of syntactical identity between expressions) on a subexpression of u :

I	$\alpha xyz \longrightarrow xz(yz)$	for $x,y,z \in V \cup \{*_1\}$
II	$\psi xyz \longrightarrow \begin{cases} y & \text{if } z \equiv x \\ x & \text{if } z \not\equiv x \end{cases}$	for $x,y,z \in V \cup \{*_1\}$
III	$*_1 x \longrightarrow *_1$	for $x \in V \cup \{*_1\} \cup \{*_2\}$
IV	$*_2 x \longrightarrow *_2$	for $x \in V \cup \{*_1\} \cup \{*_2\}$
V	$x *_2 \longrightarrow *_2$	for $x \in V \cup \{*_2\}$

B) "$u \Longrightarrow v$" ("u reduces to v") as the reflexive and transitive closure of $u \longrightarrow v$.

Hence $u \Longrightarrow v$ means that there exist $x_0, x_1, \ldots, x_n \in D$ such that $u \equiv x_0 \longrightarrow x_1 \longrightarrow \ldots \longrightarrow x_{n-1} \longrightarrow x_n \equiv v$.

We call such a sequence of immediate reductions a computation, which is a terminating computation when $v \in V \cup \{*_1\} \cup \{*_2\}$. An expression u is said to be irreducible iff there is no expression $u' \not\equiv u$ such that $u \Longrightarrow u'$. Hence u is irreducible iff $u \in V \cup \{*_1\} \cup \{*_2\}$.

Representing V , $*_1$ and $*_2$ outside D for sake of clarity, D can be divided as in the following figure 1:

(fig.1) D

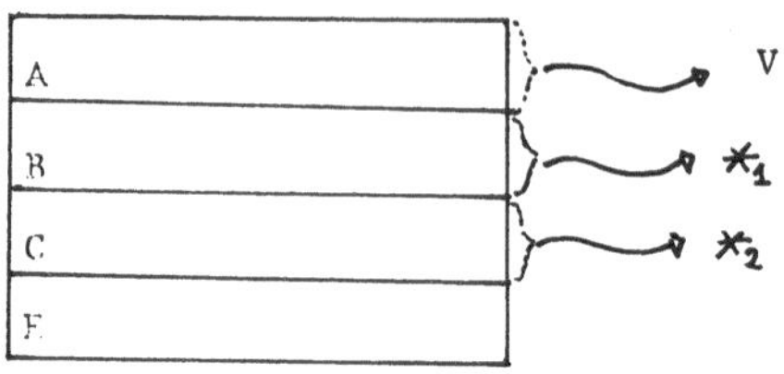

where A is the subset of D that reduces to V by the rules, B is the subset of D that reduces to $*_1$ by the rules, C is the subset of D that reduces to $*_2$ by the rules and E is the subset of D to whose elements one of the rules is always applicable.

Remark. Suppose $u \Rightarrow z$ with $u \not\equiv z$ and $z \in V \cup \{*_1\} \cup \{*_2\}$.

Rules are such that all reducible subexpressions of u are reduced to elements of $V \cup \{*_1\} \cup \{*_2\}$ and the same holds for every expression obtained by the rules within the computation starting with u and terminating with z. Hence the computation from u to z makes all im mediate reductions that it is possible to do to pass from u to z and two computations starting with u and terminating with z differ each other because of the different order according to which those immediate reductions have been done. Hence when $u \Rightarrow z$ every compu tation starting with u is a terminating computation and terminating with z. Hence aut $u \Rightarrow z \in V \cup \{*_1\} \cup \{*_2\}$ aut u is always reducible. In the second case, consider two whatever expressions v and w such that $u \Rightarrow v$ and $u \Rightarrow w$, and consider all immediate reductions done within the computation starting with u and terminating, say, with v. One of the following three cases holds:

1) they are contained in those from u to w, 2) they contain those from u to w , 3) they are contained in those obtained going on reducing v and w until two computations are constructed which start with u and have the same immediate reductions apart from the order.

The Church-Rosser property for $\Rightarrow$ is a consequence of these facts.

3. New rules compatible with R. Now we wonder whether it is possible to add rules because of which some elements of E reduce to $V \cup \{*_1\} \cup \{*_2\}$, in particular whether it is possible to add the following ones:

VI "$x \longrightarrow *_1$" for $x \in E_1$

VII "$x \longrightarrow *_2$" for $x \in E_2$

where $E_1 \subsetneq E$ and $E_2 \subsetneq E$ have infinitely many elements and will be specified in the sequel.

Because of such rules figure 1 may be transformed into figure 2:

(fig.2) D

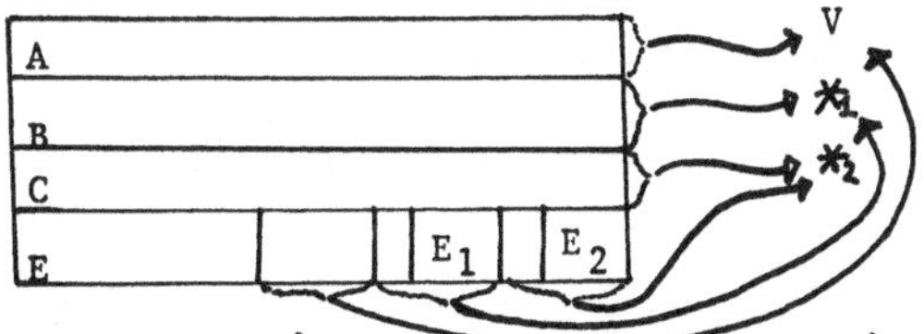

Let $R = \{I, II, III, IV\}$, $R' = R \cup \{VI, VII\}$ and write $\Rightarrow_{R}$ and $\Rightarrow_{R'}$ for the relative predicates. Let us also call a-R-reducible and a-R'-reducible every expression always reducible by rules of R and of R' respectively, in the sense that it is not possible to obtain an irreducible expression by those rules.

In [5] it is shown that there exist finitely many elements of E to which rule VI can be applied.

Now a rule not in R is compatible with R in the sense of satisfying the following condition, for every $u \in D$ and $z, z' \in V \cup \{*_1\} \cup \{*_2\}$:

Cond: If $u \Rightarrow_{R'} z$ then u is not a-R'-reducible and for every z' such that $u \Rightarrow_{R'} z'$ it holds $z' \equiv z$.

For rules VI and VII this condition means both the following ones:

Cond 0: If $u \Rightarrow_{R} z$ then u is not a-R'-reducible and for every z' such that $u \Rightarrow_{R'} z'$ it holds $z' \equiv z$;

Cond 1: If u is a-R-reducible and $u \Rightarrow_{R'} z$, then u is not a-R'-reducible and for every z' such that $u \Rightarrow_{R'} z'$ it holds $z' \equiv z$;

and since cond.0 is in fact satisfied (because when $u \Rightarrow_R z$ neither rule VI nor rule VII can be applied) in the sequel it is sufficient to consider cond 1 only.

4. Definition of E_1 and E_2. Let $x \in E$ be said stable iff there exist $y \in E$ and $y' \in E$ such that $x \Rightarrow_R y \Rightarrow_R y'$ and y occurs inside y'. For instance the expressions $\alpha ii(\alpha ii)$, $\vartheta(\vartheta\alpha)\psi$, $\vartheta(w\alpha)(\vartheta(w\alpha))$, $\vartheta(w\alpha)(\vartheta\alpha)$ (where $i \equiv \alpha(\psi\psi)(\psi\psi\psi)$, $\vartheta xy \Rightarrow_R x(\vartheta x)y$ and $wxy \Rightarrow_R xyy$) are stable, while the expressions $\vartheta\alpha\psi\psi$, $\alpha(\vartheta\alpha)x(\vartheta\vartheta x)$ are not stable.

Now we consider the following subsets of the set of stable expressions:

$$E_1 = \{ x \mid \text{ either } x \in H \text{ or } x \in H' \}$$

where

$$H = \{ x \mid x \Rightarrow_R y \Rightarrow_R y \ , \ x \Rightarrow_R y \Rightarrow_R yz \text{ with } z \in V \cup \{*_1\} \cup \{*_2\} \}$$

$H' = \{ x \mid x \Rightarrow_R y \Rightarrow_R zy \ , \ x \Rightarrow_R y \Rightarrow_R zyz' \ , \ x \Rightarrow_R y \Rightarrow_R z(yz')$ with $z' \in V \cup \{*_1\} \cup \{*_2\}$ and $z \in V \cup \{*_1\}$ is such that either $*_1$ is a fixed point of z by R or $z*_1 \Rightarrow_R h \in H \}$.

and

$E_2 = \{ x \mid x \Rightarrow_R y \Rightarrow_R zy \ , \ x \Rightarrow_R y \Rightarrow_R zyz' \ , \ x \Rightarrow_R y \Rightarrow_R z(yz')$ with $z' \in V \cup \{*_1\} \cup \{*_2\}$ and $z \in V \cup \{*_2\}$ is such that $*_1$ is neither a fixed point of z by R nor $z*_1 \Rightarrow_R h \in H \}$.

For example expressions $\alpha ii(\alpha ii)$, ϑwz, $\alpha(\alpha k(kz))i(\alpha(\alpha k(kz))i)$ with $z \in V \cup \{*_1\}$ are in H; expressions $\vartheta(\vartheta\alpha)(\psi *_1 *_1)$, $\vartheta(\vartheta\alpha)\psi$, $\vartheta(\alpha(k*_1))z$, $\vartheta\alpha(w(w(\vartheta\alpha)))(w(w(\vartheta\alpha)))$, $\vartheta(w(\alpha*_1))(\vartheta(w(\alpha*_1)))$ with $z \in V \cup \{*_1\}$ are in H'; expressions $\vartheta(\alpha(k\psi))\alpha$, $\vartheta(\alpha(kk))i$, $\vartheta(\alpha(k(k\alpha)))z$, $\vartheta(w(\alpha\psi))(\vartheta(w(\alpha\psi)))$, $\beta\psi(w(w(\beta\psi)))(w(w(\beta\psi)))$, with $\beta xyz \Rightarrow_R x(yz)$ and $z \in V \cup \{*_1\}$ are in E_2.

Remark 1 - $H \cap H' = \emptyset$ and $E_1 \cap E_2 = \emptyset$.

Remark 2 - The set of expressions which reduce to $*_1$ by R' is of course bigger than E_1. Analogously for E_2.

Remark 3 - For every $x \in E_2$ rule $x \longrightarrow *_1$ is not compatible with R. In fact $x \in E_2$ entails there exists $z \in V$ which does not have $*_1$ as a fixed point by R' and either $x \underset{R}{\Rightarrow} y \underset{R}{\Rightarrow} zy$ or $x \underset{R}{\Rightarrow} y \underset{R}{\Rightarrow} zyz'$ or $x \underset{R}{\Rightarrow} y \underset{R}{\Rightarrow} z(yz')$, hence $x \underset{R}{\Rightarrow} z *_1 \underset{R'}{\not\Rightarrow} *_1$ holds.

Remark 4 - For some elements of E_1, $x \longrightarrow *_2$ is not compatible with R. It is enough to consider $x \in E_1$ such that $x \underset{R}{\Rightarrow} y \underset{R}{\Rightarrow} *_1 y$ to obtain $x \underset{R'}{\Rightarrow} *_2$ and $x \underset{R'}{\Rightarrow} *_1$ (because $*_1 *_2 \longrightarrow *_1$).

Lemma. Rules VI and VII satisfy cond 1 when $x \in E_1 \cup E_2$.

Proof - If $x \in E_1 \cup E_2$ then x is a-R-reducible. Moreover if $x \in E_1$ then, remembering the definition of E_1, only one of the following cases can hold by rules $R \cup \{VI\}$:

1) 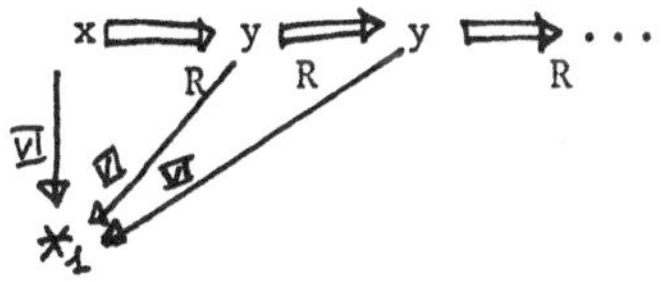

5) 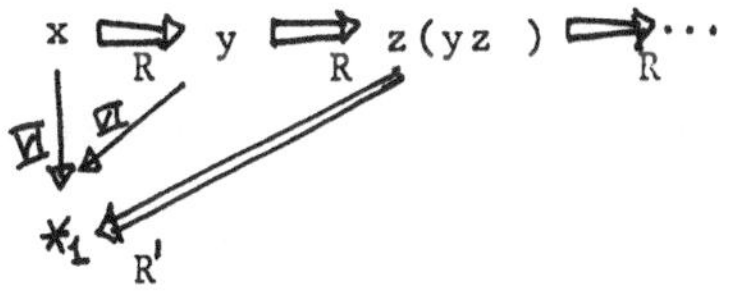

2) 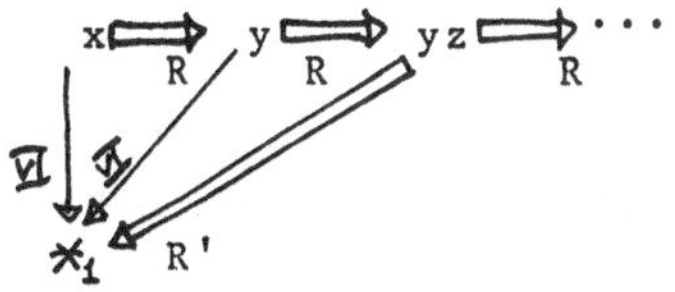

3) 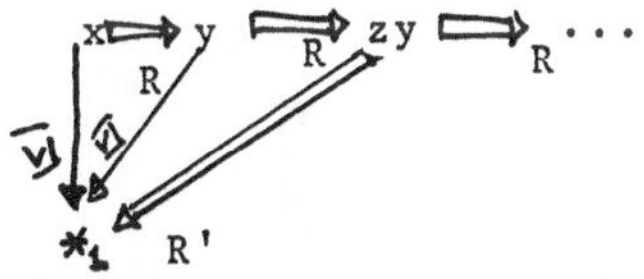

4) 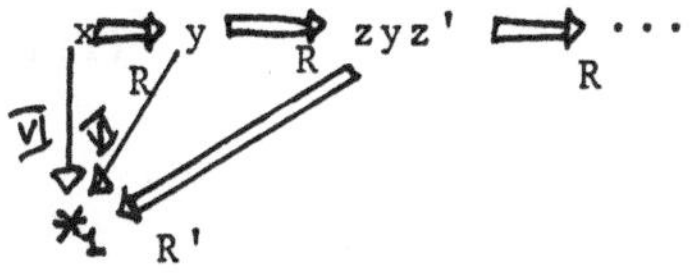

and if $x \in E_2$ then, remembering the definition of E_2, by rules $R \cup \{VII\}$ only one of the following casescan hold:

1) 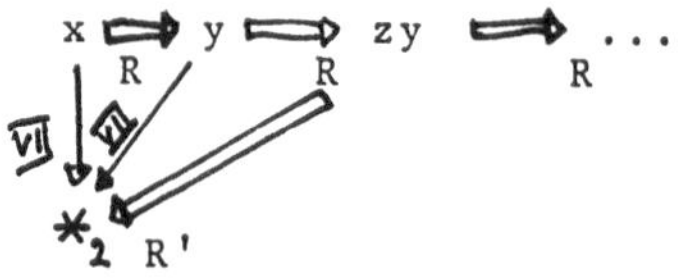

2) 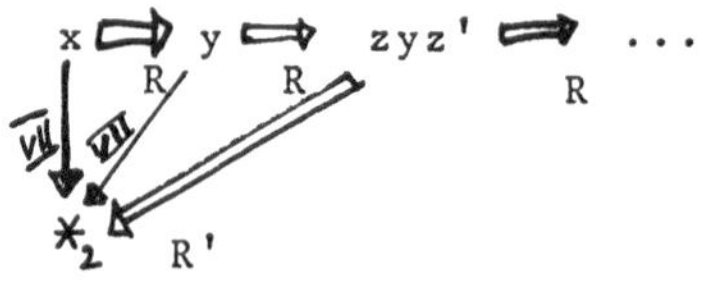

3) 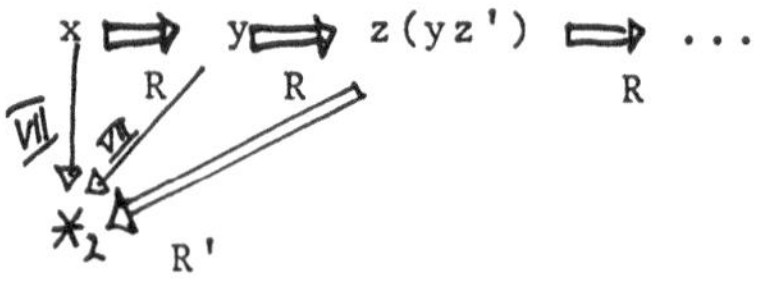

In all such cases cond1 is satisfied whatever is the order of application of the rules. ☒

Theorem - Rules VI and VII are compatible with R .

Proof - We have to show that cond 1 is satisfied. When cond 1 fails one of the following cases holds, for some $u, v \in D$ and $z, z' \in V \cup \{*_1\} \cup \{*_2\}$:

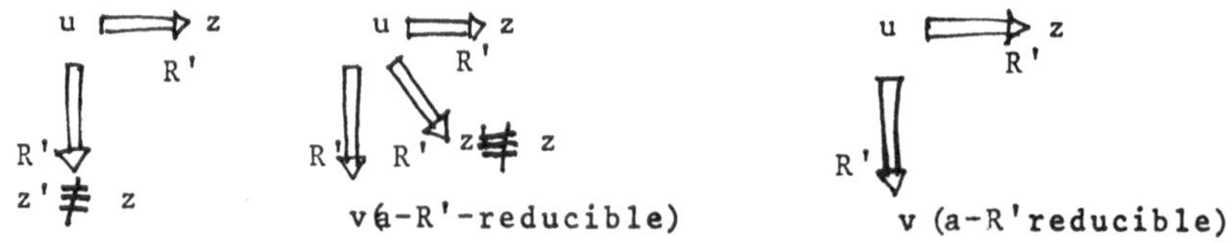

(i) (ii) (iii)

By the lemma $u \notin E_1 \cup E_2$, hence $u \in E - (E_1 \cup E_2)$. Now from $u \underset{R'}{\Rightarrow} z$ at least one rule in $\{VI, VII\}$ has been applied at least once, so u reduces by R to some expressions w_o containing some occurrences of some $x \in E_1 \cup E_2$ and no $y \in E - (E_1 \cup E_2)$. However from (i), (ii) and (iii), by changing the order of application of rules in R' on w_o the result of the computation changes too, that is z is no more obtained. This means that those occurrences of elements of E in w_o cannot be in $E_1 \cup E_2$. ⊠

Remark. The Church-Rosser property for $\underset{R'}{\Rightarrow}$ is guaranteed by Cond.

5. An uniformly reflexive structure with three indefinite elements.

Since also $x \underset{R'}{\Rightarrow} y$ is the reflexive and transitive closure of $x \to y$, let us define equality inside D as follows:

"x=y" ("x equal y") when one of the following cases holds:

i) $x \underset{R'}{\Rightarrow} z$ and $y \underset{R'}{\Rightarrow} z$ for $z \in V$

ii) $x \underset{R'}{\Rightarrow} *_1$ and $y \underset{R'}{\Rightarrow} *_1$

iii) $x \underset{R'}{\Rightarrow} *_2$ and $y \underset{R'}{\Rightarrow} *_2$

iv) x and y are such that neither $x \underset{R'}{\Rightarrow} z$ nor $y \underset{R'}{\Rightarrow} z'$ can hold for $z, z' \in V \cup \{*_1\} \cup \{*_2\}$.

Such an quality relation is a congruence relation, because it is reflexive, symmetric and transitive, and moreover it can be verified that if $x = x'$ and $y = y'$ then $xy = x'y'$.

Identifying all expressions which reduce to the same $z \in V$, all expressions which reduce to $*_1$, all expressions which reduce to $*_2$, and all expressions in iv) with $*$, and writing $x = y$

when x and y have been identified, we obtain a uniformly reflexive structure (URS, see [1] ,[2] , [4]), with domain $U_o = V_o \cup V_o^*$, where $V_o = V \cup \{*_1\}$ and $V_o^* = \{*_2, *\}$. The indefinite elements are $*_1$, $*_2$ and $*$; however since $*_1$ is manipulated by rules of R like an element of V , $*_1$ may be considered as a definite element, while $*_2$ and $*$ are not manipulated by rules of R, hence they cannot be considered as definite elements and have been put toghether in V_o^*. Moreover $*$ is an indefinite stronger than $*_2$, so that axiom-schemata of a URS, i.e. (see [5]).

ax1 - $x* = *$

ax2 - if $x,y \neq *$ then $\alpha xy \neq *$

ax3 - if $x,y,z \neq *$ then $\alpha xyz = xz(yz)$

ax4 - if $x,y,z \neq *$ then $\psi xyz = \begin{cases} y & \text{if } z = x \\ x & \text{if } z \neq x \end{cases}$

can be written in our case :

ax1 - $x* = *$

ax2' - if $x,y \notin V_o^*$ then $\alpha xy \notin V_o^*$

ax3' - if $x,y,z \notin V_o^*$ then $\alpha xyz = xz(yz)$

ax4'- if $x,y,z \notin V_o^*$ then $\psi xyz = \begin{cases} y & \text{if } z = x \\ x & \text{if } z \neq x \end{cases}$

and it can be verified that they are satisfied.

REFERENCES

[1] WAGNER E.G., Uniformly Reflexive Structures: An Axiomatic Approach to Computability, Information Sci. 1 (1969), 343-362.

[2] STRONG H.R., Algebraically Generalized Recursive Functions Theory, IBM J. Res. Develop.12 (1968),465-475.

[3] CURRY H.B. and FEYS R., Combinatory Logic, North Holland, Amsterdam, 1958.

[4] GROSS W. and VENTURINI ZILLI M., Computability and Uniformly Reflexive Structures, Proceedings of the International Computing Symposium (Venice, 1972), pp.485-495.

[5] VENTURINI ZILLI M., On different kinds of indefinite,Calcolo 11 (1974), 67-77.

ON SUBRECURSIVENESS IN WEAK COMBINATORY LOGIC

Carlo Batini - Alberto Pettorossi

Centro di Studio dei Sistemi di Controllo e Calcolo Automatici
C.N.R. - Roma

Istituto di Automatica dell'Università di Roma

ABSTRACT

In this paper weak combinatory logic as an algorithmic language is considered and various notions of structural and computational complexity are introduced. Particular attention is devoted to the definitional power of a system of combinators, that is to the concept of "subbase". Some results concerning the relations between specific subbases and their generative power are presented.

SUMMARY

1. INTRODUCTION

The main purpose of this work is to introduce the concepts of computational complexity and subrecursiveness in combinatory logic [1]. Since we are interested in the algorithmic and computational properties of combinatory logic, we think that a fruitful way of approaching these properties is to consider how limiting the power of the calculus (that is limiting the allowed definitions of "programs" or the allowed amount of "resource" used by "programs") entails limitations on the ability of manipulating "data".

Studies on computational complexity have given well established and meaningful results for several abstract machines and languages such as Turing machines, LOOP programs, rewriting systems [2]. Less considered seems to have been the definition of suitable concepts of computational complexity in combinatory logic and λ-calculus. An interesting step in this direction was made by H.R. Strong [3] who defined a measure of depth of computation in a programming language based on Wagner's URS and showed that for each partial recursive function there is an index with uniformly bounded measure of computation.

In order to carry on our investigation on complexity properties of combinatory logic in §2 and §3 we shall examine properties of structural and computational complexity of combinators in weak calculus [4] and in §4 we examine the ability of combinations in generating pure applicative combinations [1]. For reasons that will be made clear later we restricted ourselves to considering only proper combinators [1] or, in general, combinators for which it is possible to define suitable input-output relations.

2. APPROACHES TO COMPUTATIONAL COMPLEXITY IN COMBINATORY LOGIC

The two basic notions of complexity in the literature are structural and computational complexity.

We now introduce the formal definitions of some possible measures of these notions. Structural complexity is inherent to a combinator as a static well formed object in a specific base.

Definition 2.1. - *Length* (SL) of a combinator χ is recursively defined by the following rules:

i) if χ is a basic combinator, $SL(\chi) = 1$;

ii) if $\chi = (\chi_1\chi_2)$, $SL(\chi) = SL(\chi_1) + SL(\chi_2)$.

Definition 2.2. - *Depth of the parenthesis structure* (SD) of a combinatory χ is defined by:

i) if χ is a basic combinator, $SD(\chi) = 0$;

ii) if $\chi = (\chi_1\chi_2)$, $SD(\chi) = 1 + \max\{SD(\chi_1), SD(\chi_2)\}$.

A simple relation between the two measures of structural complexity is the following:

Fact 2.1. - $\lceil \log_2 SL(w) \rceil \leq SD(w) \leq SL(w)-1$ for any combinator w.

Computational complexity measures are related to the reduction of combinators to normal form in the weak calculus.

The measures we may define, depend on the computation rule choosen in the reduction process. Among the rules that guarantee to reach the normal form we will choose the standard (leftmost outermost) rule.

Definition 2.3. - *Number of steps of computation* (CT) of a combinator χ is defined by:

i) $CT(\chi) = t$ if χ reaches a normal form in t steps;

ii) $CT(\chi)$ = undefined otherwise.

Definition 2.4. - *Size of computation* (CS) of a combinator χ is defined by:

$CS(\chi) = \ell$ if $\ell = \max_i\{SL(\chi_i)\}$ where χ_i is a formula achieved during the reduction process.

Definition 2.5. - *Depth of computation* (CD) of a combinator χ is defined by:

$CD(\chi) = d$ if $d = \max_i\{SD(\chi_i)\}$ where χ_i is a formula achieved during the reduction process.

Analogously to what is made [5,6,7] for acceptable Gödel numbering of partial recursive functions we will now give the following definitions for measures of complexity in the weak calculus of combinators:

(1) $|\ |$ is a structural complexity measure if:

  i) $|\ |$ is a recursive mapping from the set of combinators to the integers;

  ii) $\forall n$ the number of combinators w such that $|w| = n$, is finite.

(2) C is a computational complexity measure for a combinator w if C is a partial recursive mapping from the set of combinators to the integers and:

  i) w has normal form implies that $C(w)$ is defined;

  ii) $C(w)$ is defined implies that "w has normal form" is decidable;

  iii) $C(w) = n$ is decidable.

It is not difficult to verify that, if the cardinality of the base is finite, SL and SD are structural complexity measures, while CT, CS and CD are computational complexity measures (*).

As we have already remarked, in general, the way the properties of complexity measures are studied is to consider how limitations on the measures result in limitations on the power of computational systems. On the other hand, for the above listed measures this does not seem to be the way of achieving interesting general results. It is certainly possible to simulate tape and time bounded Turing machines, restricted rewriting systems, primitive recursive computations, etc. in combinatory logic, and to define in this way the corresponding classes of combinators, but this approach does not give classifications of combinators well matched to the computational peculiarities of combinatory logic. These peculiarities are essentially:

i) the ability of "packing" data in such a way that "unpacking" is impossible from outside [9] and that data can be accessed in other than a sequential way;

ii) the "rightward" mechanism of operating of the calculus that makes it more similar to a tag-machine or to a non-erasing Turing machine than to other classical computing systems;

iii) the impossibility in a non typed calculus of an "a priori" distinction between programs and data and, inside a program, between primitives and constructs.

The last point is particularly interesting because the variability of the argument (or at least of the size of the argument) is at the base of the classification of the complexity of programs and functions (typically characterized by asyntotic behaviour).

3. COMPLEXITY MEASURES IN SUBBASES

Taking into account the type of properties of combinatory logic listed at the end of §2, we think that a promising way of studying the com

(*) The first property of computational complexity measures is satisfied by the norm introduced in [8] in the definition of NURS.

putational complexity of combinators is to use the concept of subbase and to analyze the computational power of various subbases. This somehow corresponds to the limitation of definitions in the formalism of recursive functions, which allows the generation of interesting subsets of partial recursive functions, such as the class of primitive recursive functions, the class of elementary functions and the classes of Grzegorczyk [10].

Definition 3.1. - A *subbase* $\mathcal{B}$ is a non-empty (possibly infinite) class of combinators $\mathcal{B} = \{\Phi_1,\dots,\Phi_n\}$.

In general we will refer to finite subbases of independent combinators.

Definition 3.2. - The *applicative closure* of the subbase $\mathcal{B}$ (denoted $\mathcal{B}^+$) is the class of all finite (applicative) combinations of Φ_i's. We wish, in the future, to refer to the subclass of $\mathcal{B}^+$ whose elements are in normal form and proper. We will indicate this class by $\mathcal{B}^+_{np}$.

We now summarize a few examples of basic results holding for particular subbases.

Theorem 3.1.-For any combinator w in $\{B,C,K\}^+$ the structural and computational complexity satisfy the following limitations:

i) $CT(w) \leq SL(w) - 1$

ii) $CS(w) = SL(w)$

iii) $CD(w) \leq SL(w) - 1$

iv) $CD(w) \leq \left\lfloor 2^{SD(w)-1} + \frac{SD(w)-1}{2} \right\rfloor$

Proof. (i),(ii) - From a theorem of Curry [1] no combinator exists in $\{B,C,K\}$ with duplicative effect. So, at every reduction step the length of the formula decreases at least by 1.

(iii). In a formula, whose length is n, appear n-1 applications.

(iv). Since SD(w) is fixed, the maximum number of basic combinators w is $2^{SD(w)}$.

In every reduction step the basic combinators may increase the depth of a formula by at most 1. Therefore, if n is the number of basic combinators used to achieve the maximum depth, we have

$CD(w) = SD(w) + n$, where n is the maximum

integer such that: $SD(w)+n \leq 2^{SD(w)} - n - 1$. (The last inequality is the (iii) written for the deepest formula achieved). Hence point (iv) follows.

Q.E.D.

Theorem 3.2. - In the subbase {B} the expotential growth of the depth is achievable. In fact:

i) $\forall w \varepsilon \{B\}^+$ such that $SD(w) \leq 2$ then $CD(w)=SD(w)$;

ii) $\forall n>2 \ \exists \ w_n \varepsilon \{B\}^+$ such that:

$$SD(w_n) = n$$
$$CD(w_n) = N + \sum_{i=0}^{n+1} b_i \quad \text{where } b_o=b_1=0 \text{ and } b_{i+2}=b_{i+1}+b_i+1$$

Proof. The point (i) is immediate.

For the point (ii) it is not difficul to see the class of combinators w_n, whose structure is recursively defined as follows:

$$w_n = d_{n-1}w_{n-1} \qquad \text{where:} \quad w_o = d_o$$
$$d_n = B\, d_{n-2}d_{n-1}$$
$$d_o = B$$
$$d_1 = B\,B,$$

has the property that all its B's with 3 arguments, in the reduction process, increase by 1 level the initial depth of the w_n. We also have $w_n = d_{n-1}(d_{n-2}(\ldots(d_o d_o)\ldots))$, and $n = SD(w_n)$.

Let b_i be the number of B's with 3 arguments in d_i. We can see $b_o = b_1 = 0$ and $b_i = b_{i-1}+b_{i-2}+1$.

Therefore: $CD(w) = n + \sum_{i=0}^{n-1} b_i$ Q.E.D.

Remark 3.1.-In the subbase {B,C} the limit of Theorem 3.2. can be improved.

For example, if $q = B^2(CBB)B\, d_4 d_5 w_6$,

we have $CD(q) = SD(q) + \sum_{i=0}^{SD(q)-1} b_i+1$.

As far as combinators without normal form are concerned, we may define SL(w)(n) and SD(w)(n) to be the length and the depth of the combinator w at the n-th reduction step and we may show some properties of these functions of n for the subbase {W} and {B,W}. In particular, we will show that, in the subbase {W}, SL(w)(n) is somehow linear, while an exponential growth is possible in the subbase {B,W}.

We will first introduce the following:

Definition 3.3. - The *number* $n_r(w)$ *of right-applied objects of* $w\varepsilon\mathcal{B}^+$, is defined by: $n_r(w) = k+1$, where $w=(\ldots((b\chi_1)\chi_2)\ldots\chi_k)$, $b\varepsilon\mathcal{B}$ and $\chi_i\varepsilon\mathcal{B}^+$, $1\le i\le k$.

Remark 3.2-The decomposition of w in right-applied objects is unique.

Definition 3.4. - The *subwords* of a combinator $w\varepsilon\mathcal{B}^+$ are recursively defined as follows:

(i) if w=(ab) where a,b$\varepsilon\mathcal{B}$ then a and b are subwords of w;

(ii) if w=(AB) where A,B$\varepsilon\mathcal{B}^+$ then A, B and the subwords of A and B are subwords of w.

It is now necessary to prove the following lemma.

Lemma 3.1. - Let w be a combinator of $\{W\}^+$. Let $\bar{w}$ be the leftmost subword of w such that $n_r(\bar{w})=3$. We will call $\bar{w}$ the leading subword of w.

(i) If $\bar{w}$ exists:

- w has not normal form;
- $\forall n$ SL(w)(n) = SL($\bar{w}$)(n)+c, where c is a constant.

(ii) If $\bar{w}$ does not exist, w is in normal form.

Proof. Point (ii) is immediate. Point (i) is proved by the following facts:

(i) In any reduction step of a word $w \varepsilon W^+$, we have: $n_r(w)(n)\le n_r(w)(n+1)$. (*)

(ii) If $w=w_1w_2$ and all W in w_1 have one argument, then in the reduction process the W's in w_1 do not change the number of their arguments.

(iii) If $w=w_1w_2$ where $n_r(w_1)\ge 3$, then the reduction process which follows, does not take into account w_2.

Q.E.D.

(*) We denote by w(n) the combinator derived by w after n reductions. Obviously w(0)=w.

For the base $\{W\}$ we have the following result:

Theorem 3.3. - For any $w\in\{W\}^+$ without normal form

$\exists\bar{n}$ finite such that, if $\Delta(n)=SL(w)(n+1)-SL(w)(n)$, we have:

(i) $\forall n<\bar{n} : \Delta(n)\geq 0$ depending on n;

(ii) $\forall n\geq\bar{n} : \Delta(n)=\bar{\Delta}\geq 0$.

Proof. Given a w $\{W\}^+$ without normal form by lemma 3.1 does exist $\bar{w}$ (defined as in lemma 3.1) and $SL(w)=SL(\bar{w})+c$.

Moreover the structure of $\bar{w}$ may be one of the following:

(1) $W W w_o$

(2) $W(Ww_k)w_o$

(3) $W \;\bar{\bar{w}}\; w_o$ where $w_o,w_k,\bar{\bar{w}}\in\{W\}^+$ and $n_r(\bar{\bar{w}})\geq 3$.

The complete case analysis of (1), (2) and (3) by applying the induction principle on the combinators' length, proves the theorem [12]. Q.E.D.

For the subbase $\{B,W\}$ we have:

Theorem 3.4. - There exists w in $\{W,B\}^+$, such that SL(w)(n) grows exponentially with n.

Proof. In a constant number of reduction steps (6)
$W(W_{(2)}B_{(1)})(W(W_{(2)}B_{(1)}))p$, where $p\in\{W,B\}^+$, reaches
$W(W_{(2)}B_{(1)})(W(W_{(2)}B_{(1)}))p'$ where $p'\in\{W,B\}^+$ and $SL(p')=2\cdot SL(p)$. Q.E.D.

4. SUBBASES AND DEFINITIONAL COMPLEXITY

A particularly interesting type of results concerning subbases are related to their generative power:

Definition 4.1. - Let $\mathcal{B}$ be a subbase; let V be an infinite ordered set of variables $\{x_1,x_2,\ldots,x_n,\ldots\}$; let V^+ be the set of all finite applicative combinations of variables.

We say that $L(\mathcal{B})$ is the *language generated by* $\mathcal{B}$ if $L(\mathcal{B})$ is the smallest subset of V^+ satisfying the property that, given any $w\in\mathcal{B}^+_{np}$, then if n is the order of w, there is $X\in L(\mathcal{B})$ such that

$wx_1 \dots x_n$ reduces to X (*).

We prove first the following lemma.

Lemma 4.1. If ξ is a proper combinator whose order is 2, then $\{\xi\}^+_{np}$ is the set: $\{\xi,\xi\xi,\xi(\xi\xi),\dots,\xi(\xi(\dots(\xi\xi)\dots)),\dots\}$.

Proof. If we suppose in $\{\xi\}^+_{np}$ a ψ exists with $n>2$ right-applied objects, then ψ is not in normal formal. Q.E.D.

The following fact can easily be verified:

Fact 4.1.

(i) $L(\{K\}) = \{x_n | n \geq 1\}$;

(ii) $L(\{W\}) = \{x_1x_2x_2\} \cup \{x_1^k | k \geq 3\}$; (**)

(iii) $L(\{B,I\})$ = the language of *complete ordered applications* that is the language generated by the production $S \to (SS)$ and by substitution from left to right of all occurrences of S in a sentential form by the variables $x_1,x_2,\dots,x_n,\dots$ in this order;

(iv) $L(\{B\}) = L(\{B,I\}) - \{\bar{X}x_n |$ if $n=1$ $\bar{X}$ is the empty word, if $n>1$ $\bar{X}$ is a word of the language of complete ordered applications with the variables $x_1,\dots,x_{n-1}\}$;

(v) $L(\{C\}) = \{x_1x_3x_2\} \cup \{x_2x_3x_1\}$.

Proof of Fact 4.1.(i). From lemma 4.1. we know the structure of the elements of $\{K\}^+_{np}$: for each $n \geq 1$ there exists only one combinator $\xi_n \in \{K\}^+_{np}$ such that $SL(\xi_n)=n$, and $\xi_n = K\xi_{n-1}$.
For $\xi_1 = K$ the corresponding X is x_1. If the fact is valid $\forall n'<n$ then

$$\xi_n x_1 \dots x_{n+1} = K\xi_{n-1}x_1 \dots x_{n+1} \geq \xi_{n-1}x_2 \dots x_{n+1} \geq x_n$$

where the last reduction is guaranteed by the induction hypothesis.

Q.E.D.

Proof of Fact 4.1 (ii). Like proof of Fact 4.1. (i). Q.E.D.

Proof of Fact 4.1. (iii) and Fact 4.1. (iv). The language $L(\{B,I\})$ is contained in the language of complete ordered applications [1].

(*) Notice that we consider for example $a_1a_2a_3$ and $((a_1a_2)a_3)$ to be the same word.

(**) Notice that we consider x_i^k and $\underbrace{x_ix_i\dots x_i}_{k}$ to be the same word.

Viceversa the language of complete ordered applications is contained in $L(\{B,I\})$, because, if X is a word of the language of complete ordered applications, and:

- $X = x_1 x_2 \ldots x_n$, where $n \geq 1$, then $B^{n-1}I$ corresponds to it;
- $X = \bar{X}\ x_j \ldots x_n$ and $\bar{w} \varepsilon L(\{B\})$ corresponds to $\bar{X}$, then $B^n I\bar{w}$ corresponds to it.

The last case is the one in which $X = x_1 X_1 X_2 \ldots X_k$ where at least X_k is a combination of 2 or more variables x_i, and possibly k=1. In this case we will now prove inductively that a combinator of $\{B\}^+$ corresponds to X. One parenthesis can be removed for instance from X eliminating the one surrounding X_k by $B_{(k-1)}$. Let us assume we succeded, in the expansion procedure, to remove p parentheses, obtaining a combination of the form:

$\xi x_1 Y_1 Y_2 \ldots Y_\ell$ where at least one Y_j is a combination of 2 or more variables, and ξ is a combination of B's.

Now we may remove one more parenthesis as the one surrounding Y_j, by the deferred combinator $B_{(j)}$.

Q.E.D.

Proof of Fact 4.1. (v). We prove first that: $\forall \xi \varepsilon \{C\}^+_{np}$ the order of ξ is ≤ 3.

In every reduction step of $\xi x_1 x_2 \ldots x_n$ where $\xi \varepsilon \{C\}^+_{np}$, there are at least 3 arguments between the first basic combinator in ξ and x_4.

Infact there will always be x_1, x_2 and x_3, because:

- on the first reduction step there are x_1, x_2 and x_3;
- if there are x_1, x_2 and x_3 on the i-th reduction step, then they are also on the (i+1)-th step, because the order of C is 3, C has not compositive effect and no x_i, where $1 \leq i \leq 3$, may be on the left of the leftmost C in a formula (ξ is proper).

Therefore in order to prove the fact 4.1.(v), we can consider only the combinations of x_1, x_2 and x_3. For $x_1 x_2 x_3$, $x_2 x_1 x_3$, $x_3 x_1 x_2$ and $x_3 x_2 x_1$ after the first expansion step the last variable, that cannot be reused, is not x_3, as it should be.

Instead for $x_1 x_3 x_2$ and $x_2 x_3 x_1$ the last variable is x_3, and they are actually computed by C and CC, respectively.

Q.E.D.

For specific subbases the completeness (meta)algorithms (such as those given in [11] for the base {S,K} and in [1] for the base {B,C,W,K}) become more interesting. In fact while in a complete base those metaalgorithms cannot always give the shortest combinator corresponding(*) to a given combination of variables, this may be accomplished in the case of a subbase which is complete only with respect to a subset of combinations. For the base {B} we have the following results:

Theorem 4.1. - For any X in $L(\{B\})$ we may obtain the shortest w in $\{B\}^+$ corresponding to X.

Sketch of the proof. We will prove that, in the construction of the combinator w that corresponds to X, a sufficient rule in order to obtain the shortest w consists in the elimination of the rightmost parenthesis that can be eliminated at every expansion step by using one B. Infact, at a given expansion step, we have in general an object ξ that can be derived from S by the following productions:

$$S \to (SS_V) \mid (S_B S) \mid S_B \mid S_V \; ; \qquad S_B \to (S_B S_B) \mid B$$

$$S_V \to (S_V S_V) \mid x_1 \mid \ldots \mid x_n \; ,$$

in which all variables x_i appear on the left side of x_j, where $j>i$.

We can now eliminate:

(a) 1 or more parentheses surrounding components like $(S_V S_V)$;

(b) 1 parenthesis surrounding a component like $(S_B S)$, where S generates at least one variable;

(c) 1 parenthesis surrounding a component like (SS_V), where S generates at least one B.

Case (a). Let us assume we reached in the expansion step the following object:

$$\xi_1 = \xi' \underset{(1)}{(t_1 t_2)} \ldots \underset{(k)}{(t_k t_{k+1})} x_\ell \ldots x_n \quad , \text{ where every } t_i$$

is either a variable or a combination of the variables $x_p, \ldots, x_{\ell-1}$, and ξ' is a combination of B's and variables $x_1, \ldots, x_{p-1}$.

(*) In the sense of [1] pag. 160.

Let us consider the class S of all the strategies that remove all explicit parentheses in ξ_1 between x_p and x_n and reach an object like: $\xi_2=\xi'' \; x_p \ldots x_{\ell-1} x_\ell \ldots x_n$, where ξ'' is a combination of B's and the variables $x_1, \ldots, x_{p-1}$.

It is possible to prove the following assertion:

For every strategy $s \varepsilon S$ that eliminates first the (1) parenthesis, there exists a strategy $\bar{s} \varepsilon S$ that eliminates first the (k) parenthesis taking a not greater number of steps than s.

Since the proof is rather long [12], we give the following example where p=1 and n=6.

STRATEGY s

no.of steps	achieved formula	removed parenthesis	introduced parenthesis (as side effect)
	$x_1 \; (x_2 \; (x_3x_4)) \; (x_5x_6)$ (1) (2) (3)		
1		(1)	-
	$Bx_1x_2 \; (x_3x_4) \; (x_5x_6)$ (2) (3)		
2		(3)	(α),(β),(γ)
	$B \; (\; (\; (Bx_1)x_2) \; (x_3x_4))x_5x_6$ (α)(β)(γ) (2)		
3		(α)	-
	$BB \; (\; (Bx_1)x_2) \; (x_3x_4)x_5x_6$ (β)(γ) (2)		
4		(2)	(δ)
	$B \; (BB \; (\; (Bx_1)x_2))x_3x_4x_5x_6$ (δ) (β)(γ)		
5		(δ)	-
	$BB \; (BB) \; (\; (Bx_1)x_2)x_3x_4x_5x_6$ (δ) (β)(γ)		
6		(β)	-
	$B(BB \; (BB)) \; (Bx_1)x_2x_3x_4x_5x_6$ (δ) (γ)		
7		(γ)	-
	$B(B(BB(BB)))Bx_1x_2x_3x_4x_5x_6$		

STRATEGY $\bar{s}$

no. of steps	achieved formula	removed parenthesis	introduced parenthesis (as side effect)
	$x_1(x_2(x_3x_4))\ (x_5x_6)$ (1) (2) (3)		
1		(3)	(ε)
	$B\ (x_1\ (x_2\ (x_3x_4)))x_5x_6$ (ε) (1) (2)		
2		(ε)	-
	$BBx_1\ (x_2\ (x_3x_4\ )\ )x_5x_6$ (1) (2)		
3		(1)	(η)
	$B\ (BBx_1)x_2\ (x_3x_4)x_5x_6$ (η) (2)		
4		(2)	(μ),(σ)
	$B\ (\ (B(BBx_1))x_2)x_3x_4x_5x_6$ (μ)(σ)(η)		
5		(μ)	-
	$BB\ (B\ (BBx_1))x_2x_3x_4x_5x_6$ (σ)(η)		
6		(σ)	-
	$B(BB)B\ (BBx_1)x_2x_3x_4x_5x_6$ (η)		
7		(η)	-
	$B(B(BB)B)(BB)x_1x_2x_3x_4x_5x_6$		

w_s=B(B(BB(BB)))B; $w_{\bar{s}}$=B(B(BB)B)(BB); SL(w_s)=7; SL($w_{\bar{s}}$)=7 .

Case (b). In this case we must compare all strategies that start with the elimination of the parenthesis of (S_BS) with those eliminating first a parenthesis of (S_VS_V), if they exist.

We can prove the same assertion of case (a) in the same way.

Case (c). As case (b).

Therefore, if we have more than 2 parentheses to be eliminated at every expansion step, we can repeat this argument for all of those; it will come out that a sufficient choice for obtaining the shortest w is to eliminate the rightmost parenthesis.

Q.E.D.

As a consequence of theorem 4.1, we may notice that, if $wx_1 \ldots x_n \geq X$ and $w'x_1 \ldots x_n \geq X'$, where $X, X' \in L(\{B\})$ and $w, w' \in \{B\}^+$:

(i) if X' has a lower number of parentheses to be eliminated(*) than X, then $SL(w') < SL(w)$;

(ii) if X' is obtained from X by moving on the left one couple of parentheses of X to be eliminated, then $SL(w') < SL(w)$.

We can also establish the following:

Theorem 4.2. - For any X in $L(\{B\})$ such that $SL(X)=n$ (**) we have that if $w \in \{B\}^+$ corresponds to X, then $SL(w)=O(n)$.

Proof. The structure of X such that $X \in L(\{B\})$ and $SL(X)=n$, in which there is the minimum number (1) of parentheses to be eliminated, is of the form:

$$\bar{X}_n = x_1 \ldots x_{n-2}(x_{n-1}x_n).$$

On the other hand, the structure of X such that $X \in L(\{B\})$ and $SL(X)=n$, in which there is the maximum number of parentheses to be eliminated and these are in the rightmost position, is of the form:

$$\bar{\bar{X}}_n = x_1(x_2(\ldots(x_{n-1}x_n)\ldots)) .$$

It can be easily verified that: if $\bar{w}_n x_1 \ldots x_n \geq \bar{X}_n$, then $\bar{w}_{n+1} = B\bar{w}_n$;

if $\bar{\bar{w}}_n x_1 \ldots x_n \geq \bar{\bar{X}}_n$, then $\bar{\bar{w}}_{n+1} = B(B\bar{\bar{w}}_n)B$;

$\bar{w}_3 = \bar{\bar{w}}_3 = B$.

Q.E.D.

(*) We suppose all parentheses to be removed are explicited.

(**) The definitions of structural complexity obviously can be extended to pure combinations .

5. BIBLIOGRAPHY

[1] Curry,H.B., R. Feys, *Combinatory logic.* vol. 1, North Holland(1974)

[2] Ausiello, G., *Computational complexity - Main results and a commentary,* Séminaires IRIA (1972).

[3] Strong,H.R., *Depth-bounded computation,* J.C.S.S., Vol. 4, n. 1 (1970).

[4] Curry, H.B., J.R. Hindley, J.P. Seldin, *Combinatory logic,* vol. 2, North Holland (1972).

[5] Blum, M., *On the size of machines,* Information and Control 2 (1967).

[6] Blum, M., *A machine independent theory of the complexity of recursive functions,* J.A.C.M. 14 (April 1967).

[7] Ausiello, G., *Abstract computational complexity and cycling computations,* J.C.S.S. Vol. 5 (1971).

[8] Barendregt, H., *Normed uniformly reflexive structures,* (in these proceedings).

[9] Böhm, C., W. Gross, *Introduction to the CUCH,* Automata Theory (ed. Caianiello) N.Y. (1966).

[10] Grzegorczyk, A., *Some classes of recursive functions,* Rozprawy Matematyczne (1953).

[11] Rosenbloom, P., *The elements of mathematical logic,* Dover (1950).

[12] Batini, C., A. Pettorossi, *Some properties of subbases in weak combinatory logic,* Rapporto dell'Istituto di Automatica - Università di Roma (1975).

SEQUENTIALLY AND PARALLELLY COMPUTABLE FUNCTIONALS

(Extended Abstract)

V.Yu.Sazonov

Inst.Mat., 630090, Novosibirsk, USSR

Developing our results [5], we define sequential and parallel functionals for a type-free model due to D.Scott [3,4] and, in particular, for the typed model [2] (which is naturally embedded in the type-free one). The sequentially computable (typed) functionals are proved to be exactly those objects, which are expressible in D.Scott's language LCF [2], and this is an answer to a question stated in [2].

NOTATION. $\lambda(C)$ is the set of λ-calculus terms [1] with constants from C and without free variables.

1. MODEL D FOR TYPE-FREE FUNCTIONALS is built up here as in [3] by means of inverse limit of recursively defined sequence $(D_n, j_n)_{n=0}^{\infty}$ with the only difference, that D_0 is not a continuous lattice, but the T_0-space $\{\perp, 0, 1, 2, \dots\}$ with the base of open sets: $\emptyset, D_0, \{0\}, \{1\}, \{2\}, \dots$. The projection j_0 is defined by the equality $j_0(f) = f(\perp)$ for all $f \in D_1 = [D_0 \to D_0]$. As in [3] the T_0-space D is homeomorphic to the space of continuous functions $[D \to D]$. We get a continuous lattice $\overline{D} = D \cup \{\top\}$ by adding the largest ("overdefined") element $\top$ to D . The binary operation in D (which exists due to homeomorphism between D and $[D \to D]$ and is naturally called application) may be extended to $\overline{D}$ by setting $x\top = \top x = \top$ for all $x \in \overline{D}$. The spaces $D_0, D_1, \dots$ and also all the spaces D_α of the typed model [2] are projections of the space D and are naturally identified with the subspaces of D (see [3] pp. 122, 125-126). Elements from D_0 represent constant functions, that is, for every $a \in D_0$ and $x \in D$ $ax = a$. For every $a, x_1, x_2, \dots \in D$ we define the value of the infinite product $a x_1 x_2 \dots \in D_0$

by:

$$\alpha x_1 x_2 \ldots = \begin{cases} x \in D_0, & \text{if } (\exists n \geqslant 0)\ \alpha x_1 x_2 \ldots x_n = x\ ; \\ \perp\ , & \text{if } (\forall n \geqslant 0)\ \alpha x_1 x_2 \ldots x_n \notin D_0. \end{cases}$$

Any object $\alpha \in D$ is corresponded thereby to some continuous function (of infinitaly many variables) from $[D^N \rightarrow D_0]$, this correspondence proving to be one-to-one. For $\alpha, x_1, x_2, \ldots \in \overline{D}$ we define additionally the value $\alpha x_1 x_2 \ldots \in \overline{D}_0 = D_0 \cup \{T\}$ to be T, when $\alpha = T$ or $x_n = T$ for some $n \geqslant 1$.

The space D is a model for λ-calculus [3,4]. That is, there exists (the only) mapping (or interpretation) $A \mapsto \overline{\overline{A}}$ from the set $\lambda(D)$ to the set D for which the following rules hold:

1) if $A = \alpha \in D$ is an atom, then $\overline{\overline{A}} = \alpha$;

2) if $A = A_1 A_2$, then $\overline{\overline{A}} = \overline{\overline{A_1}}\, \overline{\overline{A_2}}$;

3) if $A = (\lambda x. B)$, then $\overline{\overline{A}}$ is given by the equality $\overline{\overline{A}}\alpha = \overline{\overline{[\alpha/x]B}}$ valid for all $\alpha \in D$.

It is convenient here to formulate the next result, which is in fact a particular case of a much more general Theorem 4 (see below).

Theorem 1. Let $A \in \lambda(\emptyset)$ be any term, and $x_1, x_2, \ldots$ be a variables, then $\overline{\overline{A}} \neq \perp$ iff there exist $1 \leqslant i \leqslant n, A_1, \ldots, A_k \in \lambda(\emptyset)$ $(k \geqslant 0)$ so that the term $A x_1 x_2 \ldots x_n$ is convertible to the term $x_i (A_1 x_1 x_2 \ldots x_n) \ldots (A_k x_1 x_2 \ldots x_n)$ by the rules of the λ-calculus [1].

If $\overline{\overline{A}} \neq \perp$ and $x_1, x_2, \ldots$ run D, then

$$\overline{\overline{A}}\, x_1 x_2 \ldots = x_i (\overline{\overline{A_1}} x_1 x_2 \ldots x_n) \ldots (\overline{\overline{A_k}} x_1 x_2 \ldots x_n) x_{n+1} x_{n+2} \ldots \quad (*)$$

The computation of the value of the left part in $(*)$ is reduced, thereby, to the computation of the right part, which may be simpler (if we know, for example, that X_i has some fixed value in D_0).

2. STRATEGIES OF SEQUENTIAL (PARALLEL) COMPUTATION

Strategies of computation to be dealt with are essentially some way of reducing one task to the other ones, which are simpler in some sense. For example, we can naturally associate with every term $A \in \lambda(\emptyset)$ some strategy, which realizes the reduction (*).

We shall consider more general strategies. At first, we shall give an intuitive description. Strategies will form some set M and be interpreted by elements of the space $\overline{D}$ by means of some mapping $\mathcal{F}: M \rightarrow \overline{D}$. The reduction carried out by any startegy from M is analogous to the Turing reduction in the algorithm theory. Namely, the task of computation of the value of $\mathcal{F}(m)X_1X_2... \in \overline{D_0}$ (for $m \in M$ and $X_1, X_2, ... \in D$) will be reduced (according to the strategy m) to the tasks of computation of the values of some expressions of the kind

$$X_i(\mathcal{F}(m_1)X_1X_2...X_n)...(\mathcal{F}(m_k)X_1X_2...X_n)X_{n+1}^{\#}X_{n+2}^{\#}... \in \overline{D_0}, \quad (**)$$

where $m_1,...,m_k \in M$ $(k>0)$, $1 \leq i \leq n$, $\# \in \{0,1\}$ and $X^1 \stackrel{df}{=} X$, $X^0 \stackrel{df}{=} \bot$ The reduction itself will be realized as some undeterministic computation (induced by the strategy m) with an "oracle". The result of computation (in some path) may be some integer $V \in N$. The path of computation leading to the integer result is called the resulting one. In the course of computation some intermediate results may also appear. These may be: a question to the "oracle" about the value of an expression of the kind (**), or the state of undeterminicy "?". The questions to the "oracle" may be codified by strings of the kind $\# i n m_1...m_k$, where $\#, i, n, m_1,...,m_k$ satisfy the conditions stated above. The set of those strings for a given M will be denoted by the symbol [M]. We suppose the answer of the "oracle" to the question $\# in\, m_1 \dots m_k$ to be equal to the integer τ if the value (∗∗) is equal to τ and $\tau \notin \{\perp, \top\}$. If $\tau = \perp$, then the answer is "undefined", and the result of the whole computation is considered as undefined. The value $\tau = \top$ is considered to be a contradictory answer. We say that a mapping $\mathcal{F}: M \to \overline{D}$ and a strategy $m \in M$ are <u>consistent</u>, if for every $x_1, x_2, \dots \in D$ and in every path of computation (induced by the strategy m) all the answers of the "oracle" (defined by $\mathcal{F}$) are consistent (that is $\neq \top$) and all the resulting paths (if any) give the same result $v \in N$.

As the interpretation of the strategies from M in the space $\overline{D}$ we shall always consider the lowest (and consequently only one) mapping $\mathcal{F}: M \to \overline{D}$ for which the following equation is valid for all $m \in M$ and $x_1, x_2, \dots \in D$

$$\mathcal{F}(m)x_1x_2\dots = \begin{cases} v \in N, & \text{if } \mathcal{F} \text{ and } m \text{ are consistent and there exists some resulting computational path giving } v\text{;} \\ \perp, & \text{if } \mathcal{F} \text{ and } m \text{ are consistent and there exist no resulting paths;} \\ \top, & \text{if } \mathcal{F} \text{ and } m \text{ are inconsistent.} \end{cases}$$

Let us pass from the intuitive to a formal description of the strategies. Notice, that the path of computation leading to an intermediate or final result may be defined by a string ("prompting") $w = v_1 v_2 \dots v_s \in N^*$ in which all the integers $v_1, v_2, \dots, v_s$ serve successively as the answers of the "oracle" or for the removal of the undeterministic state "?". Let us denote by $\mathcal{K}(m,w)$ the result (final or intermediate) of the computation induced by the strategy $m \in M$ with the "prompting" w.

<u>Definition</u>. Let M be a set, and $\mathcal{K}: M \times N^* \to N \cup [M] \cup \{?\}$ be a partial function. The pair $(M, \mathcal{K})$ is called a <u>system of</u>

<u>strategies</u>, if for every $w \in N^*$, $v, \ell \in N^*$ and $m \in M$:

($\mathcal{K}(m,w) = v$ or $\mathcal{K}(m,w)$ is undefined) $\Rightarrow$ $\mathcal{K}(m,w\ell)$ is undefined.

A system $(M,\mathcal{K})$ is called the <u>system of sequential</u> or <u>deterministic strategies</u>, if $\mathcal{K}(m,w) \neq ?$ for all m and w. $(M,\mathcal{K})$ is the <u>discrete system of effective strategies</u>, if $M = N$ and $\mathcal{K}$ is an effective function. It is not difficult to show that there exists an <u>interpretation</u> $\mathcal{F}$ (see above) for every system $(M,\mathcal{K})$.

A system $(P,\mathfrak{A})$ is called the <u>universal system of strategies</u>, if for any other system $(M,\mathcal{K})$ there exists the only mapping (homomorphism) $\varphi : M \to P$ for which the following equation with any $m \in M$ and $w \in N^*$ is valid:

$$\mathfrak{A}(\varphi(m),w) = \begin{cases} v, & \text{if } \mathcal{K}(m,w) = v \in N; \\ \#in\,\varphi(m_1)\ldots\varphi(m_k), & \text{if } \mathcal{K}(m,w) = \#in\,m_1\ldots m_k \in [M]; \\ ?, & \text{if } \mathcal{K}(m,w) = ?; \\ \text{is undefined}, & \text{if } \mathcal{K}(m,w) \text{ is undefined.} \end{cases}$$

We shall call <u>effective strategies</u> those strategies from P which lie in homomorphic images of the discrete systems of effective strategies.

<u>Proposition.</u> There exists an universal system of strategies $(P,\mathfrak{A})$ so that P is some class of partial numeral functions from N to N, the effective strategies being just those from P which are effective functions.

We define the subset Q of <u>sequential strategies</u> in P as the union of homomorphic images of all the sequential strategy systems. Let us denote by $\mathcal{V}$ the only interpretation of system $(P,\mathfrak{A})$ in $\overline{D}$. Strategies from $R \overset{df}{=} \mathcal{V}^{-1}(D)$ are naturally called <u>consistent</u>. It is obvious that $Q \subset R$ One can prove that $\mathcal{V}(R) = D$. We shall call the functionals from $\mathcal{V}(Q) \subsetneq D$ - <u>sequentially computable</u>, and all the others - <u>essentially parallel</u>. We shall give the name <u>of effectively sequential functionals</u> for those which are the interpretations of the effective sequential strategies.

3. BASIC RESULTS

Theorem 2. There exists an universal effectively sequential functional $U \in D$ which has the following property: if $f \in D_1$ represents strategy τ (in the sense of Proposition given above), then $Uf = \mathcal{V}(\tau)$. For the typed model [2] (similarily to type-free one) there exists a family of universal functionals $\{U_\alpha\}$, where α runs the set of all types, each U_α being expressible in D.Scott's language LCF [2].

Theorem 3. The sequentially computable (typed) functionals and only they are expressible in D.Scott's language LCF.

Actually, the proof of the facts is based on Theorem 4 having also an independent value.

An operation $A \mapsto \hat{A}$, mapping every term $A \in \lambda(P)$ to some strategy $\hat{A} \in P$ may be naturally defined. Without going into particulars we notice only that this operation is based on the λ-conversion rules [1] and on computations induced by strategies from P. Moreover, if only effective (sequential) strategies occur in a term A, then $\hat{A}$ is also an effective (sequential) strategy. The restriction of the operation on terms $A \in \lambda(\emptyset)$ was essentially described above in Theorem 1.

Theorem 4. For every term $A \in \lambda(R)$ we have

$$\hat{A} \in R \quad \text{and} \quad \mathcal{V}(\hat{A}) = \overline{\overline{A^{\mathcal{V}}}},$$

where $A^{\mathcal{V}} \in \lambda(D)$ denotes the result of substituting occurences of every strategy π in A by $\mathcal{V}(\pi)$. In particular, $\mathcal{V}(\widehat{\pi\tau}) = \mathcal{V}(\pi)\mathcal{V}(\tau)$ for $\pi, \tau \in R$, and $\mathcal{V}(\hat{A}) = \overline{\overline{A}}$ for $A \in \lambda(\emptyset)$.

Also, one can easily obtain in purely sintactical way the results similar to the following ones [4] :

$$\overline{\overline{(\lambda x.\, xx)(\lambda x.\, xx)}} = \bot, \qquad \overline{\overline{(\lambda f.\, (\lambda x.\, f(xx))(\lambda x.\, f(xx)))}} = Y,$$

where $Y \in D$ is the minimal fixed point operator.

The author is debt to prof. B.A.Trachtenbrot for unfailing support and attention to this work.

REFERENCES

1. Hindley, J.R., B.Lercher,J.P.Seldin, Introduction to combinatory logic, London math.soc. Lecture note series 7 (1972).
2. Scott, D., A type-theoretical alternative to CUCH, ISWIM, OWHY, Oxford (1969).
3. Scott, D., Continuous lattices, in Toposes, Algebraic Geometry and Logic, Lecture notes in Math. 274(1972),97-136.
4. Scott, D., Lattice-theoretical models for various type-free calculi, in Proc. of the IVth Int.Congress for logic, mathodology and philosophy of science, Bucharest (1972).
5. Сазонов, В.Ю. О выразимости и вычислимости объектов в языке Д.Скотта LCF, Третья всесоюзная конференция по мат.логике, Тезисы докладов, Новосибирск (1974), 191-194.

COMPUTATION ON ARBITRARY ALGEBRAS

A. Dubinsky
Queen Mary College
University of London.

Introduction: The purpose of this paper is to investigate the connexion between the generalised automata theory of [Thatcher and Wright, 1966], [Mezei and Wright, 1965] and the algebraic theory of computation as developed in [Landin,1969].

Section 1 extends definitions used in Universal Algebra, so that algebras are equipped with multi-valued operations (called here non-determinate).

Section 2 discusses equation systems in the light of [Bekic, 1969].

Section 3 shows how the functions computed by generalised automata on the subsets of the carrier of an arbitrary non-determinate algebra are the image of the subsets computed on the initial algebra under a structure preserving mapping. In the case of finite generalised automata these latter subsets are simply the recognisable subsets and the structure is that provided by Kleene's theorem. A simple characterisation is also provided for the functions computed by infinite automata. The section starts by showing that in the finite case, Mezei and Wright's equational subsets correspond to the computed sets in [Landin,1969].

1. Preliminary definitions.

This section is devoted to a generalisation of the notion of an algebra and a partial algebra, as developed in the study of Universal Algebra (See [Grätzer,1968]) to that of a non-determinate (n.d) algebra, which is a set equipped with multi-valued operations, as in [Landin,1969]. For a given alphabet Ω, n.d. algebras form a category with morphisms generalising the homomorphisms of algebras and the strong homomorphisms of partial algebras. In contradistinction with partial algebras the category of n.d. algebras, like that of algebras, has initial objects, which are essential to a definition of a subset recognised by an automaton.

Partial algebras allows to represent logical tests as partial identities with complementary domain of definitions (e.g. [Landin, 1969]) N.d. algebras arise when combining (partial) algebras in larger entities or - and this amounts essentially to the same thing - when studying equation systems such as those in section 2.

A correspondence f from set A to set B is a subset of the cartesian product $A \times B$. The image $A'f$ of $A' \subseteq A$ under f is the set $\{b \mid (a,b) \varepsilon f \ \& \ a \varepsilon A'\}$

A k-ary non-determinate (n.d) operation $f:A^k \to A$ on set A is a triple (A^k, f, A) where f is a correspondence from A^k to A, and $k \geq 0$. One writes $a \leftarrow f(a_1,\ldots,a_k)$ when $((a_1,\ldots,a_k),a) \varepsilon f$.

If for each k-tuple $(a_1,\ldots,a_k) \varepsilon A^k$, $k \geq 0$, there is at most one $a \varepsilon A$ such that $a \leftarrow f(a_1,\ldots,a_k)$, then $f:A^k \to A$ is a partial operation on A. When there is an $a \varepsilon A$ s.t. $a \leftarrow f(a_1,\ldots,a_k)$, the partial operation $f:A^k \to A$ is said to be defined at $(a_1,\ldots,a_k) \varepsilon A^k$, and one simply writes $a=f(a_1,\ldots,a_k)$. An *operation* on A is then an everywhere defined partial operation.

Let $f:A^k \to A, g_i:A^n \to A, i=1$ to $k, k>0$. Then *$f\cdot(g_1,\ldots,g_k):A^n \to A$* is the n.d. operation (A^n,h,A) where

$$h=\{((a_1,\ldots,a_n),a) \mid x_i \leftarrow g_i(a_1,\ldots,a_n), i=1 \text{ to } k \ \& \ a \leftarrow f(x_1,\ldots,x_k) \text{ for some } x_1,\ldots,x_k \in A\}$$

$f\cdot(g_1,\ldots,g_k)$ is said to be obtained by *composition* from $f,g_1,\ldots,g_k$.

Let $(\Omega_n \mid n \geq 0)$ be a family of sets. Then $\Omega= \quad (\Omega_n \mid n \geq 0)$ is called an *alphabet*.

An n.d. Ω-algebra A_Ω is a pair (A,r) where A is a set, called the *carrier*, and r is a family $(r_f:A^k \to A \mid f \in \Omega_k$ for some $k)$ of n.d. operations on A. It is convenient to write f_A instead of r_f. These n.d. operations are called the *basic* n.d. operations of A_Ω.

The *distinguished elements* of A_Ω are those $a \in A$ such that $a \leftarrow f_A()$ for some $f \in \Omega_0$.

A_Ω is a *partial algebra* if its basic n.d. operations are partial operations. It is an *algebra* if the basic n.d. operations are operations.

N.d. algebra A_Ω is *finite* if the carrier A is finite and if, for each $a \in A$, $a \leftarrow f_A(a_1,\ldots,a_n)$ for some $n \geq 0$ and $a_1,\ldots,a_n \in A$ only for finitely many $f \in \Omega$.

$\hat{A}$, where A is a set, is the set of all subsets of A.
The *subset algebra $\hat{A}_\Omega$* of n.d. algebra A_Ω is the Ω-algebra

with carrier $\hat{A}$ and operations $f_{\hat{A}}$, for $f\epsilon\Omega_k, k\geq 0$ such that, for $A_1,\ldots,A_k \subseteq A$, $f_{\hat{A}}(A_1,\ldots,A_k) = (A_1 x \ldots x A_k)f$.

Let A_Ω, B_Ω be partial algebras. A mapping $h: A\rightarrow B$ is a homomorphism $h: A_\Omega \rightarrow B_\Omega$ if, for $f\epsilon\Omega_k, k\geq 0$ and $a_1,\ldots a_k \epsilon A$, one has $h(f_A(a_1,\ldots,a_k)) = f_B\ (h(a_1),\ldots,h(a_k))$ when $f_A(a_1,\ldots,a_k)$ is defined. Such a homomorphism is a strong homomorphism ([Grätzer, 1968])if it is the case that when $f_B(h(a_1),\ldots,h(a_k))$ is defined then $f_A(a_1,\ldots,a_k)$ is defined. When A_Ω, B_Ω are algebras, these two definitions become the usual definition a homomorphism of algebras.

Let A_Ω, B_Ω be n.d. algebras, and h a correspondence from A to B. Let $Im_h: \hat{A}\rightarrow\hat{B}$ be the mapping such that for $A' \subseteq A$, one has $Im_h(A') = A'h$. If $Im_h: \hat{A}_\Omega \rightarrow \hat{B}_\Omega$ is a homomorphism of algebras from $\hat{A}_\Omega$ to $\hat{B}_\Omega$, then the triple (A_Ω, h, B_Ω) is a morphism of n.d. algebras.

If h is a mapping while A_Ω and B_Ω are partial algebras, or algebras, then h can be shown to be a strong homomorphism, or a homomorphism of algebras, respectively.

N.d. Ω-algebras are the objects of a category Nalg(Ω) with morphisms: the morphisms of n.d. algebras.

N.d. algebra C'_Ω is a subalgebra of n.d. algebra A_Ω if C' is closed under the basic n.d. operations of A_Ω, that is, if for any $n\geq 0$, $f\epsilon\Omega_n$, $a_1,\ldots,a_n \epsilon C'$, one has that $a \leftarrow f_A(a_1,\ldots,a_n)$ implies $a\epsilon C$, $a \leftarrow f_{C'}(a_1,\ldots,a_n)$. The least subalgebra of A_Ω will have carrier $[\emptyset]_{\mathcal{S}(A)}$ = the intersections of the carriers of all the subalgebras of A_Ω

The direct product $A_\Omega \times B_\Omega$ of n.d. algebras A_Ω, B_Ω is the n.d.

Ω-algebra with carrier $A\times B$ and basic n.d. operations $f_{A\times B}: A^k \to A$, $k\geq 0$ such that $(a,b) \leftarrow f_{A\times B}((a_1,b_1),\ldots,(a_k,b_k))$ iff $a \leftarrow f_A(a_1,\ldots,a_k)$ and $b \leftarrow f_B(b_1,\ldots,b_k)$

For $n\geq 0$, the <u>n-ary polynomial symbols over alphabet Ω</u> are defined by:

1°) if $n > 0$, then $e_1,\ldots,e_n$ are n-ary polynomial symbols

if $n = 0$, then $f\in\Omega_0$ is an n-ary polynomial symbol

2°) if $k > 0$ and $f\in\Omega_k$ and $p_1,\ldots,p_k$ are n-ary polynomial symbols, then $f\cdot(p_1,\ldots,p_k)$ is an n-ary polynomial symbol.

<u>$P^{(n)}(\Omega)$</u> then denotes the set of the n-ary polynomial symbols over Ω.

The <u>n-ary polynomials on n.d. algebra B_Ω</u>, for $n\geq 0$, are then the n.d. operations $p_B: B^n \to B$ defined by:

1°) if $p=e_i$, $1\leq i\leq n$ then $p_B=e_i^n$ which is the ith <u>projection</u> on B^n, s.t. $e_i^n\ (b_1,\ldots,b_n) = b_i$

if $p = f$ where $f\in\Omega_0$, then $p_B = f_B$

2°) if $p = f\cdot(p_1,\ldots,p_k)$ where $f\in\Omega_k, k>0$ and $p_1,\ldots,p_k \in P^{(n)}(\Omega)$ then $p_B=f_B\cdot(p_B,\ldots,p_B)$ where $\cdot$ is the composition of n.d. operations.

A <u>polynomial algebra</u> P is an algebra with carrier $P^{(n)}(\Omega)$ for some $n\geq 0$ and where - for $k>0$, $f\in\Omega_k$, one has

$$f_P\ (p_1,\ldots,p_k) = f\cdot(p_1,\ldots,p_k)$$

for any $p_1,\ldots,p_k \in P^{(n)}(\Omega)$

- the distinguished elements, if any, are defined to suit each particular definition.

Let $\mathfrak{P}^{(0)}(\Omega)$ be the polynomial Ω-algebra with carrier $P^{(0)}(\Omega)$ and distinguished elements f, for all $f\varepsilon\Omega_0$. Observe that for any n.d. algebra B_Ω, the mapping $h:\hat{P}^{(0)}(\Omega)\to\hat{B}$ such that for any $S\subseteq P^{(0)}(\Omega)$, $S\mapsto\bigcup(p_{\hat{B}}()|p\varepsilon S)$, is an additive homomorphism of Ω-algebras where by additive one means that $h(S) = \bigcup(h(\{p\})|p\varepsilon S)$ On the other hand, for any homomorphism $h':\hat{\mathfrak{P}}^{(0)}(\Omega)\to\hat{B}_\Omega$ one has, for any $p\varepsilon P^{(0)}(\Omega)$, that $p_{\hat{P}^{(0)}(\Omega)}()\mapsto p_{\hat{B}}()$, while $p_{\hat{P}^{(0)}(\Omega)}()=\{p\}$ Thus, if h' is additive, one has h' = h since h' coincides with h on singletons. Hence, for any n.d. algebra B_Ω, there is exactly one morphism from $\mathfrak{P}^{(0)}(\Omega)$ to B_Ω. This is the definition of an <u>initial object</u> in a category, as defined in [MacLane,1971], and it is immediate that initial objects are unique up to isomorphism.

An <u>initial Ω-algebra</u>, written F_Ω, will be an initial object of <u>Nalg</u>(Ω) (and thus isomorphic to $\mathfrak{P}^{(0)}(\Omega)$). F_Ω is also an initial object in the obvious category of Ω-algebras.

One will assume that one is considering sets in a Universe of Sets.

2. Equation Systems over Lattice Algebras and Regularity.

If L is a complete lattice, and $(A_i|i\varepsilon I)$ a family of elements of L, one will write $\bigsqcup(A_i|i\varepsilon I)$ or $\bigsqcup_{i\varepsilon I} A_i$ for its join. When I is finite, the join will be explicitly said to be a <u>finite join</u>. (It will then be considered as a finitary operation on L). And $\sqsubseteq$ will be the partial order associated to L.

A mapping $f:L_1\to L_2$ between complete lattices will:

-<u>preserve join</u> if, for any family $(A_i|i\varepsilon I)$ of elements of L_1, $f(\bigsqcup_{i\varepsilon I} A_i) = \bigsqcup_{i\varepsilon I} f(A_i)$

-be continuous ([Bekic,1969]) if for any ascending sequence $A_1 \sqsubseteq A_2 \sqsubseteq \ldots$ of elements of L_1, $f(\bigsqcup_{i>0} A_i) = \bigsqcup_{i>0} f(A_i)$

If $L_1 = \hat{B}_1$, $L_2 = \hat{B}_2$, join = set theoretic union, for sets B_1, B_2 then continuity is the notion defined in [Landin,1969] and also the "distributivity over ω-chains" in [Mezei and Wright,1965] And $f:\hat{B}_1 \to \hat{B}_2$ is additive whenever it preserves join.

An algebra L_Ω is a lattice algebra (Wagner) if

1^0) L is a complete lattice

2^0) the basic operations are continuous

An example is the subset algebra $\hat{B}_\Omega$ of n.d. algebra B_Ω: the basic operations are additive in each argument, the polynomials continuous.

If L is a complete lattice, L^k is a complete lattice for $k>0$ with join, partial order defined "componentwise". The join $\bigsqcup_{i\varepsilon I} f_i$ of a family $(f_i : L^k \to L \mid i\varepsilon I)$ of k-ary operations on L for $k>0$ is defined by

$$(\bigsqcup_{i\varepsilon I} f_i)(A) = \bigsqcup_{i\varepsilon I} (f_i(A)), \text{ for } A\varepsilon L^k$$

For $I = \emptyset$, the join is $\lambda(X_1,\ldots,X_k)[\bot]$ where $\bot$ is the least element of L and thus the k-ary operations on L form a complete lattice with the above join.

If $\cdot$ is the composition of operations on L, iteration $*,i$- for $i>0$-is the partial operation on operations of L such that for $f:L^n \to L$ and $n \geq i$, $f^{*,i} = \bigsqcup_{k>0} f^{k,i}$

with $f^{k+1,i} = (f \sqcup e_i^n) \cdot (e_1^n, \ldots, e_{i-1}^n, f^{k,i}, e_{i+1}^n, \ldots, e_n^n)$

and $f^{1,i} = f \sqcup e_i^n$

If L is a complete lattice, a regular partial algebra of operations on L is a set of operations on L closed under composition, finite join and iteration *,i for all i>0. In particular, if L_Ω is a lattice algebra then $R(L_\Omega)$ is defined as the least regular partial algebra of operations on L containing the basic operations of L_Ω with positive arity.

Such notions allow us to characterise the least solutions of certain systems of equations which turn out to describe "computations by finite automata".

A system E of m equations

$$X_i = \bigsqcup_{j \in J_i} f_{j_L}(X_{j_1},\ldots,X_{j_{a_j}}) \sqcup \bigsqcup_{j \in J_i'} X_j, \quad i = 1 \text{ to } m,$$

where J_i is finite, f_{j_L} is a basic operation of lattice algebra L_Ω, $1 \leq j_1,\ldots,j_m \leq m$ and $J_i' \subseteq \{1,2,\ldots,m\}$ and $X_1,\ldots,X_m$ are variables, will be called a linear system of equations over lattice algebra L_Ω

Solutions are thus m-tuples of elements of L. The first component of the least solution of E is the element of L defined by E.

The techniques for solving such systems can be found in [Bekic,1969] namely his Iteration Lemma and Bisection Lemma. This leads to

Theorem 1 If Ω is such that $\Omega_0 = \{f_0,\ldots,f_{n-1}\}$, then the set of elements of L_Ω which are defined by linear systems of equations over L_Ω is precisely $\{F(f_{0_L}(),\ldots,f_{n-1_L}()) \mid F \in R(L_\Omega)\}$

Details of the proof will appear in [Dubinsky,1975].

3. Computations on arbitrary algebras.

An Ω-automaton is a pair (A_Ω, E) where A_Ω is a n.d. algebra and $E \subseteq A$. Following [Landin,1969], the subset of B computed by Ω-automaton (A_Ω, E) on n.d. algebra B_Ω is

$$C_{A,E,B} = \{b \mid (a,b) \varepsilon [\emptyset]_{\mathcal{S}(A\times B)} \text{ for some } a\varepsilon E\}$$

If A_Ω is finite then (A_Ω, E) is a finite automaton and $C_{A,E,B}$ a finitely computed subset.

A possible interpretation is that A_Ω is a "program" for "machine" B_Ω. Let A_Ω, B_Ω be partial algebras and suppose that $\Omega_k = \emptyset$ for $k>1$, and that A_Ω is finite. One then associates to A_Ω a "flowchart", which is a directed graph with labelled edges, by taking as vertices the elements of A and having an edge from a_1 to a_2 labelled f iff $a_2 \leftarrow f_A(a_1)$. The distinguished elements of A_Ω are the "start" vertices of the flowchart, the elements of E its "exit" vertices. On the other hand, the elements of B are then the "machine-states" while the basic partial operations are the "machine-instructions". The distinguished elements of B are then the initial data associated to each "start". Note that if $p\varepsilon P^{(0)}(\Omega)$, p_A can then be regarded as the composite label along a path in the flowchart when it originates at a "start" vertex. This together with the following theorem shows that the definition of $C_{A,E,B}$ reduces to the usual one made in terms of a sequence alternating "machine states" and instructions taking a "state" to the next one.

Note : if A is an Ω-automaton, one writes $C_{A,B}$

Theorem 2

$$C_{A,E,B} = \bigcup (p_{\hat{B}}() \mid p\varepsilon P^{(0)}(\Omega) \ \& \ p_{\hat{A}}() \cap E \neq \emptyset)$$

Proof: One easily establishes that

a) for any n.d. algebra D_Ω,

$$[\emptyset]_{S(D)} = \{d \leftarrow p_D() \mid p \varepsilon P^{(0)}(\Omega)\}$$

b) $(a,b) \leftarrow p_{A\times B}()$ iff $a \leftarrow p_A()$ and $b \leftarrow p_B()$

c) $\{b \mid b \leftarrow p_B()\} = p_{\hat{B}}()$

Then

$$C_{A,E,B} = \{b \mid (a,b) \varepsilon [\emptyset]_{S(A\times B)} \ \& \ a \varepsilon E\}$$

$$= \{b \leftarrow p_B() \mid p \varepsilon P^{(0)}(\Omega) \ \& \ a \leftarrow p_A() \ \& \ a \varepsilon E\} \text{ by a),b)}$$

$$= \bigcup (p_{\hat{B}}() \mid p \varepsilon P^{(0)}(\Omega) \ \& \ p_{\hat{A}}() \cap E \neq \emptyset) \text{ by c)}$$

Corollary 3: If F_Ω is an initial Ω algebra, (A_Ω, E) an Ω-automaton and B_Ω a n.d. algebra, $C_{A,E,B} = \bigcup (p_{\hat{B}}() \mid p_F() \varepsilon C_{A,E,F})$

Proof: Since F_Ω is isomorphic to a polynomial algebra,

$p_F() = p'_F()$ iff $p = p'$, for $p,p' \varepsilon P^{(0)}(\Omega)$,

hence $p_{\hat{F}}() = p'_{\hat{F}}()$ iff $p = p'$.

But then, since $C_{A,E,F} = \bigcup (p_{\hat{F}}() \mid p \varepsilon P^{(0)}(\Omega) \ \& \ p_{\hat{A}}() \cap E \neq \emptyset)$ one has $p_F() \ \varepsilon \ C_{A,E,F}$ iff $p_{\hat{A}}() \cap E \neq \emptyset$, while $p_{\hat{F}}() = \{p_F()\}$.

Theorem 4:

If $h: B_\Omega \to B'_\Omega$ is a morphism of n.d. algebras, $C_{A,E,B'} = Im_h(C_{A,E,B})$

Proof: immediate, by corollary 3 and the additivity of Im_h

Proposition 5 If (A_Ω,E) is a n.d. automaton and h the unique morphism from F_Ω to A_Ω, then $C_{A,E,F} = E\ h^{-1}$

Proof: Im_h is the unique additive homomorphism such that, for $p \varepsilon P^{(0)}(\Omega)$, $p_F^{\wedge}() \mapsto p_A^{\wedge}()$. But $p_F^{\wedge}() = \{p_F()\}$ since F is an algebra while $p_A^{\wedge}() = \{a|a \leftarrow p_A()\}$. Thus h is the correspondence from F to A such that $(p_F(),a) \varepsilon h$ iff $a \leftarrow p_A()$

Hence

$$Eh^{-1} = \{p_F()|a \leftarrow P_A() \text{ for some } a \varepsilon E\}$$
$$= C_{A,E,F}, \text{ by proof of theorem 2}$$

Corollary 6 $C_{F,E,F} = E$

Corollary 7 If B_Ω is a n.d. algebra and $h_B:F_\Omega \to B_\Omega$ the unique morphism from F_Ω to B_Ω then $C_{A,E,B} = (E\ h^{-1})\ h_B$ where $h:F_\Omega \to A_\Omega$ is the unique morphism from F_Ω to n.d. algebra A_Ω

When A_Ω is finite, $C_{A,E,F}$ will be called the subset of F recognised by (A_Ω,E). By Broposition 5, when Ω_0 is also finite, this definition coincides with that of a recognisable set in [Mezei and Wright, 1965], [Thatcher and Wright, 1966].

In fact, Theorem 4 leads to the main result (their Theorem 5.5) in [Mezei and Wright, 1965], if one takes into account the following result of [Landin, 1969]. Let A_Ω, B_Ω be n.d. algebras and let EQU $(A_\Omega \times B_\Omega)$ be the system of equations:

$$X_a = \bigcup (f_B^{\wedge}(X_{a_1},\dots,X_{a_n})|f \varepsilon \Omega_n,\ n \geq 0\ \&$$
$$a_1,\dots,a_n \varepsilon A\ \&\ a \leftarrow f_A(a_1,\dots,a_n)), a \varepsilon A$$

Then, if $(\sigma_a | a \varepsilon A)$ is the least solution of $EQU(A_\Omega \times B_\Omega)$, one has:

$$\sigma_a = \{a\}\ C' \text{ (image under correspondence } C'\text{)},$$

where $C' \subseteq A \times B$ is the carrier of the least subalgebra of $A_\Omega \times B_\Omega$

One then extends the definition of the equational subsets of B in [Mezei & Wright 1965] to n.d. algebras and one shows that they are the subsets of B defined by linear systems of equations over $\hat{B}_\Omega$. Then one establishes that such subsets are finite unions of components of the least solution of systems $EQU(A_\Omega \times B_\Omega)$ for all finite n.d. algebras A_Ω (further details will appear in [Dubinsky, 1975]). (For algebras, this is Theorem 1 in [Eilenberg and Wright, 1967]).

This discussion of Mezei and Wright's result can be summarised by:

Theorem 8 The finitely computed subsets of the carrier n.d. algebra B_Ω are precisely the equational subsets of B. And these, in turn, are precisely the images, under the unique morphism $h: F_\Omega \to B_\Omega$, of the subsets of B recognised by finite automata.

Remember that Ω_0 is the set of 0-ary operations.

By the result on equation systems and the proof of the above theorem one has:

Corollary 9 If $\Omega_0 = \{f_0, \ldots, f_{n-1}\}$, the finitely computed subsets of the carrier of n.d. algebra B_Ω are the elements of $\{F(f_{0_{\hat{B}}}(), \ldots, f_{n-1_{\hat{B}}}()) | F \varepsilon R(\hat{B}_\Omega)\}$

One assumes, in what follows, Ω to be such that $\Omega_0 = \{f_i | i \geq 0\}$.

For $n \geq 0$, let $\Omega(n)$ be the alphabet such that: $\Omega(n)_0 = \{f_0, \ldots, f_{n-1}\}$ while for $k>0, \Omega(n)_k = \Omega_k$

If B_Ω is a n.d. Ω-algebra, and $B_0, \ldots, B_{n-1} \subseteq B$ with $n \geq 0$, then let $B_\Omega(B_0, \ldots, B_{n-1})$ be the n.d. $\Omega(n)$-algebra with carrier B and 0-ary operations: $f_{i_{B(B_0, \ldots, B_{n-1})}}$ such that

$$f_{i_{\hat{B}(B_0, \ldots, B_{n-1})}} = B_i, \ i = 0 \text{ to } n-1,$$

while the remaining basic n.d. operations are those of B_Ω.

The n-ary operation on $\hat{B}$ computed by $\Omega(n)$-automaton A on n.d. algebra B_Ω is, for $n>0$, $f_{A,B}: \hat{B}^n \to \hat{B}$ such that:

for $B_0, \ldots, B_{n-1} \subseteq B$, $f_{A,B}(B_0, \ldots, B_{n-1}) = C_{A,B(B_0, \ldots, B_{n-1})}$

By Corollary 9, one immediately has (generalised Kleene's Theorem):

Theorem 10: Let $n>0$. The operations $f: \hat{B}^n \to \hat{B}$ computed by finite $\Omega(n)$-automata on n.d. algebra B_Ω are the n-ary operations in $R(\hat{B}_\Omega)$, the least regular partial algebra containing the non-0-ary basic operations of $\hat{B}_\Omega$.

For $n>0$, let $P^{(n)}(\Omega)_{\Omega(n)}$ be the polynomial algebra with carrier $P^{(n)}(\Omega)$ (the set of n-ary polynomial symbols) and distinguished elements $f_{i-1_{P^{(n)}(\Omega)}}() = e_i$ for $i = 1$ to n.

It is easy to verify that $P^{(n)}(\Omega)_{\Omega(n)}$ is an initial $\Omega(n)$-algebra.

For any $q \varepsilon P^{(0)}(\Omega(n))$ one has:

$q_{P^{(n)}(\Omega)}() = p$ for some $p \varepsilon P^{(n)}(\Omega)$

(obtained by replacing f_{i-1} by e_i in q, for $i = 1$ to n).

Hence, by Corollary 3 and the definition of $f_{A,B}$ one has the

Proposition 11:

For any $\Omega(n)$-automaton A, with $n>0$ and any n.d algebra B_Ω,

$f_{A,B} = \bigsqcup (p_{\hat{B}} \mid p \varepsilon C_{A,P^{(n)}(\Omega)})$

For n.d. algebra B_Ω and for $n>0$, let $\underline{[\hat{B}^n{}_\Omega \to \hat{B}_\Omega]} = \{f_{A,B} : \hat{B}^n{}_\Omega \to \hat{B}_\Omega \mid$ A is an $\Omega(n)$-automaton$\}$. It is thus the set of n-ary operations on $\hat{B}$ computed by $\Omega(n)$ automata (not necessarily finite) on B_Ω.

By Corollary 6, for any $S \subseteq P^{(n)}(\Omega)$, $n>0$,

$C_{P^{(n)}(\Omega),S,P^{(n)}(\Omega)} = S$

$((P^{(n)}(\Omega),S)$ is indeed an $\Omega(n)$-automaton)

Hence the above proposition leads to the

Theorem 12: For $n>0$ and any n.d algebra B_Ω,

$[\hat{B}^n{}_\Omega \to \hat{B}_\Omega] = \{ \bigsqcup_{p \varepsilon S} p_{\hat{B}} \mid S \subseteq P^{(n)}(\Omega)\}$

Proposition 13: for $n>0$ and B_Ω and algebra, $[\hat{B}^n{}_\Omega \to \hat{B}_\Omega]$ is a complete lattice with join: the join of operations from $\hat{B}^n{}_\Omega$ to $\hat{B}_\Omega$, and least element $\lambda(X_1,\ldots,X_n)[\emptyset]$, the n-ary operation with constant value $\emptyset$.

Proof: 1) If I is a set and $(S_i \mid i \varepsilon I)$ is a family of subsets of $P^{(n)}(\Omega)$ then $\bigsqcup_{i \varepsilon I} \bigsqcup_{p \varepsilon S_i} p_{\hat{B}} = \bigsqcup_{p \varepsilon \bigcup_{i \varepsilon I} S_i} p_{\hat{B}} \quad \varepsilon [\hat{B}^n_\Omega \to \hat{B}_\Omega]$

2) If $I = \emptyset$ then $\bigsqcup_{i\varepsilon\emptyset} (\bigsqcup_{p\varepsilon S_i} P_{\hat{B}}^{\wedge})$ is the join of an empty family of n-ary operations on $\hat{B}$, and thus, by the definition of the complete lattice of n-ary operations on $\hat{B}$, it is the constant operation $\lambda(X_1,\ldots,X_n)\ [\emptyset]$.

To study composition of computed operations, one will now consider $\bigcup_{n>0} [\hat{B}^n \to B_\Omega]$. The structure of that set will be obtained from that of $\Sigma_{n>0} P^{(n)}(\Omega)$ where Σ is the disjoint union of sets.

First, one has to define a partial operation on $\Sigma_{n>0} \hat{P}^{(n)}$ corresponding to the composition of operations.

Let, for $k>0$, the $\Omega(0)$ - algebra $\underline{P^{(k)}(\Omega)_{\Omega(0)}}$ be the polynomial algebra with carrier $P^{(k)}(\Omega)$ and with no distinguished elements. For any n.d. algebra $B_{\Omega(0)}$ and $X_1,\ldots,X_k \subseteq B$ there is an additive homomorphism $h:\hat{P}^{(k)}(\Omega)_{\Omega(0)}(\Omega) \to \hat{B}_{\Omega(0)}$ of $\Omega(0)$-algebras which is uniquely defined by:

$$\{e_i\} \mapsto X_i, \text{ for } i = 1 \text{ to } k.$$

<u>Convention:</u> In the rest of this paper, one will find convenient to identify $\hat{P}^{(n)}(\Omega)$ with its isomorphic copy in the disjoint union $\Sigma_{n>0} \hat{P}^{(n)}(\Omega)$. Thus the elements of the copy will also be considered as being sets of n-ary polynomial symbols.

Let the <u>complex product •</u> be the partial operation on $\Sigma_{n>0} \hat{P}^{(n)}(\Omega)$ such that

$S \cdot (X_1,\ldots,X_k)$ is defined for

$S \subseteq P^{(k)}(\Omega)$ and $X_1,\ldots,X_k \subseteq P^{(n)}(\Omega)$ when $k,n>0$, and

$S \cdot (X_1,\ldots,X_k) = h\ (S)$

where $h:\ \hat{P}^{(k)}(\Omega)_{\Omega(0)} \to \hat{P}^{(n)}(\Omega)_{\Omega(0)}$ is the additive homomorphism such that $\{e_i\} \mapsto X_i$, i = 1 to k.

$\sum_{n>0} P^{(n)}(\Omega)$ is obviously closed under $\cdot$

Proposition 14 for k,n>0 and $p \in P^{(k)}(\Omega)$ and $X_1, \ldots, X_k \subseteq P^{(n)}(\Omega)$ one has:

$$\{p\} \cdot (X_1, \ldots, X_k) = p_{\hat{P}^{(n)}(\Omega)}(X_1, \ldots, X_k)$$

Proof: By the definition of $\cdot$, and because

$$\{p\} = p_{\hat{P}^{(k)}(\Omega)}(\{e_1\}, \ldots, \{e_k\}).$$

by the definition of $P^{(k)}(\Omega)_{\Omega(0)}$, hence of $\hat{P}^{(k)}(\Omega)_{\Omega(0)}$

Corollary 15 for k,n>0 and $S \subseteq P^{(k)}(\Omega)$ and $X_1, \ldots, X_k \subseteq P^{(n)}(\Omega)$ one has:

$$\bigsqcup_{p \in S} p_{\hat{P}^{(n)}(\Omega)}(X_1, \ldots, X_k) = S \cdot (X_1, \ldots, X_k)$$

Proof: $\bigcup_{p \in S} p_{\hat{P}^{(n)}(\Omega)}(X_1, \ldots, X_k) = \bigcup_{p \in S} \{p\} \cdot (X_1, \ldots, X_k)$

$= \bigcup_{p \in S} h(\{p\})$ by above definition of h and $\cdot$

$= h(\bigcup_{p \in S} \{p\})$ by additivity of h

$= h(S) = S \cdot (X_1, \ldots, X_k)$

Lemma 16 for k,m,n>0 and for $S_1, \ldots, S_k \subseteq P^{(n)}(\Omega)$ and $S \subseteq P^{(k)}(\Omega)$ one has:

$$\left(\bigsqcup_{p \in S} p_{\hat{P}^{(m)}(\Omega)}\right) \cdot \left(\bigsqcup_{p_1 \in S_1} p_{1\,\hat{P}^{(m)}(\Omega)}, \ldots, \bigsqcup_{p_k \in S_k} p_{k\,\hat{P}^{(m)}(\Omega)}\right)$$

$$= \bigsqcup_{p' \in S \cdot (S_1, \ldots, S_k)} p'_{\hat{P}^{(m)}(\Omega)}$$

Proof: 1) Let $X_1,\dots,X_n \subseteq P^{(m)}(\Omega)$

Then, for $p\varepsilon P^{(k)}(\Omega)$,

$$p_{\hat{P}^{(m)}(\Omega)} \cdot (\bigsqcup_{p_1\varepsilon S_1} p_{\hat{P}^{(m)}(\Omega)},\dots,\bigsqcup_{p_k\varepsilon S_k} p_{k\hat{P}^{(m)}(\Omega)})(X_1,\dots,X_n)$$

$$= p_{\hat{P}^{(m)}(\Omega)}(S_1\cdot(X_1,\dots,X_n),\dots,S_k\cdot(X_1,\dots,X_n))$$

by Corollary 15 above

$= p_{\hat{P}^{(m)}(\Omega)}(h(S_1),\dots,h(S_k))$ where $h:\hat{P}^{(n)}(\Omega)_{\Omega(0)} \rightarrow \hat{P}^{(m)}(\Omega)_{\Omega(0)}$ is

the unique additive homomorphism s.t. $\{e_i\} \mapsto X_i$, i=1 to n

$= h(p_{\hat{P}^{(n)}(\Omega)}(S_1,\dots,S_k))$ since h is a homomorphism

$= h(\{p\}\cdot(S_1,\dots,S_k))$, by above proposition 14

$= (\{p\}\cdot(S_1,\dots,S_k))\cdot(X_1,\dots,X_n)$ by definition of h and $\cdot$

$= \bigcup_{p'\varepsilon\{p\}(S_1,\dots,S_k)} p'_{\hat{P}^{(m)}}(X_1,\dots,X_n)$ by above Corollary 15 again

Thus $p_{\hat{P}^{(m)}(\Omega)} \cdot (\bigsqcup_{p_1\varepsilon S_1} p_{1\hat{P}^{(m)}(\Omega)},\dots,\bigsqcup_{p_k\varepsilon S_k} p_{k\hat{P}^{(m)}(\Omega)})$

$$= \bigsqcup_{p'\varepsilon\{p\}\cdot(S_1,\dots,S_k)} p'_{\hat{P}^{(m)}(\Omega)}$$

2) But then

$$(\bigsqcup_{p\varepsilon S} p_{\hat{P}^{(m)}(\Omega)})\cdot(\bigsqcup_{p_1\varepsilon S_1} p_{1\hat{P}^{(m)}(\Omega)},\dots,\bigsqcup_{p_k\varepsilon S_k} p_{k\hat{P}^{(m)}(\Omega)})$$

$$= \bigsqcup_{p\varepsilon S}(p_{\hat{P}^{(m)}(\Omega)}\cdot(\bigsqcup_{p_1\varepsilon S_1} p_{1\hat{P}^{(m)}(\Omega)},\dots,\bigsqcup_{p_k\varepsilon S_k} p_{k\hat{P}^{(m)}(\Omega)}))$$

by the definition of join and composition of operations.

$$= \bigsqcup_{p\varepsilon S} \left(\bigsqcup_{p'\varepsilon\{p\}\cdot(S_1,\ldots,S_k)} p'_{\hat{P}^{(m)}(\Omega)} \right) \text{ by 1)}$$

$$= \bigsqcup_{p'\varepsilon \bigcup_{p\varepsilon S}\{p\}\cdot(S_1,\ldots,S_k)} p'_{\hat{P}^{(m)}(\Omega)}$$

since $\bigsqcup_{i\varepsilon I} \bigsqcup_{b\varepsilon\Phi_i} b = \bigsqcup_{b\varepsilon \bigcup_{i\varepsilon I} \Phi_i} b$, for any complete lattice L and any family $(\Phi_i | i\varepsilon I)$ of subsets of L

$$= \bigsqcup_{p'\varepsilon S\cdot(S_1,\ldots,S_k)} p'_{\hat{P}^{(m)}(\Omega)} \text{ by the proof of Corollary 15.}$$

For any n.d. algebra B_Ω, the <u>semantic mapping</u> is the mapping

<u>$m_B: \sum_{n>0} \hat{P}^{(n)}(\Omega) \to \bigcup_{n>0} [\hat{B}^n_\Omega \to \hat{B}_\Omega]$</u>

such that $S \mapsto \bigsqcup_{p\varepsilon S} p_{\hat{B}}$ for any $S \subseteq P^{(n)}(\Omega)$ and any n>0

<u>Theorem 17</u> $f_{A,B} = m_B(C_{A,P^{(n)}(\Omega)})$ for any $\Omega(n)$-automaton A and n>0. The semantic mappings are surjections. In particular, m_F is a bijection.

<u>Proof</u>: $f_{A,B} = m_B(C_{A,P^{(n)}(\Omega)})$ since

$f_{A,B} = \bigsqcup(p_{\hat{B}} | p\varepsilon C_{A,P^{(n)}(\Omega)})$

Since $[\hat{B}^n_\Omega \to \hat{B}_\Omega] = \{\bigsqcup_{p\varepsilon S} p_{\hat{B}} | S \subseteq P^{(n)}(\Omega)\}$,

m_B is onto.

For any n>0, there is an injection from $\hat{P}^{(n)}(\Omega)$ into $\hat{F}$

Hence, for $p,p'\varepsilon P^{(n)}(\Omega)$, $p_{\hat{F}} = p'_{\hat{F}}$ entails $p_{\hat{P}^{(n)}(\Omega)} = p'_{\hat{P}^{(n)}(\Omega)}$ which entails $p = p'$. Thus $m_F(S_1) = m_F(S_2)$ entails $S_1 = S_2$

For $i>0$, the complex iteration $*,i$ is the unary partial operation on $\sum_{n>0} \hat{P}^{(n)}(\Omega)$ defined for all $S \subseteq P^{(n)}(\Omega)$ with $n \geq i$ by:

$$S^{*,i} = \bigcup_{k>0} S^{k,i} \text{ where } S^{k+1,i} = (S \cup \{e_i\})(\{e_1\},\ldots,\{e_{i-1}\},S^{k,i},\{e_{i+1}\},\ldots,\{e_n\})$$

and $S^{1,i} = S \cup \{e_i\}$. Obviously $\sum_{n>0} \hat{P}^{(n)}(\Omega)$ is closed under $*,i$

Theorem 18 (Main Result) For any n.d. algebra B_Ω, $\bigcup_{n>0} [\hat{B}^n{}_\Omega \to \hat{B}_\Omega]$ is the least set of non 0-ary operations on $\hat{B}$, containing - the projections e_i^n, for $1 \leq i \leq n$

- the basic operations of positive arity of $\hat{B}_\Omega$

which is closed under - composition

- join of n-ary operations, for $n>0$

The onto mapping $m_B: \sum_{n>0} \hat{P}^{(n)}(\Omega) \to \bigcup_{n>0} [\hat{B}^n{}_\Omega \to \hat{B}_\Omega]$ preserves join. As far as finitary partial operations are concerned, it is a strong homomorphism, preserving $\cdot$, finite $\bigsqcup$ and the iterations.

In particular, m_F is an isomorphism.

Proof: 1) for $n>0$ and $B_0,\ldots,B_{n-1} \subseteq B$ there is a unique additive homomorphism of $\Omega(0)$-algebras, $h: \hat{P}^{(n)}(\Omega)_{\Omega(0)} \to \hat{B}_\Omega()$ such that $\{e_i\} \mapsto B_{i-1}$, $i=1$ to n. Hence

$$\bigsqcup_{p' \in S \cdot (S_1,\ldots,S_k)} p'_{\hat{B}}(B_0,\ldots,B_{n-1}) =$$

$$= h\left(\bigsqcup_{p' \in S \cdot (S_1,\ldots,S_k)} p'_{\hat{P}^{(n)}(\Omega)}(\{e_1\},\ldots,\{e_n\})\right)$$

since h is an additive homomorphism

$$= h\left(\left(\bigsqcup_{p \in S} p_{\hat{P}^{(n)}(\Omega)}\right) \cdot \left(\bigsqcup_{p_1 \in S_1} p_{1\hat{P}^{(n)}(\Omega)},\ldots,\bigsqcup_{p_k \in S_k} p_{k\hat{P}^{(n)}(\Omega)}\right)(\{e_1\},\ldots,\{e_n\})\right)$$

by Lemma 16.

$$= (\bigsqcup_{p\in S} p_{\hat{B}})\cdot(\bigsqcup_{p_1\in S_1} p_{1\hat{B}},\ldots,\bigsqcup_{p_k\in S_k} p_{k\hat{B}})(B_0,\ldots,B_{n-1})$$

since h is an additive homomorphism.

Thus $\bigsqcup_{p'\in S\cdot(S_1,\ldots,S_k)} p'_{\hat{B}} = (\bigsqcup_{p\in S} p_{\hat{B}})\cdot(\bigsqcup_{p_1\in S_1} p_{1\hat{B}},\ldots,\bigsqcup_{p_k\in S_k} p_{k\hat{B}})$,

which entails that m_B preserves $\cdot$ and that $\bigcup_{n>0} [\hat{B}^n{}_\Omega \to \hat{B}_\Omega]$ is closed under composition.

2) Since $\bigsqcup_{i\in I} \bigsqcup_{p\in S_i} p_{\hat{B}} = \bigsqcup_{p\in \bigcup_{i\in I} S_i} p_{\hat{B}}$, one has

$\bigsqcup_{i\in I} m_B\ (S_i) = m_B\ (\bigcup_{i\in I} S_i)$

Thus m_B when restricted to $\hat{P}^{(n)}(\Omega)$, with $n>0$, preserves join.

3) Using an inductive argument, one then proves that $\bigcup_{n>0} [\hat{B}^n{}_\Omega \to \hat{B}_\Omega]$ is closed under the iterations, and that m_B preserves the iterations. It follows easily that m_B is a strong homomorphism

4) Since every element of the set $\bigcup_{n>0} [\hat{B}^n{}_\Omega \to \hat{B}_\Omega]$ is of the form $\bigsqcup_{p\in S} p_{\hat{B}}$ for some $S \subseteq P^{(n)}(\Omega)$, $n>0$ one has the required characterisation of that set.

Suppose there is a mapping from $\{f_B | f\in\Omega_k, k>0\}$ to $\{f_{B'} | f\in\Omega_k, k>0\}$ such that $f_B \mapsto f_{B'}$, for $f\in\Omega_k, k>0$. Then- and only then - there is a mapping $m_{B,B'} = \bigcup_{n>0} [\hat{B}^n{}_\Omega \to \hat{B}_\Omega] \to \bigcup_{n>0} [\hat{B}'^n{}_\Omega \to \hat{B}'_\Omega]$ such that, for any $S \subseteq P^{(n)}(\Omega)$, $n>0$, $\bigsqcup_{p\in S} p_{\hat{B}} \to \bigsqcup_{p\in S} p_{\hat{B}'}$. This mapping is obviously onto.

Note that since F_Ω is an initial Ω-algebra, $m_{F,B}$ exists for any n.d. algebra B_Ω.

<u>Proposition 19</u> When it exists, $m_{B,B'}$ is a strong homomorphism of regular partial algebras which, for $n>0$, preserves the join of n-ary operations.

Proof: 1) $m_{B,B'}(\bigsqcup_{p\varepsilon S_0} p_{\hat{B}}.(\bigsqcup_{p_1\varepsilon S_1} p_{1_B},\ldots,\bigsqcup_{p_k\varepsilon S_k} p_{k_{\hat{B}}})) =$

$$= m_{B,B'}(\bigsqcup_{p\varepsilon S_0\cdot(S_1,\ldots,S_k)} p_{\hat{B}}) \text{ since } m_B \text{ preserves } \cdot$$

$$= \bigsqcup_{p\varepsilon S_0\cdot(S_1,\ldots,S_k)} p_{\hat{B}'}$$

$$= \bigsqcup_{p\varepsilon S_0} p_{\hat{B}'}\cdot(\bigsqcup_{p_1\varepsilon S_1} p_{1_{\hat{B}'}},\ldots,\bigsqcup_{p_k\varepsilon S_k} p_{k_{\hat{B}'}}) \text{ since } m_B \text{ preserves } \cdot$$

2) A similar argument shows that the join of n-ary operations for n>0, is preserved. The rest follows easily.

Proposition 11 entails, since join of n-ary operations is preserved for all n>0,

Proposition 20 For any $\Omega(n)$-automaton A and n>0, when $m_{B,B'}$ exists, one has: $m_{B,B'}(f_{A,B}) = f_{A,B'}$

Let $S(\Omega)$ be the least subset of $\Sigma\hat{P}^{(n)}(\Omega)$ which contains the polynomial symbols $f(e_1,\ldots,e_k)$ for all $f\varepsilon\Omega_k, k>0$,
-and which is closed under $\cdot$, and $*,i$ for all i>0 and finite (including 0-ary) union of subsets of $P^{(n)}(\Omega)$, for all n > 0.

Then $R(\hat{B}_\Omega) = S(\Omega)\, m_B$ for any n.d algebra B_Ω, while for any n.d. algebra B'_Ω one has $R(\hat{B}'_\Omega) = R(\hat{B}_\Omega)\, m_{B,B'}$ whenever there is a mapping from $\{f_B | f\varepsilon\Omega_k, k>0\}$ to $\{f_{B'} | f\varepsilon\Omega_k, k>0\}$ s.t. $f_B \mapsto f_{B'}$, for all $f\varepsilon\Omega_k, k>0$.

Theorem 21: Let B_Ω B'_Ω be n.d. algebras. Let $s_B : S(\Omega)\to R(\hat{B}_\Omega)$ be the mapping such that for $S\varepsilon S(\Omega)$, $S \to \bigsqcup_{p\varepsilon S} p_{\hat{B}}$.
Whenever there is a mapping from $\{f_B | f\varepsilon\Omega_k, k>0\}$ to $\{f_{B'} | f\varepsilon\Omega_k, k>0\}$

such that $f_B \mapsto f_{B'}$ for all $f \in \Omega_k, k>0$, let $s_{B,B'}: R(\hat{B}_\Omega) \to R(\hat{B}'_\Omega)$ be the mapping such that $\bigsqcup_{p\in S} p_{\hat{B}} \mapsto \bigsqcup_{p\in S} p_{\hat{B}'}$ for all $S \in S(\Omega)$

Then

1) $s_B: S(\Omega) \to R(\hat{B}_\Omega)$ is a strong onto homomorphism preserving $\cdot$, finite join and the iterations and it preserves join. In particular, $s_F: S(\Omega) \to R(F_\Omega)$ is an isomorphism.

2) If, for $\Omega(n)$-automaton A, with $n>0$, one has $C_{A,P^{(n)}} \in S(\Omega)$, then $s_B(C_{A,P^{(n)}(\Omega)}) = f_{A,B}$.

3) When it is defined, the mapping $s_{B,B'}: R(\hat{B}_\Omega) \to R(\hat{B}_\Omega')$ is a strong onto homomorphism of regular partial algebras which also preserves join. In particular, $s_{F,B}$ is defined for any n.d. algebra B_Ω.

4) If, for $n > 0$ and $\Omega(n)$-automaton A, one has $f_{A,B} \in R(\hat{B}_\Omega)$, then, if $S_{B,B'}$ is defined, one has $s_{B,B'}(f_{A,B}) = f_{A,B'}$

Proof: s_B and $s_{B,B'}$ are restrictions of $m_{B'}, m_{B,B'}$.

Since $R(\hat{B}_\Omega) = S(\Omega)\, m_B$, while $f_{A,B} = m_B(C_{A,P^{(n)}(\Omega)})$ one has that $f_{A,B} \in R(\hat{B}_\Omega)$ iff $C_{A,P^{(n)}(\Omega)} \in S(\Omega)$

A consequence of this, and of theorem 1 is then (this is Kleene's theorem):

The elements of $S(\Omega)$ are precisely the sets of n-ary polynomial symbols recognised by finite $\Omega(n)$-automata, for all $n>0$.

Acknowledgements: The research reported in this paper was partly done while working for IBM U.K. Laboratories under their Advanced Education program, partly while working under an S.R.C. grant. Further details will appear in a doctoral thesis to be submitted at the University of London.

I have been greatly helped by constant encouragement from Professor P.J. Landin and Dr. E.G. Wagner. I have also to thank Dr.J.W.Thatcher for his comments when I started the work reported here.

References:

Bekic, H., Definable operations in general Algebras and the theory of Automata and Flowcharts, Unpublished (1969).

Dubinsky, A., Ph.D thesis to be submitted at the University of London (1975).

Eilenberg, S., Wright, J.B., Automata in General Algebras, Information and Control, 11, 452-470 (1967).

Grätzer, G., Universal Algebra, Princeton, Van Nostrand and Co.Inc (1968).

Landin, P.J., A Program Machine Symmetric Automata Theory, in: Machine Intelligence 5, Ed. Meltzer and Michie, Edinburgh University Press, (1969, 99-120.

MacLane, S., Categories for the Working Mathematician, New York, Heidelberg, Berlin, Springer Verlag. (1971).

Mezei, J., Wright, J.B., Generalised Algol Like Languages, IBM Research Report RC.1528 (1965).

Thatcher, J.W., Wright, J.B., Generalized finite automata theory, with an application to a decision problem of second order Logic, IBM Research Report RC1713 (1966).

ON SOLVABILITY BY λI - TERMS

J.W. KLOP

Mathematical Institute, Budapestlaan-Utrecht-Holland.

0. Introduction. In [1], (2.1) the following theorem is proved: if M is a closed λI-term, then: M has a normal form ⇔ ∃λI-term F s.t. FM = I ⇔ M is I-solvable, where 'M is I-solvable' means: ∃λI-terms $N_1,\ldots,N_n$ s.t. $MN_1\ldots N_n = I$.
The proof of the main lemma (2.11) in [1] uses induction on a Σ_4^0-statement and is technically cumbersome; therefore H. Barendregt suggested to look for a simpler one. The result is the proof below, which uses the same strategy as that of Barendregt in [1]. It is not much shorter, but easier, we think.
Further, an idea in one of the corollaries in [1] is used as a hint in the construction of a generator for the λI-calculus.

1. DEFINITION. SO is the set of closed λI-terms defined by:
(i) $I \in SO$
(ii) $\vec{\alpha}_0,\ldots,\vec{\alpha}_n \in SO \Rightarrow \lambda x_0\ldots x_n.\ (x_0\vec{\alpha}_0)\ldots(x_n\vec{\alpha}_n) \in SO$ where $\vec{\alpha}_i \in SO$ is short for $\alpha_{i_0},\ldots,\alpha_{ik_i} \in SO$.
Other notations:
(1) α,β,... are variables ranging over SO,
(2) if $\vec{N}$ is the tuple $< N_0,\ldots,N_k >$, then $M\vec{N} = MN_0\ldots N_k$,
(3) $M \in' SO$ means: $M \in SO$ modulo convertibility, i.e. $\lambda I \vdash M = \alpha$ for some $\alpha \in SO$,
(4) $MN^{\sim k} = M\underbrace{N\ldots N}_{k\text{ times}}$

2. LEMMA. (i) $\alpha,\beta \in SO \Rightarrow \alpha\beta \in' SO$
(ii) $\alpha \in SO \Rightarrow \alpha I^{\sim m} = I$ for some m.
Proof: (i).
Define $(SO)_0 = \{I\}$
$$(SO)_{n+1} = (SO)_n \cup \{\lambda x_0\ldots x_m\cdot(x_0\vec{\alpha}_0)\ldots(x_m\vec{\alpha}_m)\,|\,m\in\omega, \vec{\alpha}_i \in (SO)_n\}.$$
Induction on n and simple calculations, which we omit, prove: for all n, $(SO)_n$ is closed under application, modulo convertibility. $SO = \bigcup\{(SO)_n | n\in\omega\}$, hence (i).
(ii) holds for $(SO)_0$. Let $\alpha \in (SO)_{n+1}, \alpha = \lambda x_0\ldots x_m\cdot(x_0\vec{\alpha}_0)\ldots(x_m\vec{\alpha}_m)$ where $\vec{\alpha}_i \in (SO)_n, i \leq m$. Then $\alpha I^{\sim m+1} \in' (SO)_n$, and by induction

hypothesis $\alpha I^{\sim m+1}\ I^{\sim k} = I$ for some k. Hence (ii).

3. Notation. (i) $\lambda x_0 \ldots x_n . (x_0 \vec{\alpha}_0) \ldots (x_n \vec{\alpha}_n) = < \vec{\alpha}_0, \ldots, \vec{\alpha}_n >$
(ii) if $\alpha = < \vec{\alpha}_0, \ldots, \vec{\alpha}_n >$ then $l(\alpha) = n$ and $(\alpha)_i = \vec{\alpha}_i, i \leqslant l(\alpha)$,
and dually, if $\vec{\alpha} = < \alpha_0, \ldots, \alpha_n >$ then $l(\vec{\alpha}) = n, (\vec{\alpha})_i = \alpha_i, i \leqslant l(\vec{\alpha})$.

Now we define a relation $\subset$ and a function $\cup$ on SO and simultaneously $\vec{\subset}$ and $\vec{\cup}$ on $\vec{SO}$, the set of all $\vec{\alpha} = < \alpha_0, \ldots, \alpha_n >$ where $\alpha_i \in SO, i \leqslant n$ and $n \in \omega$.

4. DEFINITION.
(1) (i). $I \subset I$
(ii). $\alpha \subset \beta \Leftrightarrow l(\alpha) \leqslant l(\beta)$ & $(\alpha)_i \vec{\subset} (\beta)_i$ for all $i \leqslant l(\alpha)$
(iii) $\vec{\alpha} \vec{\subset} \vec{\beta} \Leftrightarrow l(\vec{\alpha}) \leqslant l(\vec{\beta})$ & $(\vec{\alpha})_i \subset (\vec{\beta})_i$ for all $i \leqslant l(\vec{\alpha})$
(Note that (iii) is the 'dual' of (ii).)
(2) (i) $I \cup I = I$
(ii) $\alpha \cup \beta = \beta \cup \alpha = < (\alpha)_0 \vec{\cup} (\beta)_0, \ldots, (\alpha)_{l(\alpha)} \vec{\cup} (\beta)_{l(\alpha)}, (\beta)_{l(\alpha)+1}, \ldots, (\beta)_{l(\beta)} >$ if $l(\alpha) \leqslant l(\beta)$
(iii) dual of (ii)
Now $\subset$ is transitive, $\alpha \subset \alpha \cup \beta = \beta \cup \alpha$, and dually.

5. Notation. $M = M(x_0, \ldots, x_n)$ means: $FV(M) \subset \{x_0, \ldots, x_n\}$.
If $M = M(x_0, \ldots, x_n)$ and $\vec{\alpha} = < \alpha_0, \ldots, \alpha_a >, a \geqslant n$ then
$M\vec{\alpha} = M(\alpha_0, \ldots, \alpha_n)\alpha_{n+1} \ldots \alpha_a$.

6. MAIN LEMMA. Let $M = M(x_0, \ldots, x_n)$ be a λI-term in β-normal form.
Then $\exists \vec{\alpha}\ \forall \vec{\beta} \vec{\supset} \vec{\alpha} (M\vec{\beta} \in' SO)$
Proof:
Induction to the definition of β-normal forms (see [1], (2.8))
(i) $M \equiv x$; $x \equiv x_0$, $M = M(x_0)$. Take $\vec{\alpha} = < I >$.
(ii) $M \equiv \lambda x.N$; $M \equiv M(x_0, \ldots, x_n), x \equiv x_{n+1}, N \equiv N(x_0, \ldots, x_{n+1})$.
Induction hypothesis: $\exists \vec{\alpha}\ \forall \vec{\beta} \vec{\supset} \vec{\alpha} (N\vec{\beta} \in' SO)$.
Such an $\vec{\alpha}$ works also for M.
(iii) $M \equiv x_0 M_0 \ldots M_k$; $M_i \equiv M_i(x_0, \ldots, x_n), M \equiv M(x_0, \ldots, x_n)$.
Ind. hyp.: $\forall i \leqslant k\ \exists \vec{\alpha}_i\ \ \forall \vec{\beta} \vec{\supset} \vec{\alpha}_i\ \ (M_i \vec{\beta} \in' SO)$.
Define $\vec{\beta} = \vec{\alpha}_0 \vec{\cup} \ldots \vec{\cup}\ \vec{\alpha}_k$ and take $\vec{\gamma} = < \gamma_0, \ldots, \gamma_g >\ \vec{\supset} \vec{\beta}$ s.t.
$g \geqslant n+1$, by prolonging $\vec{\beta}$ with some I's, if necessary.
Define σ by $\begin{cases} l(\sigma) = k+1 \\ (\sigma)_i = < \gamma_{n+1}, \ldots, \gamma_g > \end{cases}$

Define $\vec{\xi} = \langle \xi_0,\ldots,\xi_z \rangle, \xi_0 = \langle \vec{\xi}_{00},\ldots,\vec{\xi}_{0z_0} \rangle$ by.

$$\begin{cases} \xi_0 = \sigma \cup \gamma_0 \\ \xi_{i+1} = \gamma_{i+1} \quad (i+1 \leq g). \end{cases}$$

So $\vec{\xi} \supset\!\!\!\rightarrow \vec{\gamma}$ and $\xi_0 \supset \sigma$. There are two cases: (1) $z_0 \leq z+k-n$, (2) else. If (1), then $M\vec{\xi} = \xi_0 M_0(\xi_0,\ldots,\xi_n)\ldots M_k(\xi_0,\ldots,\xi_n)\xi_{n+1}\ldots$ $\ldots\xi_z = (M_0(\xi_0,\ldots,\xi_n)\vec{\xi}_{00})\ldots(M_k(\xi_0,\ldots,\xi_n)\vec{\xi}_{0k})\vec{\psi}$ for some $\vec{\psi} \in' SO$. Because $\langle \xi_0,\ldots,\xi_n \rangle \vec{\xi}_{0i} \vec{\supset} \langle \gamma_0,\ldots,\gamma_n \rangle(\sigma)_i = \gamma, i \leq k$, by ind. hyp. we have $M_i(\xi_0,\ldots,\xi_n)\vec{\xi}_{0i} \in' SO, i \leq k$, hence by 2.(i), $M\vec{\xi} \in' SO$. If (2), then an equally simple calculation shows $M\vec{\xi} = \mu I$ for some $\mu \in SO$, hence also $M\vec{\xi} \in' SO$. By the same calculations we see: $\vec{\phi} \vec{\supset} \vec{\xi} \Rightarrow M\vec{\phi} \in' SO$.

7. THEOREM. If M is a closed λI-term in β-n.f., then M is I-solvable.
Proof: By 6, $\exists\vec{\alpha}\ M\vec{\alpha} \in' SO$, hence by 2, $\exists m\ M\vec{\alpha}I^{\sim m} = I$.
<u>Remark</u>: It can be proved that the $\alpha \in SO$ have the greatest possible 'solving power' in this sense: if M is a closed 'unary' λI-term in n.f., $M = \lambda x.x\vec{N}$, then M can be solved by one α: $\exists\alpha M\alpha = I$. And hence, if M is closed n-ary and in n.f., $M = \lambda x_1\ldots x_n.x_i\vec{N}$, then $\exists\alpha_1,\ldots,\alpha_n\ M\alpha_1\ldots\alpha_n = I$.

8. LEMMA. If $M_0,\ldots,M_k$ are closed λI-terms in β-n.f., then $\exists\vec{\alpha}\ \forall i \leq k\ M_i\vec{\alpha} = I$.
Proof: Take $\vec{\alpha}_i$ s.t. $\forall\vec{\beta} \vec{\supset} \vec{\alpha}_i\ \ M_i\vec{\beta} \in' SO$; $\vec{\beta} = \vec{\alpha}_0 \vec{\cup}\ldots\vec{\cup}\vec{\alpha}_k$, then $\exists m\ \forall i \leq k\ \ M_i\vec{\beta}I^{\sim m} = I$.

9. THEOREM. If $M_0,\ldots,M_k$ are closed λI-terms in β-n.f. then there is a closed λI-term K^* s.t. $K^*NM = N$ for all λI-terms N and all $M \in \{M_0,\ldots,M_k,K^*\}$.
(K^* is a "local K for $\{M_0,\ldots,M_k,K^*\}$".)
Proof: By 8, $\exists\vec{\alpha}\ \forall i \leq k\ \ M_i\vec{\alpha} = I$. Take $a = l(\vec{\alpha}) \geq 2$.
By 2, $\exists n\ \alpha_1\vec{\alpha}I^{\sim n} = I$ and $\alpha_0\alpha_2\ldots\alpha_a I^{\sim n} = I$.
Define $K^* = \lambda xy.y\vec{\alpha}I^{\sim n}x$, then $K^*NM_i = M_i\vec{\alpha}I^{\sim n}N = N$ and $K^*NK^* =$ $= K^*\vec{\alpha}I^{\sim n}N = \alpha_1\vec{\alpha}I^{\sim n}\alpha_0\alpha_2\ldots\alpha_a I^{\sim n}N = N$.

Remark: it is easy to show that there is no "local K" for all closed λI-terms in n.f.

10. The idea of a "local K" is useful in the construction of the following example of one single generator (or primitive combinator) for the λI-calculus. First two similar cases:
(1). [2] gives an example of a generator for the λK-calculus: define X = λx.x(xS(KK))K, then XXX = K and XK = S.
(2). [3] gives a pair of generators for the combinatory semigroup SCL_K (which has application X∘Y = BXY and is generated by B, C_*, C_*S, C_*K where B = λabc.a(bc) and C_* = λab.ba):
take C_* and H = λx.x(BK(C_*K))S, then
H∘C_* = C_*K, (C_*K)∘C_* = K, H∘K = C_*S, (C_*S)∘C_* = S, and S∘K = B.
(3). Define X = λx.xJAJCD, where J = λabcd.ab(adc); then
XX = XJAJCD = JJAJCDAJCD = JA(JCJ)DAJCD = A(JCJ)(AAD)JCD and
XI = JAJCD = AJ(ADC).
If A is a "local K for {C,D,A}", then
XX = A(JCJ)AJCD = JCJJCD = CJ(CCJ)D and XI = AJ(ADC) = AJD = J.
Defining C = λabc.cab
D = λxy.xIIII(yE)
E = λxy.xIII(yIIII)
we have XX = DJ(CCJ) = JIIII(CCJE) = CCJE = ECJ = CIII(JIIII) = I.
Define A = λxy.yIIIIx, then A is a "local K" for {C,D,A} and the result is XX = I, XI = J.
Because I,J generate the closed λI-terms, X does so too.

REFERENCES

[1] HENK BARENDREGT, A characterization of terms of the λI-calculus having a normal form.
Journal of Symbolic Logic, vol. 38, nr. 3, pp. 441-445.

[2] HENK BARENDREGT, A one point base for the λK-calculus, mimeographed.

[3] C. BÖHM, A two point base for the combinatory semigroup, hand-written note.

SOME PHILOSOPHICAL ISSUES
concerning
THEORIES OF COMBINATORS

Dana Scott
Oxford University

Abstract. The paper presents in an informal way several conflicting viewpoints concerning concepts, theories, models, and applications of λ-calculus and the combinators.

Introduction. During the course of the Rome Symposium it became progressively more obvious that there is a very wide variety of motivations for studying λ-calculus and that the objectives of research need not be particularly well integrated. Perhaps that is as it should be, but perhaps too we should try to be a little clearer in making our standpoints understandable so that at least some measure of comparison is possible. In particular, we can see by listing the questionable points that many of the issues and features are, as some people say, *orthogonal* to one another. This means that wild mixtures are possible, and much confusion is probable, if we do not try to state from the outset how and why certain choices are being made.

There is no pretence here to having given a "definitive" philosophy of combinators, but the author's experience and observation covers a certain amount of ground from several fields. Thus by presenting a series of contrasts at not too great length, he hopes he will at least provoke people to ask themselves questions about the proper placing of their results and the direction of their work. Certainly one point seems definite: there is as yet no combinatory *logic*. This is much too absolute a term. The combinators appear in many rôles in many theories, but they have not given us an *ultimate* foundation either for logic or for a theory of computation. They may never provide such foundations, but this is no criticism. They are interesting enough to merit study, and they deserve the favour of careful distinctions. Magic may have its place, but it is not here. There is more than sufficient evidence that they have a job to do - and if you do not fully understand how they do it, that is your fault not theirs.

In order to impose some slight organization of ideas, we shall divide the discussion into four parts: Concepts, Theories, Models, Applications. We intend here no hierarchy nor any indication of an historical progression. As a matter of fact people usually see the (rather formal) theories first, motivated by possible (and possibly unrealized) applications. It is only later (if ever) that they ask what concepts the theories were meant to formalize. The answer is not always so easy; sometimes models can help and often they fail - as they maybe hard to construct. It is also not easy to present the conceptual (pre-systematic) ideas apart from the theories and models in a way that is natural and makes the systematization seem inevitable. But that is usual in mathematics, and many people have good intuitions without being really able to explain them. This does not mean, however, that we should not _try_ to give the necessary explanations. It is also true that many people have only vague ideas, and still they churn out endless numbers of "results"; but that is no reason to condone such activity. As the λ-calculus has now entered a "second period" of fairly wide-spread interest, we really ought to take some time for serious reflection.

§ 1. Concepts. The conceptual basis for the λ-calculus is the notion of _function_, and the combinators are certain very general functions that can be used to define new functions from old. So far so good, but we cannot stop here. Just because we employ functional notation and discuss free and bound variables, it does not mean we know what we are doing. There are somany different kinds of functions, we are definitely obliged to say more. If we do not, then (just as with the pioneers of the subject) we should not be surprised to find that either our theories are too weak to be really useful or too strong to be consistent. This brings us to our first major contrast:

1.1 Typed vs. Untyped. The completely type-free logic is a will-O-the-wisp. This has been shown time and time again; yet the vision remains. And formal theories are quite as bad as drugs in keeping such illusions alive. The cold light of day reveals, however, perfectly solid ground. The notion of a type (as in type theory) is a sound concept. Functions do indeed have domains of definitions and ranges of values, and these ideas can be specified without having to say exactly which function is under consideration. Types are a way of making precise a portion of our ideas of separating functions into kinds; but this is only a start, since there are other properties besides domains and ranges that often need detailing.

A possible point of confusion concerns the relations between types. For example, one type may be more inclusive than another. Indeed, after becoming familiar with the usual finite types - often separated into a fairly rigid, noninclusive hierarchy - we may find it reasonable to introduce infinite types which include all the finite levels. This is done extensively in Zermelo-Fraenkel set theory, and for a certain kind of function concept this was done in the author's D_∞-models for λ-calculus. The point of confusion comes in because at these infinite levels we may think that we have achieved the goal of being type free. This is not the best way to explain the situation, however. Usually - with the aid of good notation - all we have done is to push the types into the background. They can always be brought out again. When we write:

$$D_\infty = D_\infty \to D_\infty$$

we do not mean that D_∞ is "type free". All we mean (after we say what "→" means) is that D_∞ is a highly self-contained (or reflexive) domain and that certain operations do not take us outside the domain. The trick of course is to obtain a sufficiently high degree of "self-containedness". An absolute notion is impossible - unless one redefines what he expects of "absolute".

The moral then is to accept types, yet to realize how flexible the concept can be. Certain features of the type-free paradise are possible, but there are always limits. Speak not of the untyped but of the infinite types.

Having come to terms with natural limitations (which is not to say that one has a grasp of their scope), it is good to recall at this point a rather more specific contrast that has to do especially with functions:

1.2 Explicit vs. Recursive. Perhaps the terminology here is too specific. The point will become clear if we keep the finite types in mind. Even just with pure function types (all the type symbols generated from atomic ones by "→" alone), there are plenty of typed combinators. For example there are all the typed versions of I,S, and K . The question is: are there more? The answer is: no - if all we demand is closure under explicit definability - but it is yes if we also expect recursive definitions (as given, for example, by typed versions of the Y combinator). The problem is not one of what types to use, but rather one of what closure conditions to require.

It need hardly be stressed, but the same type symbols may have different interpretations in different theories. Constant use of the "type-free" theories can make us insensitive to this obvious fact. The type-free theories also have the drawback that certain combinators - like the paradoxical combinator Y - give certain advantages "for free". Thus Y gives us recursion, which in a typed theory must be added as a new primitive. What else comes "for free" in the untyped theory? No one has ever provided a complete analyses; and maybe there is no final answer, since stronger and stronger theories could be possible.

In discussing closure conditions, we should not forget the conditional expression popularized by McCarthy. In giving recursion equations it is very convenient, and in some formulations it is needed a primitive as it is not definable in terms of the pure combinators. Another question about "truth-value" combinators will come up in 1.10 below.

1.3 Total vs. Partial. As soon as recursive definitions of any generality enter, the question of partial functions will arise. Often in the theory of types the underlying concept is that of total function. There is always a conflict which it is difficult to resolve. In the case of Kleene's recursion theory of higher-type functionals this conflict has often been remarked. In this admitedly infinitistic theory partial functions have a very secondary rôle which the (still unpublished) theory of Platek was meant to improve. Just how the traditional λ-calculus enters into these theories has only been partly explored (by Kleene and by Platek) and the work is not well known.

The type theories of de Bruijn and Martin-Löf (see the paper of de Vrijer in this symposium) also put the emphasis on total functions. The author's other paper in this symposium tries to suggest some connections with the "type-free" systems, but these proposals are rather tentative.

The author's lattice-theoretic models were concieved to accomodate partial functions from the start and the idea was heavily influenced by Platek's earlier work (there are also connections to Davis and Nerode). But the new point in the construction of the D_∞ model was that the spaces of finite type, the D_n , could be made cumulative - because we were using partially ordered spaces of partial functions. The trick of the partial ordering was that monotone well-defined (that is, total) functions give a model for a concept of partial function. There is a warning to take into account here to which we shall return in 1.4: the partially-ordered spaces which are so good for the fixed-point theorem may not model all aspects of how a function can be partial.

The issue is how to make clear the distinction between total and partial, and there is still very much left to say on the conceptual level.

1.4 Stable vs. Unstable. It was difficult to know what to call this contrast. It could be called "Curry vs. Kleene". It has to do with another aspect of partial functions. With Curry, application is always "meaningful" in the sense of always being allowed. With Kleene, the application $\{e\}(n)$ between numbers (here e is a Gödel number) is necessarily from time to time undefined. Curry's application is stable while Kleene's is unstable (this is meant to be descriptive not critical!)

The "modern" form of Kleene's theory is given with URS (see Barendregt's paper in this symposium). It is clear that Kleene carried over much of his experience with λ-calculus to these other structures, for example in his proof of his recursion theorem. Kleene also defined an analogous application $\{\alpha\}(\beta)$ for number-theoretic functions (see the Kleene-Vesley book). Troelstra has used the idea, but it does not seem to have been studied "abstractly" like the URS.

The question is: what really is the connection between the two kinds of theories of combinators? Could the two theories have a common generalization? Or is one "better" than the other?

1.5 Strict vs. Inclusive. This contrast might also be called "Church vs. Curry". Church seemed to feel (as a consequence of the paradoxes?) that only expressions in normal form were meaningful. Curry on the other hand was willing to face all combinators. In his lecture at the Symposium Barendregt explained the situation very neatly interms of the distinction between the λ-I-calculus and the λ-K-calculus. Church favoured the λ-I-calculus, and in this context, his feeling about normal forms can be made to look most reasonable. (The work Barendregt mentioned is to be published elsewhere.)

There is a definite conceptual point here: In the λ-I-calculus a function is meant to strictly depend on all its arguments. The question of dependence in logic has been very extensively discussed by Anderson and Belnap. The latter (in unpublished work) has tried to relate the λ-I-calculus to the logic (in the way Curry-Howard- et al. related combinators of the λ-K-calculus to intuitionistic logic) and he has discussed models. But the work seems as yet inconclusive. What is a strict function? Is it one which is undefined (or $\perp$) for undefined arguments? That seems like a necessary condition, but is that all we should say? There is a point to the λ-I-calculus, but more needs to be done to explain it.

1.6 Pure vs. Mixed. We might also call this "(η) vs. (ξ)". In detailing one's concept of function, is one going to be moved to be so pure that absolutely everything is taken as being (uniquely) a function? That is what the (η)-rule means:

$$(\eta) \qquad \lambda x.y(x) = y$$

Everything y <u>is</u> a function (namely: the function $\lambda x.y(x)$). Well, we know that this axiom is <u>consistent</u> (in pure λ-calculus), and it holds in the D_∞-models (where, starting with the injection of D_o into D_1, everything is made to be a function). But is this axiom conceptually desirable? Might we not want to mix in some other sorts of elements among the functions?

It of course depends on what we want to do with the λ-calculus. We should, however, not confuse this point with that of 1.4. We can coherently make the application $y(x)$ stable even when y is not a function. In the author's graph model $(P\omega)$ we find instead of (η):

$$(\eta^*) \qquad y \subseteq \lambda x.y(x)$$

and this inclusion had an interesting connection with closure operators. There seems to be no trouble in "coercing" objects y to closely related functions $\lambda x.y(x)$, but why identify them? In an "unstable" theory we might even want $y(x)$ to be "meaningless" when y is not a function. Remember: being type-free does not mean having no type distinctions; rather it means having a large pot of accumulated elements with enough structure for a tasty stew. In cooking there are reasons to mix flavours, so why not in function theory? In any case, one should be conscious of what choices are possible.

1.7 <u>Absolute vs. Relative</u>. The combinators are great fun, but why is so much effort spent on pure λ-calculus? What reason is there to suppose that everything can be defined solely in terms of abstraction and application? As a matter of fact not everything can be so defined. Thus, should not more attention be given to <u>relative</u> definability, to definability in terms of given primitives? The automata people do it; the recursion-theory people do it; the logicians do it; and the λ-people should too. If λ-calculus is ever to have any applications of any real importance, then it will be necessary. (Needless to say, there is no difficulty in talking about relative computability in the D_∞ and $P\omega$ models.) And, by the way, what ever became of the δ-rules that people used to discuss? Is there not some point to them?

1.8 Equational vs. Ordered. This point will come up again in Section 2 when we discuss theories, but there is a question of conceptual justification. Why spend all your thought on equations between combinators? The lattice-theoretic models have shown that there are very pleasant theories of partial orderings between λ-terms. The ideas connected with approximate normal forms, which were reported on in several papers in this symposium, have worked out very well and they essentially use inclusion and limits. In fact, how can we understand the fixed-point operator without limits? Of course, the study of partially ordered systems is more special than pure λ-calculus, but is it any less interesting? Might there not be other relations between terms that would also make sense? There is no reason that all useful properties can be expressed as equations - unless we add new operations (like $\cap$ or $\cup$ for lattices). More imagination about λ-calculus would be very welcome.

1.9 Compatible vs. Inconsistent. Even if a partial ordering is granted, there is a vexing question about forming unions of elements. Should $a \cup b$ (or, if you like: $a \sqcup b$) always make sense. Just as in the case of Curry vs. Kleene, where we might have reasons for having $x(y)$ undefined, we might feel that certain pairs of elements are inconsistent with each other. Do we want $a \cup b$ always to exist even when a and b somehow conflict?

In order to make the theory cleaner, Scott used lattices where all sets of elements had joins - all elements were compatible with each other. This gives in particular his horrid "top" element $\top$. Now it is Scott vs. practically everyone else, since no one really likes $\top$. One reason to have a full lattice was the extension theorem for continuous functions. If D is a continuous lattice then any continuous function $f : X \to D$ can be extended to a continuous function $\bar{f}: Y \to D$ where Y is any superspace with $X \subseteq Y$ as a (topological) subspace. Indeed this property characterizes the continuous lattices, and it seemed very reasonable if we want rich models for partial functions. For example any partly defined continuous $f : A \to B$ where $A,B \subseteq D$, can be found as a restriction of an everywhere defined $\bar{f} : D \to D$ if D is a continuous lattice. If $D = D \to D$, we can see why it is nice to have one type of element $u \in D$ represent all manner of continuous functions on all subspaces of D . But all this argument shows is consistency. These lovely continuous lattices are possible, but what of other structures?

Several hours work did not give good definitions for the category of CPO's. We shall return to this question under a discussion of models. The point is from the conceptual side to say all you mean. It is not enough just to have the ideas of a partial ordering and of limits of chains because this is all that is needed for fixed-points. If one is to have a satisfactory theory of functions, functionals, objects both partial and total, then one has to face other questions about existence of elements. The $a \cup b$ question is one, and the extension question (which is closely related) is another having to do with completeness or fullness of the models. Much more investigation is needed in this area.

1.10 Deterministic vs. Nondeterministic. It is hard to know what to say on this score because current ideas are so vague. We certainly have a feeling for the difference between the two kinds of computation; we can give examples of nondeterministically definable functions; but we cannot yet say that anyone has proposed very clear theoretical distinctions. Thus a union $a \cup b$ or a symmetric or in a boolean condition $p(x) \vee q(x)$ is nondeterministic(in that we have to try out both sides and cannot say in advance which will finish first). But this surely is not the whole story? There are other notions of search which will be needed (as in the theory of search computability of Moschovakis).

There is a great temptation to equate parallel computation with non-deterministic ideas, but much care is needed here. With parallelism there is sharing of data, and there is a need to resolve conflicts (inconsistencies) which does not seem to come up in simple nondeterminism. Despite a large number of papers in theoretical computer science on this problem area, the author remains very unsatisfied. When a really good answer is reached, however, he is sure that it will affect the λ-calculus, possibly with the introduction of new primitives that are as yet unrecognized.

1.11 Intensional vs. Extensional. In 1.10 we already spoke of computations. Up to this point we really only were questioning the property of computability (or definability). The latter is an extensional notion. It may turn out that the problems of parallelism and nondeterminism have only intensional solutions: the notions only make sense with regard to the way in which values are found. It may also be that problems of normal forms and Church-Rosser Theorems can only be discussed with particular computation rules in mind. If it turns out that way, fine! But there is so much confusion between use and mention, terms and denotations, form and content in writings on λ-calculus that it is often very difficult to know when you are standing on solid ground. Thus as a final plea on conceptual clarity the author entreats people to make the necessary distinctions. It does not matter which side you are on as long as you know where you stand.

§ 2. Theories. The points to be brought up in the next two sections are much more technical, though some of them have already been touched upon. The main suggestion here is that people should be aware of the very wide variety of ways in which a theory can be formalized, since they are often far from equivalent.

2.1. Constants vs. Variables. There was a small movement at one time to ban the variable. Some philosophers still discuss it. The elimination of variables from logic is possible, but the price one has to pay in algebraic operators is very high indeed. Without casting our vote one way or the other, we can see there can be differences in point of view which make variables - at least free variables - natural. For example an extensionality rule like:

$$\frac{\vdash \tau(x) = \sigma(x)}{\vdash \tau = \sigma}$$

(where x is a "new" variable not free in σ or τ) does not mean the same as the ω-rule:

$$\frac{\vdash \tau(\xi) = \sigma(\xi) \quad \text{all terms } \xi}{\vdash \tau = \sigma}$$

(where the terms are all closed terms). For one thing the ω-rule is infinitistic and gives a highly nonelementary theory as was pointed out by Barendregt in his lecture. On the other hand the rule with the variable has a quite different import with regard to a model, since the hypothesis is meant to hold for all values of x in the model, not just for those that are definable by closed terms.

We are trying to say that variables are good or bad, but it does seem that principles can be expressed with their aid that cannot be formulated without them. When the main emphasis has been on the pure combinators, variables have often been pushed to the sidelines and sometimes not given enough attention.

2.2. Equational vs. Conditional. It is the case that there are many curious and surprising equations that hold among the combinators, but these are of the status of completely general truths. In making theories and in proving theorems that are to have specialized applications, we often need rather more "local" truths. As a very minor example consider the obvious conditional:

$$u \circ u = u \vdash u \circ (u \circ u) = u \ ,$$

where $a \circ b = \lambda x.a(b(x))$. If the function u has its square equal to itself, then so with the cube (and every other positive power). All through algebra there are such conditionals to be found. There is no way to express this fact by a single equation. Many other examples can be found in the work of de Bakker and Milner. Thus interesting as the equations are, λ-calculus is more than just an equational theory.

2.3. Free vs. Quantified. The same concepts may be involved in many different theories. During the symposium Barendregt exhorted us to study combinatory algebras as algebras just as we do in mathematics with other structures. It is a good question whether anyone outside the "club" will want to listen to us, but that is a different problem. The point to be taken here is that the same structure can be studied in many ways. We know this in logic as regards first- and second-order arithmetic and in mathematics where, say, the reals also have a topology as well as an algebra. The point is a very obvious one, but sometimes it seems to be forgotten in studies of λ-calculus.

One reason for difficulties may be that on the level of theories (rather than models) there is generally no unique way to pass from the lower-order theory to the higher one. In 2.2 we mentioned conditional equations, and it may well be the case that the set of all valid (or all desired) equations does not determine which conditional equations should be valid. (Of course, some are determined just from the conditional laws of equality). It is probably even worse when we go from free-variable theories to quantified theories.

As an example of a quantified, first-order theory, we have the axioms (α), (β), (ξ) mentioned by the author in his other paper for the symposium. It is in the extensionality axiom (ξ) that the quantifier enters. Off hand the author does not know whether there is a simple set of "combinatory" axioms that axiomatize the equations provable in this theory. He does not know an equivalent theory in terms of S and K (without λ). He has no idea how to axiomatize the quantifier-free part of this theory. All in all, it seems fair to say that the (first-order) quantified theory of combinators is an untouched area.

2.4. Algebraic vs. Inductive. Even if we do not care to pass to quantified theories, there are distinctions to make at the quantifier free level. Most formulations of λ-calculus are very algebraic. But there is no reason whatsoever to believe that the laws of reduction are the only laws for proving terms equal. Take the paradoxical combinator, for example:

$$Y = \lambda u.(\lambda x.u(x(x)))(\lambda x.u(x(x))) \quad .$$

It is defined in pure λ-language. In Scott's models this Y is the least fixed-point operator. From the work of Böhm it is known that there are many definitions of similar combinators that cannot be proved equal to Y by conversion alone. In the models all these combinators are equal. Many of the desired equations can, however, be proved by using Scott's Induction Rule. For instance, an interesting equation is:

$$Y(\lambda f\ \lambda x.g(x)(f(x))) = \lambda x.\ Y(g(x))\ .$$

In Curry-style combinators we could write:

$$BYS = BY$$

The author has never attempted a proof, but it seems very likely that the methods of Böhm could be used to show that this cannot be proved by reduction. The proof by induction is very simple.

The conclusion is that if we really believe in the combinators, we may want to add an induction rule to our algebraic theories. This seems very reasonable if we want the combinators to help us with recursion theory. The reduction rules give us "for free" the recursion _equations_. But no one who has ever done any recursion theory would ever imagine that the equations were enough. Recursive _definitions_ must always be joined with inductive _proofs_. It is just these natural proof methods that are usually missing from ordinary combinatory algebra. (The same could be said about the theory of URS.) It does not really seem right to call λ-calculus really a _calculus_ if the induction rule (and some theory of limits) is missing. As far as it goes λ-algebra is fine, but it does not go far enough.

2.5. _Finitary vs. Infinitary_. Even though it is concerned with infinite processes, the induction rule is a finitary rule - at least in the sense that it gives a recursively enumerable set of theorems. We have already mentioned Barendregt's ω-rule which is essentially an infinitary rule. No doubt there could be infinitary versions of the induction rule. Going further, Fitch takes a highly infinitary view of combinatory "logic". This is discussed in the author's other paper as rather being a _truth definition_. (The distinction between syntax and semantics can sometimes blur when we pass to infinitary systems. The reason is that stronger rules make more notions absolute.) We have, therefore, enough evidence to show that it is interesting to think about infinitary rules in the λ-calculus. It is another very underdeveloped area.

2.6. Classical vs. Non-classical. Here is an even less developed area. First there is three-valued logic that seems very appropriate to λ-calculus (see the author's other paper). Just how do theories work out in this logic? Next, remember that λ-calculus has quite reasonable interpretations restricted to computable functions. Thus we can certainly look at the theory from a constructive point of view. Constructivism is not the same as Intuitionism, but many of the aims are the same. Models of λ-calculus (or URS) can certainly be defined intuitionistically. Thus (especially in quantified theories) there is a question of how to give intuitionistic formalizations. Is there going to be a marked difference between classical and intuitionistic versions of λ-calculus? What about free-choice sequences and Brouwer's theorem on continuity of functions? There ought to be something to say on this score.

§ 3. Models. As it was unavoidable not to speak of models earlier, we shall be very brief here.

3.1. Terms vs. Values. If one's view of λ-calculus is very algebraic (even: equational), then after one has found a formally consistent theory, then it is not at all surprising to find that there are term models of the theory. But you get no more out of the term models than what you put into the theory. Term models do not by themselves provide any conceptual analysis nor do they give much indication of how to find applications of the theory. Model constructions that can be explained in advance of theory formalizations have a chance of doing both. It is not at all clear that "meaning as use" is an adequate philosophy. This is not to disparage formalism; but there is more to mathematics than mere formalism. Even if one denies this, there is still the question of how to relate λ-calculus to other parts of mathematics (where functions are certainly considered important). Term models are precious little help for this purpose.

3.2. Constructive vs. Nonconstructive. The problem was really raised in 2.6: discuss the constructive aspects of λ-calculus models. On the other hand there may be interesting non-constructive models, too (see 3.5 below). Are term models constructive if the equality relation is not decidable? There may very well be a point that the URS version of combinators is more constructive (more intensional). In the $P\omega$-model, for example, there are many submodels. The least one is the class of recursively enumerable sets (constructive?) and every enumeration degree gives another model (see the authors Kiel paper). Which of these are constructive? More models can be found using the theory of retracts and the fixed-point construction. The constructive aspects have not been very much discussed (though work of Constable and Egli is oriented more in this direction).

3.3. CPO's vs. Lattices. This is a more technical point. We touched on it in 1.9 and 1.10. There is a strong feeling with many people that chain-complete partially ordered sets are enough and that continuous lattices are too much. The reason for taking continuous lattices was given in 1.9. There is a serious question here of how to judge the level of generality. Many CPO's can be completed to continuous lattices in a natural way (or at least embedded in them). Thus model constructions with CPO's may turn out just to be submodels of continuous lattices. So now who is to say which point of view is more general? Of course there may be an argument that (certain) CPO's are more economical models (having just the right kind of elements) and this would be understandable. But the study has not gone foward far enough yet to be really conclusive. (As mentioned in 1.9 there is also a problem of defining further completeness properties of CPO's, another interesting question.)

3.4. Inverse Limits vs. Graph Models. Many people were sorry to see the author give up talking so much about the D_∞ construction. One reason was that inverse limits were rather tiresome to explain to people who had never thought about such algebraic constructions, and the necessary computations were quite long to give in complete detail. The advantage of the graph model $P\omega$ was that its existence required no proof. The whole definition could be given in half a page. This does not mean that a person could understand the import of the construction in a flash, but at least he was not tired out even before he started.

Another advantage of the graph model is that it can be connected with ordinary recursion theory very easily. This is a little harder to see with inverse limits (whether for lattices or CPO's). Since, for the author at least, one of the reasons to develop λ-calculus is to have a simple and flexible way of presenting involved recursive definitions, it seems reasonable to relate things to the standard theories as quickly as possible so people can see how the new theory is a useful generalization.

The inverse limits are hardly lost, however. The graph model $P\omega$ is universal for all separable continuous lattices: they are all retracts of $P\omega$. (All T_0-spaces with a countably-based topology are also subspaces of $P\omega$ as well.) Thus $P\omega$ is a very rich model, and the author has shown (in the Kiel paper) that a calculus of retracts makes their definitions easy to find. Further, the retracts introduced by fixed-point constructions can often be related to inverse limits. So the theory of these spaces is not really abandoned and they are even put in a wider context of recursively defined concepts.

3.5. Continuous vs. Monotone. Ordinary (finitary) recursions lead to continuous functions. The reason is, roughly, that the recursion is solved by finite iterations taken to a simple infinite limit. If we ever hope to be able to compute such functions, then we should keep to this continuous level. However, there are more infinitistic concepts: transfinite recursion for example.

In set theory, let H_ω be the set of hereditarily finite sets (all members of members of ... of members are finite). Let H_{ω_1} be the set of hereditarily countable sets. (H_{ω_1} is uncountable, while H_ω is countably infinite.) We can form graph models (almost exactly by Plotkin's original suggestion) by defining:

$$F(X) = \{y \mid \exists x \subseteq X.\ \langle x,y\rangle \in F\}$$

$$\lambda X.\tau = \{\ \langle x,y\rangle \mid y \in \tau[x/X]\}\quad .$$

Here the capital letters range over the power set (PH_ω or PH_{ω_1}) and the small letters over the small sets (H_ω or H_{ω_1}). The first model is our familiar model of continuous functions, while the second is the model of σ-continuous functions (continuous for unions of directed sets closed under countable sups). Such functions are monotone but only weakly continuous. PH_{ω_1}, however, is a very interesting λ-calculus model in which the Y-operator now does transfinite fixed points (no closure after ω in general).

We could go to higher cardinalities even to infinity in the strongest sense: let the X's be arbitrary classes and the x's just sets. We still have an interpretation of λ-calculus in, say, Gödel-Bernays set theory. (Note: no need for bound class variables.)

In the other direction we can find a host of submodels (even of

PH_{ω_1}). All of what is sketched here could be done over admissible sets. (Instead of arbitrary subsets, we should use just the Σ_1-subsets to have a λ-calculus model.)

This seems to indicate there is a vast field for investigations of λ-calculus related to, say, descriptive set theory. There is certainly more to λ-calculus than meets the eye, even if we agree that too much has already met the eye in the published literature.

§ 4. Applications. It is difficult for the author not to write just about his personal interests. It is to be hoped that the other contributions to this symposium will provide a counterbalance. In concluding this paper, those applications that have especially concerned the author and have provided the motivation for him will be quickly reviewed.

The main interest that led to the discovery of the lattice-theoretic models was the need to find a basis for Strachey's theory of programming language semantics. The situation seems reasonably satisfactory at the moment and many people are working at giving language definitions in this style. A very great deal remains to be done to make the approach really practical, however, and by now the field is too large to survey here even superficially. The main effect of the λ-calculus models seems to have been - in very general terms - to give people more imagination about recursive definitions on abstract structures.

The relative success with programming language semantics suggests finding more connections with classical recursion theory. Though there have been many papers on abstract recursion theories of various sorts, no very unified view has come out. Possibly one lack is a good discussion of functionals, and here is a place that λ-calculus might find an application. Another place to look for connections is a study of degrees other than the Turing degrees. The so-called enumeration degrees of sets of integers are very easily defined in the $P\omega$ model and might merit more study. In another direction we could look for more connections with notions of computational complexity. Before starting on such projects, it would be well to consider the gains expected. It is unfortunate, but much of recursion theory leads a very isolated existence.

A very brief indication of how λ-calculus could be interpreted in set theory was mentioned in 3.5. This also might provide some applications. It is doubtful, however, that λ-calculus will lead to a particularly surprising alternative foundation for set theory as was once hoped.

In logic, λ-calculus is tied very closely to the theory of types. Infinite and transfinite types look interesting and this also brings us to category theory which is very close to type theory. Indeed the λ-calculus models give us new cartesian-closed categories, and if one only knew what conclusions to draw, one could find applications here. There is also proof theory which has been closely connected to λ-calculus. We would be grateful for a few new ideas at this point, however.

OPEN PROBLEMS

The following is a list of open problems sent in by various participants of the conference. They are roughly divided accordingly to their relation to

- I *Pure λ-calculus*
- II *Models*
- III *Illative theory*

Some of the problems do not seem to be too hard. None of them has the status of a "classical" open problem, which in fact the theory is lacking until now.

Henk Barendregt.

I. Pure λ-calculus.

1. Add to the λ-calculus new constants D, D_1, D_2 with reduction rules $D_i M_1 M_2 \geqslant M_i$, $i=1,2$, $D(D_1 M)(D_2 M) \geqslant M$ (surjective pairing).
   - Does the extended system satisfy the Church-Rosser theorem ?

   The problem originally was posed by C. Mann in 1972. To simplify matters, Hindley in 1973 posed the related question whether the theory extended with a δ and the reduction rule $\delta MM \geqslant M$ satisfies Church-Rosser. Quite a lot of effort has been spent on the problem.

2. Let f map λ-term into λ-terms (possibly containing free variables).
f is representable iff $\exists F \forall M\ f(M) = FM$.
f and f' are dual iff $\forall M,N\ f(M)N = f'(N)M$.
   - Are dual maps representable ?

   If attention is restricted to closed terms the answer is NO.

3. As S-term is an element of the least class of combinators containing S and closed under application.

- Is the equality between S-terms decidable ?

There are S-terms without a normal form, e.g. AAA, with A = SSS.

4. The ω-rule is:
FZ = F'Z for all closed Z ⇒ F = F'.
- Is the set of theorems of the λ-calculus + ω-rule a complete Π^1_1 set ?

For the theory containing $\mathcal{H} = \{\Omega=\Omega' \mid \Omega, \Omega'$ *unsolvable*$\}$ *the answer is YES.*

5. A term F is surjective iff for all closed Y there exists an X such that FX = Y.
- If F is surjective, is there a G such that F∘G = I ?

6. The range of a combinator F, RaF, is defined as {FM|M closed}.
- Is there a sequence $F_0, F_1, \ldots$ such that $\mathrm{RaF}_0 \subsetneq \mathrm{RaF}_1 \subsetneq \ldots$.

7. Show that the equation BYS = BY cannot be proved in the λ-calculus without induction. Are there other such simple equations involving Y that can be proved by suitable induction rules ? (For example Böhm's sequence of Y operators is already known).

8. The graph of a combinator M is the set $G_M = \{N \mid M \geqslant N\}$ with the relation ⩾. What properties characterize the set of graphs of combinators ?

II. Models.

1. The λI-calculus with extensionality and all terms without normal form equated is consistent. Does there exist a mathematical model for this theory ?

2. Are there interesting equalities/orderings holding in Pω or D_∞, which cannot be proved by Scott's Induction Rule ?

3. Is the equality in all D_∞ (whatever the initial projections) maximal ?

4. Consider λ-calculus based on axioms (α),(β),(ξ). If D is a model, a subtype is just a subset $A \subseteq D$. For any subtypes A and B define

$$A \to B = \{f \in D \mid \forall x \in A . f(x) \in B\}.$$

A combinator (λ-term without free variables) is said to have functionality, if there is an expression in subtype letters and "→" such that in all models and for all subtypes the combinator belongs to the compound type. For example:

$$S \in [A \to [B \to C]] \to [[A \to B] \to [A \to C]]$$

is true in all models. Is it decidable whether a combinator has functionality ? Does a combinator with functionality always have a normal form ? Is there always some unique most general functionality for a combinator with it ?

5. The same problems can be posed when we add as a type formation construct the cartesian product:

$$A \times B = \{< x,y > \mid x \in A \wedge y \in B\}.$$

Does it make any difference which pairing function is used ? We could also consider sums:

$$A+B = \{0\} \times A \cup \{1\} \times B$$

where 0 and 1 are conveniently chosen distinct combinators.

6. Consider the Graph Model Pω for the λ-calculus and the URS of Kleene with continuous function application {α}[β]. In each of these models we can consider the "types" given by definite classes. These types form a category in the mathematical sense of category theory.

Question: are these categories equivalent ? We could also consider this problem with regard to "variable" types (properties of Σ and Π).

7. What is the relation between the URS-theory of combinators and the standard theory where application is always defined ? Is

there a uniform way of converting a *URS* into a λ-calculus model ?

8. By the ordinal of a model for λ-calculus we understand the sup of the closure ordinals for all monotone, first-order inductive definitions. Are the ordinals of λ-calculus models just the admissible ordinals ?

III. Illative theory.

1. The basic axioms (α),(β),(ξ) for λ-calculus are to be taken as first-order axioms (particularly in the case of (ξ)).
What is the equivalent (first-order!) theory using application, S and K as primitives rather than λ ?

2. What is a proper intuitionistic version of λ-calculus (or *URS* theory) ? And is there any difficulty in showing constructively that models exist ? Is the Church-Rosser theorem intuitionistic ?

3. Given a functionality-type τ, characterize the set of combinators M such that $\vdash \tau M$ (cf. Given a formula A of implicational propositional logic, characterize the set of its derivations.)